AeroDynamic: Inside the High-Stakes Global Jetliner Ecosystem

AeroDynamic: Inside the High-Stakes Global Jetliner Ecosystem

Kevin Michaels

Ned Allen, Editor-in-Chief
Lockheed Martin Corporation
Bethesda, Maryland

Published by
American Institute of Aeronautics and Astronautics, Inc.
12700 Sunrise Valley Drive, Reston, VA 20191-5807

Cover photo: 787 production line in Everett, WA; Copyright © Boeing.

American Institute of Aeronautics and Astronautics, Inc., Reston, Virginia

1 2 3 4 5

Library of Congress Cataloging-in-Publication Data

On file.

ISBN 978-1-62410-402-2

CONTENTS

PREFACE xi

CHAPTER 1 A NEW ECOSYSTEM 1

References 5

CHAPTER 2 THE EUROPEAN DREAM 7

The Early Years of the Jet Age 7
The 1960s: Accelerated Innovation 10
The 1970s: Bigger, Higher, and Faster 12
Competitive Dynamics of the 1970s and US Deregulation 19
The 1980s: Rise of Airbus 23
Twin-Aisle Refresh 28
Late-1980s Growth 30
Unducted Fans? 33
European Dream 34
References 34

CHAPTER 3 MOONSHOTS 37

1990: An Eventful Year 37
The Early 1990s Bust 39
A New Super-Jumbo? 43
And Then There Were Two 45
Mega-Merger 47
The Changing Nature of Airlines 48
The Late 1990s Boom 51
European Restructuring 53
The A380: The First Moonshot 54
A Second Moonshot: The 787 56
Moonshot Number 3: The A350 XWB 59
Global Airline Restructuring 62

Moonshot Delays . . . 64
Another Airline Crisis . . . 66
References . . . 68

Chapter 4 The Regional Revolution . . . 71

The 1980s: Turboprop Dominance . . . 71
Canadian Revolution . . . 73
Embraer: The Unlikely Competitor . . . 77
Exit Fokker . . . 80
Regional Jet Demand Growth . . . 82
Regional Jet Aeropolitics . . . 85
The Fee-for-Service Dream House . . . 86
British Aerospace Responds . . . 87
70-Seaters: The Next Wave . . . 87
More Attrition . . . 90
Regional Jet Transition . . . 93
Bombardier's 747 . . . 94
References . . . 96

Chapter 5 Modern Miracles . . . 99

Jet Engine Origins . . . 100
Pratt & Whitney's Emergence . . . 103
Changing Fortunes . . . 106
Trent: More than a River . . . 110
Brian Rowe's Big Gamble . . . 111
The Rolls-Royce Force of Conviction . . . 113
A New Aeroengine Market: Regional Jets . . . 116
Single-Aisle Roller Coaster . . . 118
Large-Thrust Fireworks . . . 119
Revolutionary Technology . . . 123
Module Reuse: A Key to Profitability . . . 124
Exit Pratt & Whitney? . . . 125
References . . . 125

Chapter 6 Gearheads . . . 127

No Experience Necessary . . . 127
Breakthrough: The Mitsubishi Regional Jet . . . 132
Getting Real: The Bombardier C-Series . . . 134
CFM's LEAP Ahead . . . 137
Enter the Dragon . . . 140
The Shot Heard 'Round the World: The A320neo . . . 142
In a Corner: Boeing's Launch of the 737 MAX . . . 143

Regional Reengining: The E2 147
Pratt & Whitney: Rolls-Royce Rapprochement? 148
A Changed Landscape 149
References 150

CHAPTER 7 RETRENCHMENT **153**

Restructuring Airlines 153
Twin-Aisle Delays 161
Reengining Twin-Aisles 162
Moonshot Debacles 164
The A350 XWB: An Airbus Bright Spot 166
New Regional Jet Competition 168
Single-Aisle Hijinks 171
The Mother of All Jetliner Battles 173
The Airbus–Bombardier Surprise 177
References 179

CHAPTER 8 SUPPLY CHAIN 2.0 . . . . . **183**

The Way They Were: Early Jetliner Supply Chains 184
Bombardier and Embraer: Tier 1 Pioneers 186
Harnessing Globalization: Low-Cost Countries 189
Tier 1 Expands 191
Boeing's Supply Chain Nightmare 192
The More for Less Era 196
Rebalancing 198
Leaning Operations 200
Globalizing Engineering 204
References 208

CHAPTER 9 BRAINIACS **211**

The Early History of Avionics 211
Avionics in the Jet Age 214
Goodbye Third Pilot 216
Supplier Consolidation 219
Avionics in the Bustling 1990s 222
Brussels Bombshell 226
Momentum Shift 228
Thales Rising 233
GE's Return to Avionics 235
Brainiacs 236
References 237

CHAPTER 10 SKIN AND BONES 239

Jetliner Materials: The Early Years 240
Enter Composites 244
Material Innovation Golden Age 247
The 787 Shocker 251
Emergence of Major Tier 1 Aerostructures Suppliers 255
Material Supplier Competition 257
Revolution from Below 261
New Frontiers 266
References 267

CHAPTER 11 INTERIOR DESIGNERS 269

Jetliner Interiors: Early History 269
The Unlikely Consolidators 272
Interiors in the New Millennium 276
Supplier Consolidation 278
Internet Cafes in the Sky? 279
Interiors in the 2010s 283
Operations Headaches 285
Evolving Connectivity 287
The Interiors Segment Comes of Age 290
References 292

CHAPTER 12 UNSUNG HEROES 295

Aircraft Equipment: The Early Years 295
Aircraft Equipment in the 1990s 302
2000s: "Moonshots" Drive Innovation 307
The Mega-Systems Award 311
Rising China 314
2010s: Consolidation Continues 315
References 319

CHAPTER 13 LIFEBLOOD 321

Maintenance, Repair, and Overhaul: A Large and Crucial Sector 321
MRO Origins 323
Airline Maintenance Alliances 324
The 1980s: Aftermarket = Afterthought 326
The Influence of ETOPS 328
The 1990s: The Birth of the MRO Sector 330
MRO in the New Millennium 338
PMA Parts: The OEM Alternative 340

Return to Growth 342
Return on Assets 343
References 348

CHAPTER 14 FOUR CLUSTERS AND A FUNERAL **351**

The Funeral: Southern California 351
Hecho En México 355
Singapore: Asia's Übercluster 359
Morocco: The Unlikely Cluster 364
The Southeastern United States: Aerospace Manufacturing Hotspot 368
Four Clusters and a Funeral: The Takeaways 375
The New Landscape of Jetliner Production 376
References 380

CHAPTER 15 HORIZONS **383**

Reference 386

INDEX **387**

ABOUT THE AUTHOR **403**

SUPPORTING MATERIALS **404**

Preface

AeroDynamic traces the rise and transformation of the jetliner business from the end of the Cold War to the present. This $300-billion industry is not only the pinnacle of high technology and advanced manufacturing, but also is among the largest export industries for many advanced economies and a key enabler of military aerospace and national defense.

Although high-profile aircraft manufacturers like Airbus and Boeing receive the lion's share of attention in the media and aerospace publishing, this book approaches the industry from an *ecosystem* perspective. There are hundreds of suppliers behind each shiny new aircraft that rolls off the assembly line; indeed, the major aircraft manufacturers account for just 40% of economic activity. *AeroDynamic* will shine the light on this industry's entirety. Individual chapters are dedicated to the history of the avionics, aircraft systems, aerostructures, and interiors segments. In addition, several themes are explored including the changing nature of airlines, the role of globalization, shifting original equipment manufacturer (OEM) supply chain strategies, the rise and fall of aerospace clusters, and the growing importance of the aftermarket.

The breadth of the subject matter was a challenge in researching this book. Each chapter could rightfully be the subject of an entire book. I have been an aerospace management consultant for most of my 30-plus-year career, which has enabled me to work in just about every nook and cranny in the jetliner ecosystem. I often drew upon this knowledge and the personal networks developed along the way to supplement the research and analysis. Errors and omissions are mine and mine alone.

Kevin Michaels
August 2018

Chapter 1

A New Ecosystem

The day of 6 September 1993 was eventful, like many days in Moscow in the early 1990s. A constitutional crisis gripped the Russian capital as President Boris Yeltsin engaged in conflict with the Supreme Soviet, the nation's parliament, three days after he attempted to illegally sack his vice president. The economy was in freefall as it struggled to transition to free market capitalism. The most important event that day, however, at least from the perspective of the aerospace industry, was the opening of the Boeing Technical Research Center. Just 18 months after the collapse of the Soviet Union and end of the Cold War, a US aerospace giant was opening its first major design center outside of the United States, and just 500 m from the Kremlin.[1]

One month earlier, a group of Chinese toured a McDonnell Douglas aircraft factory in Columbus, Ohio. They were eager to purchase massive computer-controlled manufacturing machinery and move it back to China to support production of MD-90 aircraft in China, known as the *Trunkliner* program. China had already produced 28 MD-82s, assembling aircraft kits shipped to Shanghai since 1986. Demand for aircraft was taking off as a result of economic reforms ushered in by former Premier Deng Xiaoping; gross domestic product (GDP) growth that year was a whopping 14%. But this was different. China wanted to make key components for Western aircraft rather than simply assembling kits. Despite an offer from some of the plant's employees to buy the machines to protect their jobs, McDonnell Douglas struck a deal to sell 19 of the machines to China. Months later, jump-suited Chinese workers crated the massive machines and loaded them into trucks. In March 1994, the plant was shuttered [1].

[1]The Free Library, "Boeing Opens Technical Research Center in Moscow," http://www.thefreelibrary.com/BOEING+OPENS+TECHNICAL+RESEARCH+CENTER+IN+MOSCOW-a013248637 [retrieved 29 Dec. 2015].

These investments, although modest, were harbingers of a geopolitical sea change that would transform the aerospace industry and its ecosystem. Prior to the end of the Cold War and Chinese economic liberalization, the market for jetliners was divided between Western and Soviet zones of influence. Airlines in the USSR, India, Eastern Europe, Central Asia, and parts of Africa typically purchased Soviet equipment from the likes of Tupolev, Ilyushin, and Antonov, whereas their counterparts in Europe, Asia, the Middle East, and the Americas purchased from Boeing, Douglas Aircraft, British Aerospace, Fokker, Lockheed, and an upstart from Europe called Airbus. In the late 1980s, the Soviet Union accounted for 25% of the worldwide civilian and 40% of the worldwide military aircraft production. Now, the market was unified, and for the first time, the customer base was truly global.

Around the same time, a step change in aircraft technology was afoot. European manufacturer Airbus, an unlikely consortium founded in 1970 that few Americans expected to survive against its mighty incumbents Boeing, McDonnell Douglas, and Lockheed, had a hit on its hands with the 150-seat A320. This new aircraft, a competitor to the venerable 737 and MD-80, was loaded with advanced technology. It was the first aircraft to utilize a fly-by-wire flight system in which mechanical linkages to the aircraft's control surfaces were replaced with an electronic interface. The A320 was also the first jetliner where composites exceeded 10% of the aircraft's structural weight. In 1991, the A320 became the first Airbus aircraft to exceed 100 units per year and gave the firm a foothold in the high-volume sweet spot of the industry—150-seat, single-aisle aircraft. Moreover, Airbus planned to quickly follow with the A330 and A340 twin-aisles, which would enter service in 1993 and give it a complete aircraft family. Airbus would survive, and it would use technology as a major differentiator.

Boeing responded with a major technology leap of its own with the 777, launched in 1990. Like its Airbus counterparts, it would incorporate fly-by-wire technology and advanced composites. It would become the first aircraft to enter service with 180-min Extended-Range Twin-Engine Operational Performance Standards (ETOPS) certification, which meant that it could fly routes 180 min from the nearest diversion airport. This would ultimately transform the economics of long-haul, transoceanic travel with twin-engine aircraft like the 777 and A330 replacing three- and four-engine aircraft like the MD-11 and 747. ETOPS would also be a catalyst for improved aircraft engine and systems reliability as well as maintenance practices.

The early 1990s also brought major changes to the way aircraft were designed and produced, shaped by globalization. Boeing's 777 was the world's first all-digital design. This meant that new aircraft designs were conceived in three-dimensional computer models rather than on paper or Mylar. Engineers could design and assemble a "virtual" aircraft, avoiding the need for costly mock-ups. Design digitization, coupled with the spread of the Internet, created the possibility of dispersed global design teams. No more would engineering teams need to be co-located in major design centers in places like Seattle and Long Beach; geopolitical shifts meant that aircraft development teams could leverage human capital collaborating around the globe. New engineering centers began to flourish in places like China and India to complement Boeing's foray into Russia.

Beyond engineering, supply chains were also transformed as the aerospace supplier base globalized and aircraft manufacturers adopted Tier 1 supply chain models (Fig. 1.1) to enable them to work with fewer, more capable companies. This meant that the fundamental design of aircraft systems would be led by major suppliers like Goodrich, Honeywell, and Messier rather than internal aircraft original equipment manufacturer (OEM) engineering teams. Tier 1 suppliers would also take on much greater supply chain responsibility, managing thousands of lower tier suppliers. This development paved the way for a dramatic consolidation of aircraft suppliers that gained momentum in the 2000s and continues to the present. Globalization played an important role in shaping these new-age supply chains as manufacturers and suppliers opened new facilities in emerging economies as a means of reducing costs and opening new markets. Complex and capital-intensive work would remain with suppliers in advanced economies while labor-intensive work could be moved to low-cost countries. This led to the formation of new aerospace manufacturing clusters in far-flung locations including China, Mexico, Morocco, and Poland.

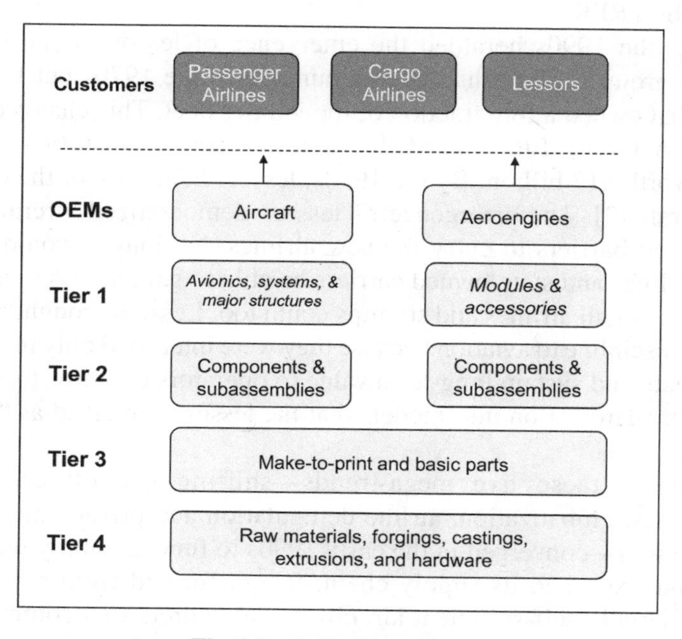

Fig. 1.1 Jetliner ecosystem.

A fourth major change in the early 1990s was deregulation and liberalization of the global airline industry. The United States deregulated air travel in 1978, allowing airlines to compete based on price; this enabled a new breed of airline known as low-cost carriers (LCCs) to emerge. People Express and Southwest Airlines were pioneers of the low-cost model in the United States. Southwest famously leveraged a single aircraft type, the Boeing 737, and the associated operational efficiencies to drive down costs and open air travel to a broader group of consumers. The new environment meant that airlines competed more on price, which led them to scrutinize their cost structures and outsource activities that were formerly conducted in-house, including maintenance, training, and catering. Also significant was liberalization through the adoption of a single market by Europe, increasing competition and paving the way for Europe's own low-cost airline revolution, led by Ryanair and easyJet. Asia would follow with its own LCC revolution in the 2000s. Coupled with liberalization was airline privatization. In the United Kingdom, Margaret Thatcher's government privatized British Airways in 1987, with major European airlines soon following her lead. Governments began to side with the interests of air travel consumers rather than coddling state-owned and national airlines. The net effect was a significant decrease in airline ticket prices, which stimulated air travel demand. In 1980, global air travel was 1,035 billion revenue passenger kilometers (RPKs)—a measure of activity that combines the number of passengers with kilometers flown. By 1990, this figure nearly doubled to 1,854 billion RPKs.

Finally, the 1990s heralded the emergence of lessors as an important customer group. Lessors had been around since the 1970s but were niche players that owned a tiny fraction of the jetliner fleet. This changed in 1989 when GPA Group Ltd. placed the largest orders in history—some 700 aircraft worth $17 billion. By the 1990s, lessors held 20% of the orders for new aircraft [2]. The emergence of lessors democratized aircraft finance and reduced barriers to entry for new airlines. No longer would only the largest airlines and state-owned carriers be able to gain access to new fleets of jetliners; small airlines and startups could too. Lessors brought a stronger financial discipline to aviation because they were interested only in assets that could create and sustain long-term value to operators and investors. Greater aircraft standardization into models that the lessors perceived as "winners" followed.

Combined, these five mega-trends—shifting geopolitics, enabling technologies, globalization, airline deregulation and privatization, and the rise of lessors—converged in the early 1990s to fundamentally redefine the jetliner business and its supply chain. It transformed from a fragmented group of largely self-sufficient aircraft manufacturers to a complex global network of operators, lessors, OEMs, suppliers, and service companies.

It became a global ecosystem. Some 25 years after Boeing's 1993 foray into Russia:

- Air travel demand more than tripled, exceeding 4 billion passengers and 7 trillion RPKs, with Asia-Pacific the largest and most important market for jetliners.
- Aircraft lessors would emerge as major jetliner customers; they would own 40% of the jetliner fleet and account for half of all orders.
- Two new aircraft production duopolies emerged: Airbus and Boeing for large jetliners and Embraer and Bombardier for smaller, regional jetliners.
- The aircraft value chain, once dominated by the United States, dispersed around the globe thanks to globalization, with new clusters developing in far-flung locations including Singapore, Morocco, and Mexico.
- Aircraft OEMs embraced Tier 1 supply chain strategies, which set the stage for a new cadre of sophisticated aeroengines, aerostructures, and aircraft systems suppliers to develop.
- Aircraft maintenance, repair, and overhaul (MRO) emerged as a distinct aerospace sector with $70+ billion in annual activity.
- Amazingly, aircraft safety improved by an order of magnitude despite the industry's overall growth.

Understanding how this new global jetliner ecosystem formed requires retracing history to the start of the jet age.

REFERENCES

[1] Mintz, J., "Sale of Aircraft Machinery to China Shows Perils of Exporting Technology," *Washington Post*, June 7, 1998, A08.

[2] Weiner, E., "GPA Group Gives Boeing a Record $9.4 Billion Order," *New York Times*, 18 April 1989, https://www.nytimes.com/1989/04/19/business/gpa-group-gives-boeing-a-record-9.4-billion-order.html [retrieved 5 February 2017].

Chapter 2

The European Dream

The jet engine revolutionized air travel around the world like no other technology. Compared to airplanes powered by piston engines, jet-powered planes, or jetliners, could fly at tremendous speeds, climb faster, and fly higher. Passengers also preferred the relatively quiet and vibration-free operation of jets.

The jetliner age began in 1949 in the United Kingdom with the flight of the de Havilland DH 106 Comet. Manufacturers from numerous countries joined the chase, yet by the 1970s, jetliners were largely a US affair, with nearly 90% of Western jetliners produced by Boeing and Douglas Aircraft. Underpinning its dominance of civil aviation was a massive military-industrial complex that was fighting the Cold War. Longtime aviation journalist Pierre Sparaco captured the prevailing sentiment when he wrote, “It is a keystone of essential Americana, creating a broadly-held feeling among Americans that any challenge to the supremacy of this vital asset stems from an insolent provocation” [1].

Yet by the late 1990s, Airbus—a new four-nation European consortium created in 1970—achieved virtual parity with US giant Boeing. Against all odds, it vanquished the jetliner businesses of McDonnell Douglas and Lockheed.

How did this happen?

The Early Years of the Jet Age

Following the end of World War II, significant aircraft engineering and production capability remained in just a handful of countries, including the United Kingdom and the United States. The United States, which built most of the long-range bombers and transport aircraft, would seemingly have a leg up in producing jetliners; however, it was the United Kingdom that led the way. The United Kingdom had two important advantages: aeroengine technology and planning. The United Kingdom pioneered commercial jet engines, and as early as 1942, British authorities under Lord Brabazon began planning for leadership in civil aircraft. The United Kingdom at the time had dozens of aircraft and aeroengine manufacturers. These advantages gave birth to the de Havilland DH 106 Comet (Fig. 2.1). The aircraft would carry 36 passengers and fly at an unheard of 420 kt at an altitude of 34,000 ft. In December 1945, just four months after the end of World War II, state-owned British Overseas Aircraft Corporation (BOAC) placed the world's first jetliner order for eight Comets. The aircraft first flew on 27 July 1949, powered by four somewhat crude de Havilland Ghost jet engines. The Comet would become the world's first production jetliner, but it wasn't the first to fly. That honor would belong to the Vickers Viking, a turboprop aircraft that demonstrated jet propulsion in April 1948 with Rolls-Royce Nene aeroengines. The jet-powered Viking would never go into production.

Another jetliner prototype flew just 13 days later, on the other side of the Atlantic. This project wasn't American—it was Canadian. The Avro C102 was well-suited for busy routes in the Eastern United States and attracted the interest of Trans World Airlines owner Howard Hughes. However, delays on Avro's CF-100 advanced fighter program—perceived as a more lucrative investment during the Korean War, led to the termination of the project in 1951.

The Comet entered service with BOAC on 2 May 1952 carrying 36 passengers from London to Johannesburg. The 23-hour, five-stop journey

Credit: Aviation Photography of Miami

Fig. 2.1 The first jetliner: the de Havilland Comet.

cost £175 one way and £315 round trip.[1] It was an immediate hit. Ten months later, disaster struck when a Canadian Pacific Airlines Comet crashed during takeoff, killing all 11 on board. Two months later, another went down after takeoff from Calcutta, killing 43. Then in January 1954, a Comet dived into the Mediterranean off the island of Elba, killing 35. After extensive investigation, a new type of test was devised—a cabin pressurization test. When the Comet's fuselage was pressurized for the 3060th time, a break appeared in the corner of one of its windows [2]. The investigation concluded that premature metal fatigue was caused by the windows' rectangular shape. The Comet would be grounded and redesigned with circular windows, a stronger fuselage, and more powerful Rolls-Royce Avon engines. But the damage to the Comet's reputation was done—it would not re-enter service until 1958, opening the door for new suppliers.

Surprisingly, the first commercially successful jetliner would come from the Soviet Union. The Tupolev 104, which was based on a bomber design, began revenue service with Aeroflot in 1956 on its Moscow–Omsk–Irkutsk route. Some 200 would be produced by 1960, and between 1956 and 1958 it was the world's only operating jetliner.

Meanwhile, the jetliner business attracted interest from France, which lost much of its aircraft production capability during the war. After returning from a trip to the United States and United Kingdom in 1945, French aerospace executive Henri Ziegler warned that "no country in Europe, and France in particular…has the economic and financial fabric sufficient to compete alone against the two superpowers. The future of our aeronautics industry depends on European cooperation" [3]. Ziegler's warnings aside, in 1951 a civil aircraft committee was organized by the French government and issued a specification for a medium-range jetliner. Numerous proposals were received from an eager French aeronautics sector before it settled on Sud Aviation's SE 210 Caravelle. The world's first jetliner specifically designed for short and medium routes, it entered service in 1959 with Air France and Scandinavian Airline System. France appeared to be on its way.

In the United States, Douglas Aircraft had a commanding position in commercial aviation based on a succession of piston-engine aircraft (DC-2, DC-3, DC-4, DC-5, and DC-6), while Boeing focused primarily on the military market. Two years after Boeing's B-47 Stratojet long-range bomber flew in 1947, the airframer began to look at building a commercial jetliner. It rolled out a four-abreast 367-80 in 1954 and bet $16 million—about two-thirds of the company's total assets—on the project [4]. Its confidence would soon be rewarded when it won an order from the US Air Force for a military tanker version of the aircraft under the designation KC-135. Douglas lost the tanker

[1]"On This Day: 2 May," BBC News, http://news.bbc.co.uk/onthisday/hi/dates/stories/may/2/newsid_2480000/2480339.stm [retrieved 28 Nov. 2016].

competition in part because it was slow to embrace jet-powered aircraft, but in 1955 it decided to modify its losing tanker design by widening the fuselage and launched a civil jetliner—the DC-8. Boeing responded by widening the fuselage of the 367-80, which became known as the 707.

Pan American Airways was the "chosen instrument" of the US government for international aviation, and its legendary CEO Juan Trippe was interested in nonstop jet service across the North Atlantic. With the promise of the Comet fading, Trippe turned to domestic suppliers. In October 1954, Pan Am ordered 20 707s and 25 DC-8s on the same day. By 1958, both BOAC and Pan Am offered jet service across the North Atlantic—BOAC with an improved Comet 4 and Pan Am with the 707. The following year, Pan Am would usher in nonstop service with an extended-range 707-320.

The DC-8 trailed the 707, but entered service with United Airlines and Delta Air Lines in 1959. Adding to the competition was San Diego–based Convair, which delivered the Convair 880, then the world's fastest jetliner, in 1960.

THE 1960S: ACCELERATED INNOVATION

Soon, airlines around the world followed the example of Pan Am and BOAC, and replaced their piston-engine aircraft with jets at a rapid pace. By 1961, international airlines like Aeroflot (Soviet Union), Air France, Alitalia (Italy), Air India, Lufthansa (Germany), KLM (Netherlands), Sabena (Belgium), SAS (Scandinavia), Swissair, El Al (Israel), and JAL (Japan) connected the planet with long-haul, transoceanic routes.

Aircraft manufacturers responded by developing designs for new missions. One target was jetliners suited for shorter routes and smaller airports. Boeing seized the opportunity by engaging large airlines to understand their needs. United Airlines wanted a four-engine aircraft for its high-altitude hub in Denver. American Airlines wanted a more efficient two-engine model. Eastern Airlines wanted a third engine for its overwater flights to the Caribbean, because at that time twin-engine commercial flights were limited by regulations to routes with 60-min maximum flying time. Boeing responded with a breakthrough design—the 727 (see Fig. 2.2)—which featured three Pratt & Whitney JT8D turbofan jet engines. Two of the engines were mounted on the aft fuselage like the Caravelle, while the third engine was blended into the vertical stabilizer with a novel *S*-shaped inlet duct. It also boasted a unique *T*-shaped empennage—a first for Boeing. To enable shorter takeoff and landing distances Boeing introduced a triple-slotted flap into its wing. Another breakthrough was the addition of an auxiliary power unit (APU), a small gas turbine in the aircraft's tail that allowed electrical and air-conditioning systems to run independently of a ground-based power supply or power from one of the main engines. Initial sales of the 727 were brisk. Four major US carriers—American, Eastern, Trans World Airlines, and United—took delivery of 727s

Credit: Wikimedia

Fig. 2.2 The revolutionary Boeing 727.

within a four-month period in 1964. That year, Boeing delivered 95 727s, nearly half of the 206 jetliners entering service that year. It would go on to produce more than 800 727s by 1970. Former GE Aviation executive Fred Herzner call the 727 "the holy grail—the first single aisle aircraft developed for multi-use operations."[2]

Douglas Aircraft created its own aircraft for short to medium routes, the DC-9. It featured two aft-mounted Pratt & Whitney JT8D engines. Removing engines and pylons from the wings improved airflow at low speeds and enabled lower takeoff and approach speeds as well as reduced structural weight. It also facilitated a reduction in ground clearance, making the aircraft more accessible to baggage handlers and passengers. It would enter service with Delta in 1965, and like the 727 saw its sales explode. Douglas would deliver 69 in 1966 and by 1967 reached production parity with the 727 at 155 units. In 1968, it produced 202 DC-9s, a record for jetliner deliveries that would stand for another 23 years. It would later produce a Series 20 variant optimized for short field operations for Scandinavian Airlines System, which began operating the new aircraft in 1969.

Although sales of Douglas aircraft were brisk, the company lacked the planning, production infrastructure, and supply chain capability to manage the ramp-up. Cash penalties for late deliveries ensued. At the same time, Douglas utilized "program accounting," which allowed it to assume that a program would be profitable over the long term, and then take any cash loss

[2]Fred Herzner (former Chief Engineer – GE Aviation), interview with author, 7 Dec. 2016.

on early production and move it into a "deferred production cost" category in inventory. This had the effect of masking the losses on the income statement even though cash flow was weak. Facing the prospect of producing DC-8 and DC-9 aircraft worth billions at a loss, the company put itself up for sale in 1966 [5]. McDonnell Aircraft Corporation won the bid and took over Douglas Aircraft in April 1967. The new company would be called McDonnell Douglas.

Back in Seattle, Boeing was not resting on its 707 and 727 laurels; in 1965 it launched a new aircraft, the 737, optimized for short and thin routes. Lufthansa become the launch customer in 1965 with an order for 21 of the new aircraft. To minimize development costs, the 737 would share the nose section and the fuselage cross section of the 727. Unlike the 727, its two JT8D engines were wing-mounted. Optimized for 75–100 passengers, the 737's gross weight was about half that of the 727. United Airlines became another big customer when it placed orders for 40 of a larger version, the 737-200. The 737 would soon become a strong seller, with over 100 deliveries in 1968 and 1969.

Although sales of Boeing and Douglas jetliners surged, the fortunes of their competitors flagged (see Fig. 2.3). The Comet, the instigator of the jetliner revolution, re-entered service in 1958 with an improved design but would sell less than 100 before shuttering production in 1967. Convair would sell just 102 of its 880 and 990 (a stretched 880) aircraft before exiting the market in 1965 with massive losses. Several new European aircraft took to the skies in the mid-1960s, including the British Aircraft Corporation's BAC 1-11, the Vickers VC-10 (a modified military transport), and the Hawker Siddeley HS 121 Trident, but all were modest sellers. The most popular non-American jetliner was the Caravelle, which gained a global following as a reliable short- to medium-range aircraft. It would deliver 267 units by 1970.

Although it originated in the United Kingdom, the jetliner business was dominated by Boeing and Douglas as the 1960s ended. In 1968, a record 741 jetliners were delivered globally. Just 61 of these were from European manufacturers. The decade ended on an ominous note as jetliner deliveries plunged. By 1970, deliveries would be less than half of the 1968 peak. This would cause major issues for suppliers in subsequent years.

The 1970s: Bigger, Higher, and Faster

A convergence of events in the 1960s would lead to bold new aircraft designs that would stretch the boundaries of capacity and speed. Boeing lost a major US Air Force cargo aircraft contract to Lockheed Martin's C-5 in 1965, and believed there was a market opening for a jetliner that would hold more than 300 passengers. Pan Am's Juan Trippe was interested in such an aircraft and held discussions with Boeing to build it. In the same year,

OEM	Model	1954-58	1959	1960	1961	1962	1963	1964	1965	1966	1967	1968	1969	1970	TOTAL
de Havilland	Comet	41	18	20	14	13	2	2	1		1				**112**
Boeing	707	8	77	91	80	68	34	38	61	83	118	111	59	19	**847**
	727						6	95	111	135	155	160	114	55	**831**
	737										4	105	114	37	**260**
	747												4	92	**96**
Sud Aviation	Caravelle		18	39	39	35	23	22	18	18	20	15	11	9	**267**
Douglas	DC-8		21	91	42	22	19	20	31	32	41	102	85	33	**539**
	DC-9								5	69	153	202	122	51	**602**
British Aircraft	BAC 1-11								34	46	20	26	40	22	**188**
Others		0	0	14	33	33	16	29	24	18	11	20	21	14	**233**
	TOTAL	**49**	**134**	**255**	**208**	**171**	**100**	**206**	**285**	**401**	**523**	**741**	**570**	**332**	**3975**

Source: Teal Group
"Others" includes F-28; Trident; VC-10, & Convair880/990; excludes Soviet aircraft

Fig. 2.3 Jetliner deliveries: 1954–1970.

Boeing executive Joe Sutter was transferred from the 737 development team to manage the design study for a new airliner, already assigned the model number 747. At the time, it was widely thought that the 747 would eventually be superseded by supersonic transport aircraft, but its cargo capability would be very important. After studying alternatives, including a baseline double-decker design from the losing military transport bid, Boeing settled on a design in which the cockpit was placed on a shortened upper deck so that a freight-loading door could be included in the nose cone. This design feature yielded the 747's distinctive "bulge" (see Fig. 2.4).

In one of the jetliner industry's most important transactions ever, Pan Am ordered 25 747s in April 1966 for \$531 million. They each would carry 378–490 passengers in a 10-abreast configuration, and would be powered by four high-bypass Pratt & Whitney JT9D turbofans with 41,000 lb thrust each. The fuel efficiency of the new aeroengines had a profound impact on operating economics. Direct operating costs were estimated to be 63 cents per seat mile—30% less than the 707 [6]. There was no guarantee that these improvements would translate into reduced fares, because airline pricing at the time was regulated in the United States by the Civil Aeronautics Board and by its equivalent in other countries. In other words, fares for international flights were determined by intergovernmental negotiations. Within a year, Boeing would bring in major orders from Continental, Northwest, and United.

Huge investments would be required to bring the 747 to the market. One was a massive, 200-million-ft^3 assembly plant—the world's largest enclosed space—in Everett, Washington, at a cost of \$200 million. This put the company in debt right from the start. Some \$1.2 billion was invested in the 747 development program over a four-year period. At the time development

Credit: Eduard Marmet

Fig. 2.4 A gamechanger: the Boeing 747.

commenced in late 1965, the total shareholder's equity for Boeing was $372 million. Boeing literally bet the company on the 747, with its development costs equaling more than three times the company's value [7].

While Boeing was gaining traction with its giant new aircraft, American Airlines president Frank Kolk sent a specification to Boeing, McDonnell Douglas, and Lockheed for a new airplane, much smaller than the 747, that would better fit American's route network and development plans. It would be a wide-body, twin-aisle aircraft capable of carrying 250 passengers for 2100 miles and operating from New York's La Guardia Airport's runways. There was a catch: rather than a four-engine design like the 707 and 747, it would be powered by two huge turbofan engines [8]. What became known as the "Kolk machine" would ultimately transform the jetliner business. A flurry of activity would follow, including decisions by Lockheed and McDonnell Douglas to develop the L-1011 and DC-10 (Fig. 2.5), new wide-body airplanes that were smaller than the 747.

Lockheed viewed Kolk's request as an opportunity to get back into the jetliner business. It had been absent since the L-188 Electra, and was experiencing problems with its military transport projects. The venerable company that had developed the U-2, C-130, and SR-71 military aircraft would now attempt to gain a foothold in the jetliner business. Its solution

Credits: Los688 (DC-10), Eduard Marmet (L-1011), M Radzi Desa (A300)

Fig. 2.5 Twin-aisle contenders: DC-10, L-1011, and A-300.

was the L-1011, a 256-seat wide-body with three engines. The aircraft would feature sophisticated flight control systems, advanced avionics (including auto-land capability), and new Rolls-Royce RB211 turbofans. After major development delays with the engines (more on this in Chapter 5), Rolls-Royce went into receivership, and suddenly Lockheed and the L-1011 program were in trouble. Lockheed had already racked up $480 million in losses on four military projects, and with $900 million sunk into the program it could not afford another major cash drain [9]. Moreover, the British government refused to finance production of the troubled engines unless the US government assured them of Lockheed's survival. With some 60,000 jobs at risk, the US government, with President Nixon's backing, guaranteed $250 million in loans to Lockheed. The L-1011 eventually got back on track and entered service with Eastern Air Lines in 1972. The path was challenging, but Lockheed was in the jetliner game.

McDonnell Douglas also responded to the Kolk request for proposals (RFP) with the DC-10, a 285-seat, three-engine wide-body. This was the company's first jetliner since the merger of McDonnell and Douglas in 1967. Like Boeing with the 747, the company made a massive bet on the program—more than $1 billion at the time, which was three times the company's shareholder equity [10]. The aircraft offered a choice of engines—the General Electric CF-6 or the Rolls-Royce RB211. In February 1968, it secured the largest jetliner order in history with American Airlines: $800 million for 25 DC-10s [11]. This bested Pan Am's 1966 launch order for 747s. United would also follow with a huge order. The DC-10 entered service in August 1971 on an American Airlines flight from Los Angeles to Chicago. The opening battle between the DC-10 and L-1011 went to McDonnell Douglas. Lockheed would respond by winning major orders from Delta, TWA, and Eastern.

Where did this leave the Europeans? In the United Kingdom, distress set in following the government-sponsored Plowden Report, which in 1965 argued that its aircraft industry had a structural disadvantage compared to the United States, and that collaboration between European companies was its key to survival. Ironically, a few months before the report was issued, Britain's minister for aviation proposed a European "Air bus" project. Soon, an Anglo-French working party published a specification for a 200- to 225-seat, 800-nm-range aircraft. West Germany joined the initiative, and in 1967 united with France and the United Kingdom in supporting the concept, labeled the "A300." Participants included Sud Aviation and Aérospatiale from France, West Germany's Arbeitsgemeinschaft Airbus, and the United Kingdom's Hawker Siddeley. As former Airbus president Roger Béteille recalled, the Kolk machine had a major influence on the A300's twinjet concept and gave the Airbus concept credibility [12]. Despite this, the United Kingdom withdrew from the project in April 1969, effectively ceding control to France and West Germany. Airbus

Industrie was formally established in April 1970, with Aérospatiale (a merger of SEREB, Sud Aviation, and Nord Aviation) and the West German company Deutsche Airbus (a grouping of Messerschmittwerke, Hamburger Flugzeugbau, VFW GmbH, and Siebelwerke ATG) each owning a 50% share. The new company moved ahead with the A-300, and technical director Roger Béteille created a novel supply chain concept that emphasized division of labor. France would make the cockpit, flight controls, and part of the fuselage. West Germany would produce most of the fuselage. Hawker Siddeley—although no longer an owner—would remain part of the team as a "preferred subcontractor" for the aircraft's wing. Other contributors included the Netherlands (flaps and spoilers) and Spain (horizontal stabilizers).

The initial version of the A300 was for a 300+-seat aircraft powered by a new Rolls-Royce RB207 powerplant. With lukewarm interest in the aircraft's large capacity and Rolls-Royce consumed with the L-1011 program, Airbus decided to pursue a 250-seat version of the concept, the A300B. The twin-engine aircraft would use two CF6-50 powerplants made by General Electric, taking a risk on an aeroengine original equipment manufacturer (OEM) known primarily for its military aircraft prowess. Airbus believed that a high level of technology should be built into the A300 to give it the edge over competing aircraft. It was the first jetliner to make significant use of composites. The A300B's two-engine configuration and advanced materials gave it a significant edge versus its three-engine US competitors. It was the first twin-engine wide-body aircraft. It also boasted a 222-in.-wide fuselage, which enabled eight-abreast passenger seating and was also wide enough for two LD3 cargo containers to be loaded side-by-side. In November 1971, Airbus received its first order for six A300Bs from Air France. Lufthansa, Germanwings, and Indian Airlines followed with their own orders. In 1972, Airbus became a three-country venture when Spain's Construcciones Aeronauticas SA (CASA) took a 4.2% stake in the consortium and was awarded the contract to build the A300's horizontal stabilizer. Despite the progress, Airbus faced a significant obstacle in entering the market because the airline industry in the early 1970s was dominated by US airlines. In 1973, the world's top seven airlines, as measured by revenue passenger kilometers (RPKs), were in the United States; the airlines comprising this group were United, American, Eastern, TWA, Delta, Pan Am, and Northwest Orient (see Fig. 2.6). Broadening the prism, 10 of the top 15 global airlines were based in the United States, with Air Canada, Air France, BOAC, Lufthansa, and Japan Airlines as outliers [13]. The impact of "aeropolitics" was real: US airlines preferred equipment from US manufacturers, and Airbus had an edge with European airlines. Still, Airbus *did* break through, but with the world's 13th and 14th largest airlines. Its first production model, the A300B2, entered service in 1974 with Air France.

Rank	Airline	Country	RPKs (000)
1	United	US	648,955,000
2	American	US	475,565,000
3	Eastern	US	464,410,000
4	TWA	US	462,100,000
5	Delta	US	372,696,000
6	Pan Am	US	365,100,000
7	Northwest Orient	US	186,213,000
8	Air Canada	Canada	182,571,000
9	BOAC – International	UK	182,040,000
10	Air France	France	167,291,000
11	Lufthansa	W. Germany	162,173,000
12	Western	US	152,338,000
13	Braniff International	US	149,328,000
14	Allegheny	US	136,634,000
15	Japan Airlines	Japan	135,638,000

Source: International Civil Aviation Organization
RPK = Revenue passenger kilometers

Fig. 2.6 World's largest airlines in 1973.

While the battle for early twin-aisle supremacy raged, another competition unfolded for supersonic jetliners. At the time, they were viewed as the next logical step in the evolution of air transport, with projected demand for hundreds of aircraft. The competition was between the Concorde, a UK–France collaboration, and the Boeing 2207. The Concorde was a technological marvel that pioneered many new technologies including an analog fly-by-wire system where wires replaced mechanical linkages to flight control surfaces, and brakes made of carbon-carbon rather than metal. The first prototype flew in 1969. Meanwhile, Boeing's supersonic program, which would have had twice the capacity of the Concorde's 128 seats, was scrapped when it could not secure development support from the US Congress. The Soviet Union unveiled its own initiative, the Tupolev TU-144, which was the first supersonic transport to fly and to reach Mach 2.0. With orders from more than a dozen airlines and Boeing out of the competition, the future for the Concorde appeared bright until a disastrous series of events in the early 1970s, including a severe global recession, an oil shock, and the crash of the Tu-144 at the 1973 Paris Air Show. Noise restrictions related to the Concorde's sonic boom also severely curtailed its ability to serve lucrative business routes, including New York–London.

Ultimately, the Concorde would sell just 14 aircraft to Air France and British Airways, and would end production in 1980. The TU-144 would suffer an even worse fate: just 55 passenger flights were completed before safety concerns led to its premature retirement. The future, as it turned out, wasn't supersonic jetliners.

Competitive Dynamics of the 1970s and US Deregulation

The early 1970s were difficult for airlines and aircraft OEMs. Air travel demand slowed just as new aircraft models became available. Thus, jetliner deliveries fell from 332 in 1970 to 238 by 1972. The 727 and DC-9 were bestsellers among single-aisles, with the 747 and DC-10 leading twin-aisle sales. Jetliner demand picked up slightly in the mid-1970s before collapsing again. In 1977, jetliner sales were just 215 units. The weak sales reflected the sorry state of the global economy in the 1970s. It also was a byproduct of the heavy hand of air travel regulation by national governments. In most of the world, governments set the prices of air travel and determined the airlines and frequencies for international flights. Foreign ownership of airlines was forbidden in most countries. Limited competition and high airfares meant that air travel was limited to business travelers or wealthy leisure customers. In the United States, a government agency called the Civil Aeronautics Board (CAB) regulated most aspects of air travel. The agency had a reputation for being slow and bureaucratic. Requests by airlines to change fares or routes could take months and even years.

In 1977, US President Jimmy Carter appointed Alfred E. Kahn, a professor of economics at Cornell University, to chair the CAB. Kahn believed that flexible pricing would benefit both US consumers and the airline industry; in 1978, he partnered with the Carter Administration and the US Congress to pass the Airline Deregulation Act. Thus, the power to set airfares shifted from the CAB to airlines. Price competition ensued as existing carriers cut prices to attract new customers. It also reduced industry entry barriers, opening the door for new carriers to enter the market or expand. Southwest Airlines achieved lower costs than legacy airlines by flying an all-737 fleet and simplifying operations. It became known as a low-cost carrier (LCC). The concept was a hit with passengers, because its low airfares allowed a broader set of customers to fly. Led by charismatic CEO Herb Kelleher, Southwest soon began an expansion campaign beyond its Texas home base. Another LCC, People Express, entered the fray in 1981. Across the Atlantic, LCC pioneer Freddie Laker set up a no-frills 707 service between London and New York for £37.50 in the summer and £32.50 in the winter. Laker called his Laker Airways service a direct challenge to governments to liberalize air travel and reduce airfares, and for major airlines to become more creative [14].

Legacy US airlines were put on notice that lower costs would be important success factors. Deregulation would prove a seminal act in the US airline industry that would significantly expand air travel and jetliner demand. By 1979, jetliner deliveries would reach 411 units, with Boeing accounting for 70% of sales.

This left the remaining 30% for the other three aircraft manufacturers. McDonnell Douglas muddled through the late 1970s with DC-9 sales falling to just 22 units in 1977 and 1978 before jumping to 39 in 1979. Its DC-10 sales, which bottomed out at 14 in 1977, would more than double to 35 by 1979. And it launched an update to the DC-9, the MD-80, which first flew at the end of the decade. The aircraft would feature more-efficient JT8D-200 powerplants, a stretched fuselage, and significant aircraft component commonality with the DC-9. It would be certified as a version of the DC-9 to hold down development costs.

The 1970s were particularly brutal for Lockheed. After initial success in the mid-1970s, L-1011 deliveries would dwindle to just eight by 1978. This was hardly the necessary volume to sustain a jet transport business. Its fortunes needed to change—and fast.

Across the Atlantic, sales of the A300 were also disappointing. By mid-1976, the aircraft garnered just 33 orders and 23 options. Production rates were just one per month, and its financial results were predictably poor. The A300B's development costs reached 4.5 billion francs (approximately $900 million), which were mostly covered by reimbursable government loans. In addition, 2.5 billion francs ($500 million) covered the production learning curve of the first batch of aircraft [15]. It did not receive a single order between the end of 1975 and mid-1977. Desperate, it convinced US-based Eastern Airlines to accept an offer it couldn't refuse: it loaned it four of its "white tail" A300s (finished aircraft lacking a customer) for six months, free of charge. After a very successful trial, Eastern ordered 23 A300s in 1978. Like the earlier incident with Western Airlines, aeropolitics once again raised its head. Eastern CEO Frank Borman, a former Apollo astronaut, was accused of being un-American, even though his airline's weak financial condition meant that it probably couldn't have purchased new aircraft without Airbus's generous terms [16]. Ironically, another American astronaut would prove later to be one of the saviors of the A300.

Another A300 challenge was its size: some operators did not have enough route density to justify the relatively large A300. Swissair and Lufthansa pushed for a smaller version. This led to consideration of a 220-seat A300B10 option, which featured a shortened A300 fuselage, new wings, and new engines—the General Electric CF6-45B2 or the Pratt & Whitney JT9D-7R4. There was a major debate as to whether to pursue this aircraft or a 150-seat single-aisle aircraft to compete with the 737 and DC-9. In the end, the B10 option was selected with input from a summit between French President

Valéry Giscard d'Estaing and German Chancellor Helmut Schmidt. The new program, renamed the A310, was launched in July 1978. Initially, two variants were proposed with an anticipated development cost tab of $850 million [17]. The A310-100 would be a short-range version (2100 nm) tailored to the needs of Lufthansa. The second variant would have medium-range (2920 nm) capability. The short-range version was eventually dropped. Swissair was the launch customer, and orders followed from Air France, Lufthansa, and later Iberia.

Boeing ended the 1970s on a roll (Fig. 2.7). It booked a whopping 461 orders in 1978—more than twice those of its competitors combined. Its corporate profit reached 5.9%, its highest since 1941, and Lockheed and McDonnell Douglas appeared to be in trouble. At year end, *Dun's Review* was prescient in predicting that soon Boeing would be the only US company producing jetliners, probably with just one company as its competitor: Airbus Industrie [18].

Boeing leveraged this momentum to develop two new aircraft programs that would fill critical product portfolio voids: a twin-aisle smaller than the 747 to address the new competition, and a 727 replacement. The first would be a twinjet layout, like the A300, which would focus on mid-sized, high-density markets that were not appropriate for the larger 747. The new 220-seat aircraft, the Boeing 767 (see Fig. 2.8), would take advantage of advancements in civil aerospace technology, including reliable, high-bypass-ratio turbofans; advanced avionics; and new materials. The 767 would feature seven-abreast, 2-3-2 seating. Initially three variants were envisaged: a 767-100 with 190 seats, a 767-200 with 210 seats, and a three-engine 767MR/LR version with 200 seats intended for long-range routes.

The second aircraft, the 200-seat 757, evolved from 727 replacement studies completed in the mid-1970s. It would offer dramatic fuel efficiency improvement from two high-bypass-ratio turbofans—particularly important since the 1973 oil shock—and new wings, systems, and materials. Its high thrust-to-weight ratio would allow it to service "hot and high" airports like Denver, Salt Lake City and La Paz, Bolivia. Boeing Commercial Airplanes president E. H. "Tex" Boullioun made British Airways a prime target for the aircraft and stated his preference for the Rolls-Royce RB211 engines if it became a launch customer. He also offered British Aerospace a 20% share of the program. British Aerospace was a nationalized aerospace champion created by the 1977 merger of British Aircraft, Hawker Siddeley Aviation, Hawker Siddeley Dynamics, and Scottish Aviation. His efforts paid off at the 1978 Farnborough Air Show, when British Airways became the 757's launch customer with an order for 19 aircraft. Eastern Airlines quickly followed, buying 21. Both airlines selected 37,000-lb-thrust Rolls-Royce RB211-535C powerplants, marking the first time that a Boeing airliner was launched with engines produced outside the United States. The other option

Type	OEM	Model	1970	1971	1972	1973	1974	1975	1976	1977	1978	1979	TOTAL
Single-Aisle	Boeing	707	19	10	4	11	21	7	9	8	13	6	**108**
		727	55	33	41	92	91	91	61	67	118	136	**785**
		737	37	29	22	23	55	51	41	25	40	77	**400**
	McDonnell Douglas	DC-8	33	13	4								**50**
		DC-9	51	46	32	29	48	42	50	22	22	39	**381**
	Other Models*		45	39	36	31	23	46	46	30	18	13	**327**
Twin-Aisle	Airbus	A300					4	9	13	16	15	24	**81**
	Boeing	747	92	69	30	30	22	21	27	20	32	67	**410**
	Lockheed	L-1011			17	39	41	25	16	11	8	14	**171**
	McDonnell Douglas	DC-10		13	52	57	47	43	19	14	18	35	**298**
Supersonic	Aérospatiale/BAC	Concorde						1	7	2			**10**
		TOTAL	**332**	**252**	**238**	**312**	**352**	**336**	**289**	**215**	**284**	**411**	**3021**

Source: Teal Group
*Other Models includes F-28; Trident; VC-10; Caravelle; Mercure; BAC 1-11; excludes Soviet aircraft

Fig. 2.7 Jetliner deliveries in the 1970s.

Credits: Wikimedia (767) Makaristos (757)

Fig. 2.8 Boeing's one-two punch: the 767 and 757.

would be the Pratt & Whitney PW-2037, which Delta Air Lines selected in an order for 60 757s in 1980. Sales forecasts for the 757 varied between 500 and 1000 [19].

These two projects, Boeing's first in 15 years, would cost $2 billion to develop [20]. To partially offset these costs, Boeing would enter risk-sharing partnerships for 30% of the 767 with European and Asian suppliers. It also eventually dropped the short- and long-range versions of the 767, the former to reduce overlap with the 757.

Boeing's bold moves left an unanswered question in the United Kingdom: Should British Aerospace join Boeing on the 757 or become part of Airbus? The French were incensed that British Airways (BA) launched the 757 program. The British government, after all, owned both BA and British Aerospace. Why couldn't it direct BA to buy European? Fortunately, British airline entrepreneur Freddie Laker saved the day when he ordered 10 A300B4s for his low-cost Skytrain airline. This removed an important obstacle, and the British government put up a reported £50 million in a repayable loan towards the A310's development costs. In January 1979, British Aerospace took a 20% stake in Airbus Industrie. France and Germany's shares fell to 37.9% each, with the rest held by CASA of Spain. The following year, Airbus announced plans to build an improved version of the A300B4, to be called the A300-600 series. It had a longer fuselage than the earlier A300s, and increased capacity to 266 passengers. It also featured a forward-facing crew cockpit. Airbus ended the decade with a solid four-nation consortium, a budding family of twin-aisle aircraft, and a growing order book.

THE 1980S: RISE OF AIRBUS

It was clear from the outset that the 1980s would be a break from the past—beginning with economic policy. Margaret Thatcher became the United Kingdom's prime minister in 1979 and immediately focused on deregulation and privatization. British Petroleum (1979) and British Aerospace (1981) were two of the first major public companies returned to the private sector.

Thatcher's ideological soulmate, Ronald Reagan, assumed the US presidency in 1981. Early in his term, he fired 11,000 striking air traffic controllers who ignored his order to return to work. In carrying out his threat, Reagan also imposed a lifetime ban on rehiring the strikers. The world was changing—two economic liberals and defense hawks were in charge—but at the same time, the global economy was headed for a sharp recession. A slowdown in orders meant that the market for large jetliners was too small for four competitors. In December 1981, Lockheed announced that it was stopping production of the L-1011 after $2.5 billion in losses to focus on the defense business. It planned to phase out production in the mid-1980s. In the same year, Airbus Industrie passed both McDonnell Douglas and Lockheed to become the second largest seller of jetliners [21].

With Lockheed withdrawing from the jetliner market and the 757, 767, and MD-80 preparing to enter service, the strategic focus in Seattle and Europe shifted to single-aisle aircraft. Boeing wanted to increase the capacity and range of its 737 while maintaining commonality with previous variants. Development studies began in 1979, and by 1980 it released preliminary specifications at the 1980 Farnborough Air Show. The new aircraft, the 737-300, would have a 104-in. stretch of the fuselage, updated aircraft systems, and fuel-efficient, high-bypass-ratio aeroengines. The low clearance of the 737's fuselage (derived from the 707) caused challenges, including how to put the larger bypass ratio engines under the 737's wings. The problem was solved by placing the engines ahead of (rather than below) the wing, and by moving engine accessories to the sides (rather than the bottom) of the engine pod, giving the 737 a distinctive, noncircular air intake with a flat bottom. Boeing would revisit this design challenge in the decades ahead. The aeroengine supplier was a surprise. Rather than sticking with incumbent Pratt & Whitney for upgraded JT8D engines, Boeing selected the 20,000-lb thrust CFM56-3 engine from CFM International, a joint venture between General Electric and France's Snecma. Two factors helped to sway Boeing to shift to the new competitor. The first was Boeing's bet that new, stringent noise standards developed by the International Civil Aviation Organization (ICAO) would eventually take hold; the second was the need for greater fuel efficiency. The higher bypass ratio of the CFM56 was vastly superior to the JT8D. This would ultimately lead to a sea change in the aeroengine segment, as discussed in Chapter 5. In March 1981, after Southwest Airlines and USAir ordered 10 aircraft each, with options for 20 more, Boeing formally launched the aircraft.

While Boeing began development of its updated single-aisle, Douglas delivered its first MD-80, a JT8D-209–powered MD-81, to launch customer Swissair in September 1980. Douglas also delivered its first MD-82, a more powerful variant with JT8D-217s, to Republic Airways in 1981. It would deliver five MD-80s in 1980, and ramp up to 62 in 1981 and 33 in 1982.

Meanwhile in Europe, Airbus debated whether to expand its product range by going larger (favored by the Germans) or creating a single-aisle aircraft to challenge the 737-300 and MD-80 (favored by the French). Eventually, Airbus coalesced around creating a new single-aisle. Several options were under consideration in what was called the Joint European Transport (JET), with capacity of 131, 162, and 188 seats. In February 1981, the program was designated the A320. In refining its design, Airbus worked with Delta Air Lines on a 150-seat aircraft required by the airline. At the 1981 Paris Air Show, Air France committed to 25 A320s with an option for a further 25.

As new single-aisle concepts emerged, Boeing's new 767 gained momentum. It rolled out at Boeing's Everett facility in August 1981, and the first delivery to United Airlines took place a year later. Boeing's dual product initiative continued when the 757 rolled out from its Renton facility in early 1982 in a Hollywood-style debut. The 176-passenger 757 was priced at $32–$35 million, and at the time had 136 firm orders from seven airlines. The 216-seat 767 had 173 orders from 17 airlines [22]. Eastern Air Lines and British Airways put the 757 in commercial service in 1983; it gained popularity for short- and mid-range domestic US routes, shuttle services, transcontinental US flights, and European charter operations.

For its part, Airbus also introduced the A310-200 in 1983, 11 years after its first A300B. The aircraft entered service with Swissair the following year. With the A310 weighing 21,000 kg less and carrying 80 fewer passengers than the A300, Airbus marketed it as an "introduction to widebody operations" for developing airlines. Airbus also announced a longer-range version of the aircraft, the A310-300.

Despite the optimism surrounding the new aircraft programs, storm clouds gathered in the global economy as oil prices doubled in the late 1970s and early 1980s. The price of crude oil, less than $15 a barrel in 1978, approached $40 per barrel in 1982. At the same time, the US Federal Reserve increased interest rates dramatically to combat rampant inflation; the US Federal Funds rate approached 20% in 1981. Predictably, global gross domestic product (GDP) growth fell from 2.0% in 1980 to 0.5% in 1982, and major economies lapsed into recession. In the United States, the situation was complicated by the aftereffects of its 1978 deregulation and competition from low-cost carriers, which challenged incumbent carriers used to a regulated "cost-plus" environment. This meant that US airlines could no longer buy as many jetliners as desired, and then pass these costs along as higher ticket prices. Aircraft deliveries also plummeted from 446 in 1980 to 304 in 1982. The 747 was particularly affected, with deliveries falling from 73 to 26 over the same two-year span. McDonnell Douglas's deliveries were halved. And in 1983, Airbus logged zero net orders [23]. Just 335 jetliners were delivered in 1983.

Despite the economic malaise, Dutch OEM Fokker threw its hat into the ring of the jetliner market in 1983. Fokker had produced the F28, a 65- to 79-seat regional jet, since 1969, but the aircraft became economically obsolete. Fokker decided to stretch the fuselage significantly, redesign the wing, and nearly double the capacity to 107 passengers. It retained the rear fuselage–mounted engines and T-tail configuration, like the DC-9. Subsequently, marketed as the Fokker 100, it would feature modern avionics and new-generation Rolls-Royce Tay turbofans. It was a big gamble by a manufacturer known mostly for regional aircraft.

In 1984, the economic malaise passed. The Thatcher–Reagan deregulation made an impact, and global GDP growth hit a robust 4.5%. Optimism returned, and the jetliner business was poised for an upcycle after a nasty four-year period. Boeing rolled out its third new aircraft in three years in January 1984 when it unveiled the 737-300. The new variant, which seated 126 passengers, would enter service with USAir in November that year. This may have been a catalyst for Airbus to finally get moving on the A320. Not much had happened since Air France's 1981 orders, mostly due to squabbling between the partners and their governments. The French were strongly in favor of the new aircraft, while the Germans were circumspect. British Aerospace needed financial assistance from the British government to participate, but the government was ideologically opposed to state aid. The logjam was broken when Margaret Thatcher understood that the A320 would happen with or without the United Kingdom, and agreed to loan British Aerospace £200 million in February 1984. Finally, the A320 development program gathered momentum.

Airbus made several crucial decisions in differentiating its aircraft from Boeing. The first was the fuselage diameter of 13 ft—7.5 in. wider than the 737 and 727. Although heavier, it provided greater passenger comfort in six-abreast seating. Its 111.2-ft (33.9-m) wing had better aerodynamic efficiency than the 737 and MD-80 because of the higher aspect ratio. It integrated composites into the aircraft's nacelle—the most ambitious use of this emerging material of any jetliner. Unlike the 737-300, it would eventually offer an engine choice: the CFM56-5 (a variant of the 737's engine) or the V2500 from a new consortium called International Aero Engines. Perhaps the most important design decision was to incorporate a fly-by-wire flight control system into the aircraft. In this approach, the movements of flight controls are converted to electronic signals, processed by flight control computers, and transmitted by wires to determine how to move the actuators at each control surface. Under the leadership of Bernard Ziegler, Airbus also incorporated full flight envelope protection into its flight-control software, which prevented pilots from stalling the aircraft. Another breakthrough was that pilots controlled the aircraft by sidesticks, eliminating the traditional yokes in the cockpit. A full discussion of the A320's avionics is in Chapter 9.

There was more good news for Airbus in 1984. In September, Pan American World Airways agreed to a deal worth more than $1 billion to acquire or lease 28 A300B4s, A310s, and A320s between 1987 and 1990. This was a breakthrough for Airbus, because Pan Am historically had been one of Boeing's best customers, having acquired 208 jetliners from Boeing in the prior 25 years, including the introduction of the 747 in the early 1970s. The strategic implications of this deal were also significant. In the mid-1980s, US airlines accounted for 50% of jetliner demand, and Airbus's only significant order in the region was the sale of 34 A300s to Eastern Airlines [24].

The year 1985 brought a change in leadership when Jean Pierson, previously director of Aerospatiale's Aircraft Division, succeeded Bernard Lathiere as president of Airbus Industrie (see Fig. 2.9). He brought a more aggressive style and said that Airbus Industrie was targeting 30% of the world market for commercial aircraft. He also made it clear that the next big project would be a long-range aircraft [25]. In the same year, an obscure US marketing executive from Piper Aircraft joined Airbus North America. His name was John Leahy.

While Airbus worked on its new single-aisle, McDonnell Douglas continued to tweak its MD-80 family. It rolled out another variant, the MD-83, to launch customer Alaska Airlines in 1985. That year it would deliver a healthy 71 aircraft, and announced it would produce a shorter fuselage MD-87 variant, which would seat 109–130 passengers. Its momentum was palpable. American Airlines operated 44 MD-80s with options for 100; other important customers were Alitalia, Pacific Southwest Airlines, Scandinavian Airlines System, and TWA. It had 556 orders and options in hand—a six-year backlog—as it

Credit: Airbus

Fig. 2.9 Airbus CEO Jean Pierson.

competed head-to-head with the 737-300 [26]. The California-based OEM also made a major move in an important new market: China. In 1984, it signed a letter of with the Shanghai Aircraft Industrial Corporation for coproduction of 25 MD-80s in China. This agreement was a result of McDonnell's focus on China, then a closed economy, which included a DC-9 demonstration tour in 1979 and aerostructures subcontract work. In 1986, it launched its last variant of the MD-80, the MD-88, which was basically the MD-82/83 with an updated cockpit. Delta would be the launch customer.

The mid-1980s brought an important update to aircraft operations for long-haul flights—particularly for the new large twin-engined twin-aisle aircraft. Beginning in 1936, a pilot or operator had to prove that there were suitable landing fields at least every 100 miles along their route. With the onset of the jet age, the US Federal Aviation Administration (FAA) updated the regulation to mandate that operators had to ensure a landing area within 60 minutes of their route. In 1964, the rule was waived for three-engine aircraft, paving the way for the L-1011 and DC-10. In contrast, ICAO had a more liberal 90-minute rule for twin-engine aircraft, and countries outside the United States often followed the ICAO standards. Airbus A300, for example, had been operated across the North Atlantic, the Bay of Bengal, and the Indian Ocean under a 90-minute ICAO rule since 1976.

Aeroengine reliability advanced to the point where the FAA and ICAO concluded that a properly designed two-engine aircraft was reliable enough for long-haul transoceanic flights. In 1985, the FAA was first to approve Extended-range Twin-Engine Operational Performance Standards (ETOPS) guidelines spelling out conditions for allowing overwater flights with a 120-minute diversion period. This was sufficient for most transatlantic flights. The first ETOPS rating was awarded to Trans World Airlines in May 1985 for the 767 service between St. Louis and Frankfurt. The impact went beyond twin-aisles. Aloha Airlines, for example, could now operate 737s between Hawaii and the mainland United States—more than 2000 miles. In 1988, the FAA amended the ETOPS regulation to allow extension to a 180-minute diversion period subject to stringent qualifications, making most of the Earth's surface available to ETOPS flights. ETOPS created the potential for airlines to downsize aircraft for long-haul routes; no longer were three- and four-engine jetliners a necessity. It would also set the stage for future large twin-engine designs.

TWIN-AISLE REFRESH

By the mid-1980s, Boeing's venerable 747 had achieved 600 deliveries but was losing momentum. It delivered just 40 747s in 1984 and 1985 combined, and the order book was weak. Orders fell off considerably, and the production horizon was bleak. A major update leveraging efficient aeroengines, updated systems, and improved economies would be needed. In October 1985, the

747-400 was officially launched when Northwest Airlines ordered 10 of the new model. It planned to deploy it on its growing trans-Pacific network and take advantage of the aircraft's 8000-nm range, 60-ton payload, and 22% lower fuel consumption per seat mile versus Boeing's legacy 747s [27]. Long-haul flag carriers Singapore Airlines, Cathay Pacific, KLM, Lufthansa, and British Airways announced orders several months later; they were followed by United Airlines, Japan Airlines, and Air France. Several of these customers worked on an advisory council to advise Boeing on the design. This led to upgrades like cathode ray tube (CRT) displays, carbon brakes, and three engine choices: the Pratt & Whitney PW4056, the GE CF6-80C2, and the Rolls-Royce RB211-524G/H.

Meanwhile, Airbus decided to get serious about updating its twin-aisle product range. In January 1986, it received supervisory board approval to offer two new twin-aisles—the two-engine A330 and the four-engine A340—for simultaneous development. The goal was to produce aircraft for two different market segments—the former for high-capacity routes with the latter for long-haul operations—but with a high degree of aircraft systems and avionics commonality. These designs had originated from studies on variations of the A300 family in the 1970s, the TA9 and TA11, with TA standing for "twin-aisle." Development costs were reduced by the two aircraft programs using the same fuselage and wing, with projected savings of $500 million. One factor driving the split was customer preference; twinjets were favored in North America and quad-jets were desired in Asia; European operators had mixed views [28].

The 262-seat A330 (two classes) had a range of about 2000 nm less than the 7450-nm range of the 300-seat A340. Both would share a similar fuselage, tail, wing sections, fly-by-wire controls, and cockpit. Pilot training and certification were also common for both types. The primary difference would be in the engines. This would allow Airbus to certify the aircraft nearly simultaneously. The A330 would offer a choice of engines, while engines for the A340 were to be determined. After 14 months, Airbus managed to secure 104 orders from nine airlines including Air France and Lufthansa. Fourteen months later, Airbus announced the formal launch of the A330 and A340 programs at the 1987 Paris Air Show.

To differentiate the A340, Airbus took the unconventional step of offering a high-bypass "geared turbofan" architecture derived from the IAE V2500 as the primary engine option in January 1987. Although several customers signed preliminary contracts for this variant, the International Aero Engines board decided in April 1987 to stop the development of the new SuperFan engine, which was deemed a "paper engine" that was too risky. This forced Airbus to partly redesign the A340. It also settled on the CFM56-5C, a variant of its A320 engine, as the single engine choice on the aircraft.

The A330/340 launch essentially filled out Airbus's product range, allowing it to cover 150-seat single-aisles up to 300-seat twin-aisles. These two aircraft would be the "last large investment" requested from Airbus to achieve a strategy of offering a "complete family of aircraft," said Jean Pierson at the time [29].

While Airbus made its twin bets, McDonnell Douglas examined its DC-10 options. It started to search for a DC-10 derivative as early as 1976. Several updates were considered in subsequent years, and the aircraft gradually lost competitiveness. It averaged 11 deliveries per year in the 1983–85 timeframe, and no orders were received in 1984. It was kept alive by a US Air Force tanker contract for 60 KC-10s. After considering a white sheet aircraft, President John McDonnell made the decision to instead offer the MD-11, an enlarged version of the DC-10 with new powerplants. "Our investment is only a quarter of what it would be on an all-new airplane," he boasted at the time [30]. In December 1986, McDonnell Douglas launched the MD-11 with commitments for 52 firm orders and 40 options from 10 airlines (Alitalia, British Caledonian, Dragonair, FedEx Express, Finnair, Korean Air, Scandinavian Airlines System, Swissair, Thai Airways International, and VARIG) and two leasing companies (Guinness Peat Aviation and Mitsui). This decision came after exploring with Airbus the possibility of merging the MD-11 and A340 while producing the A330 in Europe. Airbus wasn't interested because it had its own new jetliners to worry about. To build these new aircraft, Airbus partner Aerospatiale constructed Europe's largest factory to house the A330/340 final assembly line.

Late-1980s Growth

The robust economic growth continued into the late 1980s and drove a strong upcycle in jetliner deliveries, which rebounded to 409 aircraft in 1986—a level not seen since before the recession (see Fig. 2.10). The momentum would continue as deliveries broke through 500 for the first time in 1988.

After booking 252 orders for the 737-300 in 1985, Boeing launched the 737-400 in 1986 on the strength of orders from Guinness Peat Aviation (GPA), a lessor, and Piedmont Airlines. It was a 737-300 stretch with more powerful CFM engines and capacity for 146 passengers. The following year, Boeing announced a smaller variant, the 737-500, with capacity of just 108. US low-cost carrier Southwest Airlines would be the first customer. An order boom ensued.

In October 1986, Airbus bagged a historic order when Northwest Airlines bought 100 A320s in a deal valued at $3.2 billion. This was a blow to Boeing, because Northwest had been a loyal customer and its747-400 launch customer. Airbus now had three US customers for its new single-aisle; the others were Pan Am and lessor GATX. By the end of 1986, Airbus had nearly 400 orders for the A320. It was also a great year for Boeing, which announced that it achieved record sales with nearly $19 billion in orders for 336 aircraft—smashing the previous record of $14.9 billion set the previous year. Orders included 82 747s—including 49 of the new 747-400—and 212 737s [31].

Type	OEM	Model	1980	1981	1982	1983	1984	1985	1986	1987	1988	1989	TOTAL
Single-Aisle	Airbus	A320									16	58	**74**
	Boeing	707	3	2	8	8	8	3	4	9	0	5	**50**
		727	131	94	26	11	8						**270**
		737-1/200	92	108	95	82	60	32	21	24	7		**521**
		737-3/4/500					7	83	120	137	158	146	**651**
		757			2	25	18	36	35	40	48	51	**255**
	McDonnell Douglas	DC-9	18	16	10								**44**
		MD-80	5	61	34	51	44	71	85	94	120	117	**682**
	Other Models*		16	14	12	27	29	30	36	28	36	66	**294**
Twin-Aisle	Airbus	A300	39	38	46	19	19	16	10	10	17	24	**238**
		A310				17	26	26	19	21	28	24	**161**
	Boeing	747-1/2/300	73	53	26	22	16	24	35	23	24	4	**300**
		747-400										41	**41**
		767			20	55	29	25	27	37	53	37	**283**
	Lockheed	L-1011	24	28	14	6	4	2					**78**
	McDonnell Douglas	DC-10	41	25	11	12	10	11	17	10	10	1	**148**
Supersonic	Aérospatiale/BAC	Concorde	4										**4**
		TOTAL	**446**	**439**	**304**	**335**	**278**	**359**	**409**	**433**	**517**	**574**	**4094**

Source: Teal Group
*Other Models includes F-28; Trident; VC-10; Caravelle; Mercure; BAC 1-11; excludes Soviet aircraft

Fig. 2.10 1980s jetliner deliveries.

In the Netherlands, the Fokker 100 wrapped up its development program and was certified in November 1987. Its leadership was confident that the 100-seat market would be among the fastest growing segments. They reasoned that deregulation and increased emphasis on full-fare business travel would force airlines to offer more frequent service with smaller aircraft [32]. Swissair took the first F100 deliveries in early 1988 after financial and production difficulties delayed the schedule by nearly one year. Other early customers included KLM and USAir. In 1989, American Airlines placed the largest order to date for 75 F100s. Fokker's thesis on growing 100-seat demand was looking solid.

Serious incidents tarnished the reputations of the 737 and the A320 in 1988. In May, an Aloha Airlines 737-200 suffered the loss of a section of its upper fuselage at 24,000 ft on a routine interisland flight in Hawaii. This called into question the maintenance procedures of single-aisles used in high-cycle environments. Airbus suffered a worse fate when an A320 carrying passengers crashed in front of a crowd of several thousand while flying at ground level at the Habsheim Air Show. This was also the very first public demonstration of any civilian fly-by-wire aircraft. Three people died in the accident, and the official investigation concluded it was mostly crew error for flying too low. Although suspicions about the fly-by-wire technology persisted in some quarters, the A320 successfully entered service with Air France (see Fig. 2.11). In November 1989, Airbus launched the A321, a 22-ft, 9-in (6.94-m) stretch of the A320. Its industrial footprint morphed with the new program: For the first time, Germany would be the site of final assembly. The A321 would be assembled in Hamburg while the other Airbus aircraft, including the A320, would remain in Toulouse. This

Credit: Adrian Pingstone

Fig. 2.11 The Airbus A320.

was an important and overdue update to the partnership. In making this decision, Airbus turned down a surprising offer from Lockheed to assemble the A321 in the state of Georgia.

A major accident befell a DC-10 the following year when United Airlines Flight 232 crashed in Sioux Falls, Iowa, killing 111 out of 296 passengers and crew. The cause was failure to detect a fatigue crack in the stage 1 fan disk of its GE CF6 engine. This would lead to wholesale changes in manufacturing technology to engine disks and will be discussed in Chapter 10. Although the fault was with the engine manufacturer, the crash didn't help the reputation of Douglas or the aircraft, because the FAA had previously grounded the DC-10 several times.

UNDUCTED FANS?

While delivering record numbers of single-aisle aircraft, OEMs also evaluated projects featuring an unducted fan (UDF), which was a modified turbofan engine with the fan placed outside the engine nacelle on the same axis as the compressor blades. By integrating large, slow-moving fan blades outside the engine, UDFs offered a step-function change in fuel efficiency but were unproven. High fuel prices in the 1970s led to more OEM investment, and two UDF concepts emerged in the 1980s. The first was General Electric's GE36, a UDF with a push-prop configuration. The other was the Pratt/Allison 578-DX.

The GE36 was flight tested on a Boeing 727-100 in 1986 with promising results. Boeing then proposed a new short- to medium-range 7J7 featuring the UDF. Touted as the successor to the 727, it would have carried 150 passengers and would enter service in 2002. The 7J7 was also unprecedented in its foreign content, with Japan having 25% industrial workshare. Potential customers, however, were concerned about the economics and noise of the unproven UDF engines. Boeing quietly cancelled the 7J7 in 1987.

McDonnell Douglas was intrigued by the UDF concept and the potential to re-engine the MD-80; several flight tests were conducted with a GE36. Douglas planned on two new variants featuring a UDF: a 114-passenger MD-91X and a 155-passenger MD-92X with an estimated $300- to $500-million development cost. Delta Air Lines, American Airlines, GPA, and several European airlines expressed interest, but also acknowledged that fuel prices were too low to make the new aircraft viable [33]. Money was also an issue, because McDonnell Douglas was funding the MD-11 development program at the time. It was delivering more than 100 MD-80s per year in the late 1980s; why make the MD-80 obsolete? These factors, coupled with management's conservatism, led to the shelving of the UDF initiative in 1989. Instead, it focused on a new variant, the MD-89, which would feature fuel-efficient V2500 engines. It would later become the MD-90 when Delta Air Lines place an order for 50 in November 1989.

EUROPEAN DREAM

The decade ended with the industry's largest order ever from Irish lessor GPA worth nearly $17 billion at list prices. Boeing's share was 172 aircraft valued at $9.4 billion; it was also a milestone for Airbus ($4.3 billion) and McDonnell Douglas ($3.1 billion). This was a harbinger for the growing influence of aircraft lessors as a major customer group, because they now had 20% of aircraft orders. GPA was banking on airlines to retire their older jets due to high maintenance costs and an impending noise mandate. Critics charged lessors with "hoarding" delivery positions and predicted mass cancellations in the event of an economic downturn [34].

This transaction brought Airbus's 1989 total to 421 orders, worth $34 billion—a spectacular year by any standard. After 20 years, Airbus created a full family of aircraft, helped push the mighty Lockheed out of the jetliner market, and introduced technology innovations including fly-by-wire, two-engine twin-aisle aircraft, use of composites in primary structures, and the two-person cockpit.

Airbus wasn't going away. After 35 years of Europe being fragmented and marginalized, it had achieved the European dream of a unified and powerful jetliner manufacturer.

REFERENCES

[1] Sparaco, P., *Airbus: The True Story*, Editions Privat, Toulouse, France, 2006, p. 7.
[2] Ibid., p. 20.
[3] Ibid., p. 27.
[4] Ibid., p. 30.
[5] "Douglas Invites Merger Bids," *Aviation Week & Space Technology*, 19 Dec. 1966, p. 24.
[6] Gregory, W., "Pan American Order for 747 Opens New Era in Airline Jet Transport Equipment," *Aviation Week & Space Technology*, 18 April 1965, pp. 38–39.
[7] Spitz, W., Golaswekski, R., Berardino, F., and Johnson, J., "Development Cycle Time Simulation for Civil Aircraft," NASA/CR-2002-201658, 2001, pp. 2–7.
[8] Newhouse, J., *The Sporty Game*, Alfred A. Knopf, New York, 1982, pp. 122–123.
[9] "The Lockheed Bailout Battle," *Time*, 9 Aug. 1971, http://content.time.com/time/magazine/article/0,9171,903076,00.html [retrieved 28 Dec. 2016].
[10] Spitz et al., pp. 2–7.
[11] "American Orders 25 'Airbus' Aircraft," *St. Petersburg Times*, 20 Feb. 1968, p. 12-A.
[12] Kingsley-Jones, M., "Airbus at Thirty: Family Planning," *Flight International*, 2 Jan. 2001, https://www.flightglobal.com/news/articles/airbus-at-thirty-family-planning-124235, [retrieved 17 Dec. 2016].
[13] Greenslet, E., *The Airline Monitor*, June 2016, p. 4.
[14] "Laker's Low Fare Challenge," *Flight International*, 8 July 1971, p. 42.
[15] Sparaco, p. 134.
[16] Kemp, K., *Flight of the Titans*, Virgin Books Ltd, London, 2006, p. 98.
[17] Ropelewski, R., "Airbus to Develop Two A310 Versions," *Aviation Week & Space Technology*, 4 Sept. 1978, pp. 109–112.
[18] Bauer, E. E., *Boeing: The First Century & Beyond,* TABA, Issaquah, WA, 2006, pp. 237–238.
[19] Sparaco, p. 151.

[20] Kemp, p. 124.
[21] Greenwald, J., "Catch a Falling Tristar," *Time*, 21 Dec. 1981, http://content.time.com/time/magazine/article/0,9171,925159,00.html [retrieved 28 Dec. 2016].
[22] O'Lone, R., "Boeing Rolls Out Its 757 Twinjet," *Aviation Week & Space Technology*, 18 Jan. 1982, p. 24.
[23] Sparaco, p. 182.
[24] Daniels, L., "Lift for Airbus from Pan Am," *New York Times*, 14 Sept. 1984, https://www.nytimes.com/1984/09/14/business/lift-for-airbus-from-pan-am.html [retrieved 22 April 2017].
[25] Sparaco, p. 203.
[26] Shifrin, C., "McDonnell Douglas Increases MD-80 Production," *Aviation Week & Space Technology*, 19 Aug. 1985, p. 29.
[27] Shifrin, C., "Boeing Launches Long Haul 747-400 with Northwest Order," *Aviation Week & Space Technology*, 28 Oct. 1985, p. 33.
[28] Norris, G., and Wagner, M., *Airbus A340 and A330*, MBI, St. Paul, MN, 2001, pp. 22–23.
[29] Lenorovitz, J., "Airbus Industrie Launching Production for New A330/A340 Simultaneously," *Aviation Week & Space Technology*, 24 Feb. 1986, p. 45.
[30] Greenhouse, S., "Dicey Days at McDonnell Douglas," *New York Times*, 22 Feb. 1987, https://www.nytimes.com/1987/02/22/business/dicey-days-at-mcdonnell-douglas.html. [retrieved 28 Dec. 2016].
[31] "Boeing Records Biggest Year Ever," *Aviation Week & Space Technology*, 22 Dec. 1986, p. 35.
[32] "Europeans Introduce New Aircraft to Strengthen Market Share," *Aviation Week & Space Technology*, 10 March 1986, p. 230.
[33] Dornheim, M., "Douglas Advances MD-91X Target Launch Date by Half-Year," *Aviation Week & Space Technology*, 25 May 1987, p. 33.
[34] Weiner, E., "GPA Group Gives Boeing a Record Order for $9.4 Billion," *New York Times*, 19 April 1989 https://www.nytimes.com/1989/04/19/business/gpa-group-gives-boeing-a-record-9.4-billion-order.html [retrieved 28 April 2017].

Chapter 3

MOONSHOTS

1990: AN EVENTFUL YEAR

The good times of the late 1980s quickly ended in the new decade. The United States found itself staring at the beginnings of a recession, and in August 1990, Iraq invaded Kuwait and began the Gulf War. Much like the Iran–Iraq war, which contributed to soaring oil prices a decade earlier, this military engagement severely cut oil supplies, sending the price of a barrel soaring in just a few months. Jet fuel, priced at $0.54 per gallon in July 1990, saw its price more than double to $1.20 per gallon by October.

The global jetliner fleet entering the 1990s was just over 9100 aircraft; more than 80% of these aircraft were made by Boeing or McDonnell Douglas. The top five airlines by fleet size were all based in the United States: American Airlines (490), United Airlines (427), Delta Air Lines (389), Northwest (323), and Continental (323). Three of the five next largest were also American. The United States was ascendant in global aviation (Fig. 3.1).

Despite the deteriorating environment, aircraft original equipment manufacturers (OEMs) racked up more than 850 orders in 1990—a strong year by any standard—and delivered more than 600. Airbus accounted for 95, including 58 of its new A320. McDonnell Douglas delivered 139 MD-80s, and Finnair inaugurated MD-11 revenue service in December 1990 with a flight from Helsinki to Tenerife. Delta Air Lines became the first US MD-11 operator that same month. Boeing remained the dominant supplier, however, with more than 380 deliveries.

Boeing also responded to the challenge of Airbus's new A330 and A340 twin-aisles, which were fast progressing through their respective development programs. It considered several 767 stretch configurations, including a partial double deck, but eventually settled on a white-sheet design loaded with new technology to meet the Airbus challenge. It received authority to offer the

Rank	1990		2000		2010	
1	American Airlines	490	American Airlines	704	Delta Air Lines	728
2	United Airlines	427	United Airlines	636	American Airlines	638
3	Delta Air Lines	389	Delta Air Lines	623	United Airlines	613
4	Northwest	323	Northwest	430	Southwest Airlines	547
5	Continental Air Lines	323	US Airways	401	FedEx Express	385
6	CAAC*	264	Continental Air Lines	377	US Airways	354
7	US Airways	248	FedEx Express	354	China Southern	354
8	British Airways	219	Southwest Airlines	312	Continental Air Lines	348
9	TWA	210	British Airways	262	Lufthansa	273
10	Eastern Air Lines	160	UPS Airlines	253	China Eastern	264
11	Federal Express	157	Air France	231	Air China	262
12	Lufthansa	152	American Eagle	229	Ryanair	256
13	Pan Am World Airways	133	Lufthansa	226	Air France	250
14	Alitalia	132	TWA	201	Skywest	237
15	Air France	129	Iberia	185	British Airways	226

Sources: Flight International, company websites
* Civil Aviation Administration of China, predecessor to most current Chinese airlines
Excludes Aeroflot

Fig. 3.1 Largest jetliner fleets: 1990, 2000, and 2010.

new aircraft in December 1990 [1]. The aircraft would be a twin-aisle with a capacity of 300–365 passengers and a range of up to 6,000 nm. Rather than depending on four engines for long-range flights, like the 747-400 and A340, Boeing would depend on two massive turbofans to power the aircraft. Boeing would target 180-minute Extended-Range Twin-Engine Operational Performance Standards (ETOPS) for the aircraft when it entered service, an industry first. The 777 also included fly-by-wire flight controls and a composite empennage to match similar innovations in new Airbus aircraft. The design phase for the new twinjet was different from Boeing's previous commercial jetliners. For the first time, eight major airlines—All Nippon Airways, American Airlines, British Airways, Cathay Pacific, Delta Air Lines, Japan Airlines, Qantas, and United Airlines—had a role in the development. Boeing created a new organization, the New Airplane Division, to oversee its development and named Phil Condit as its leader. United Airlines launched the program in October 1990 with one of the most significant orders in Boeing's history: 34 777s and 30 747-400s worth $11 billion. Options brought the potential value of the deal up to $22 billion—the largest in Boeing history [2].

The end of 1990 brought an important regulatory shift as governments took on the issue of noise reduction. The United States passed the Airport Noise and Capacity Act, which segmented aircraft into three groups—Stage 1, Stage 2, and Stage 3—based on noise generation. Although Stage 1 aircraft (the oldest and noisiest) were mostly out of service in the United States, the law mandated that Stage 2 aircraft, which included the 727, DC-9, and early versions of the 737 and 747, be phased out of service by 31 December 1999. Around the same time, the International Civil Aviation Organization agreed to a global phaseout of Stage 2 aircraft over a seven-year period from 1995 to

2002. Thus, operators of the roughly 4800 Stage 2 aircraft in the global jetliner fleet would soon face a decision to either retire their aircraft or fit them with costly "hush kits" to reduce aeroengine noise. The 727, with an active fleet of more than 1600 aircraft, was particularly vulnerable. Fortunately for Boeing, 286 Stage 2 747s were excluded from the ICAO mandate [3].

THE EARLY 1990S BUST

The year 1991 began with a decisive resolution to Iraq's occupation of Kuwait. Forty-two days of bombing gave way to a 100-hour ground assault by more than 30 countries that expelled Iraq from Kuwait and destroyed most of Iraq's war-fighting capability. The fighting was over by 28 Feb. 1991, but the global economy continued to decelerate. Air travel demand fell more than 2% that year. In the United States, the deteriorating environment coupled with the aftershocks of the 1978 Airline Regulation Act—including new low-cost competition—challenged major airlines. Eastern Airlines, the world's 10th largest airline, ceased operations in January 1991 after 65 years. And in December of that year, Pan American terminated operations after struggling with bankruptcy restructuring throughout the year. Pan Am was arguably the world's best-known and most influential airline—and the company that popularized international air travel and launched the 747. Two other US airlines—Continental and TWA—went through Chapter 11 restructures in 1990 and 1992, respectively. These events were a harbinger of a new, more competitive era in air travel.

Airbus began assembly of the first A340 fuselage in early 1991 in Toulouse. Fuselage sections were transported by Airbus's Super Guppy transport aircraft from Hamburg and Saint Nazaire, and wings arrived from the United Kingdom. Airbus rolled out the A340-300 in October 1991, and it flew later that month (see Fig. 3.2). Boeing, recognizing the competitive threat of the A320, began to study a next-generation 737.

Meanwhile, the MD-11 encountered performance problems with its initial customers in not achieving its advertised range with its new PW4460 and CF6-80C2 engines; the shortfall was nearly 1000 km. This led Singapore Airlines to cancel its orders and options for up to 20 MD-11s worth $3 billion and instead purchase A340s. The airline had been counting on the MD-11 to operate its 11,500-km Singapore–Paris route [4].

The news for McDonnell Douglas was devastating, and underscored its competitive disadvantage for long-haul flights. It concluded that it needed a new aircraft model, a double-decker it called the MD-12X. In late 1992, it decided to separate its military division and the loss-making C-17 military transport from its profitable jetliner division to attract the estimated $4 billion in investment for the new aircraft [5]. It signed a memorandum of understanding with Taiwan Aerospace Corporation to produce the new aircraft, and envisaged a coalition of Asian equity partners—an Asian Airbus

Credit: Wikimedia

Fig. 3.2 The A340 first flew in October 1991.

of sorts. The design grew into a 511-passenger, 9200-nm-range jumbo that would be larger than the 747 and could comfortably handle fast-growing trans-Pacific routes. At the time, the Asian air travel market was just 26% of the size of the US market, but by 2010 it was expected to reach 93% [6]. McDonnel Douglas unveiled the new jumbo in early 1992, and it initially was well-received by airlines while several final assembly location candidates in the United States were evaluated.

Unfortunately, the weak airline environment took hold, and Taiwan Aerospace Corporation (TAC) waffled on taking an equity stake while floating the idea of instead setting up a leasing company to buy 20–25 MD-12s—but only after McDonnell Douglas secured 30 orders. Industry observers worried that TAC's aggressive stance could scuttle the project and push the aircraft OEM into a strategy of harvesting its jetliner business [7]. The largest problem was the dearth of MD-12 orders; there were none. In July 1992, McDonnell Douglas announced it was putting the program on hold and transferring employees to other programs, including the promising C-17 military transport. In a letter to employees, Douglas Aircraft President Robert Hood, Jr. attributed the decision to the recession and underscored the company's commitment to the MD-12. "The industry will come back and we will come back with it," he said [8]. The reality was that airlines began to question McDonnell Douglas's ability to fund and execute an ambitious new program. "I didn't have any problem with the aircraft," said American Airlines Chairman Robert Crandall, "The problem was the financial condition of McDonnell Douglas. It is not in any shape to start a new aircraft program" [9]. McDonnell Douglas would never return to the MD-12 program (Fig. 3.3).

Boeing continued its 777-development program and named Phil Condit as president in August 1992. Despite 441 deliveries that year, it anticipated

Credit: Wikimedia

Fig. 3.3 McDonnel Douglas's MD-12 was its attempt to remain competitive; it was cancelled.

hard times and began studying the "lean production" methods of the Japanese automotive industry.

Airbus experienced ups and downs in 1992 as well. The year began with a crash of an Air Inter A320 in Strasbourg, Germany that resulted in 85 fatalities. This was the third in a series of crashes caused, at least in part, by what was believed to be pilots' unfamiliarity with the sophisticated computer system of the Airbus A320. Although crash investigators did not pinpoint a single cause of the tragedy, Airbus ultimately modified the autopilot interface to prevent pilot confusion. On the positive side, Airbus won an order for 100 A320s from United Airlines, the largest US airline and historically a loyal Boeing customer. There was more good news: The A330 completed its first flight, and Airbus delivered 158 aircraft that year including 111 A320s. Overall, there were 730 jetliner deliveries in 1992. It was the last strong year before airline losses and a global recession caught up with the jetliner business.

The year 1993 opened with an important step in European aviation liberalization, the dismantling of bilateral air travel restrictions among European Union (EU) members. Known as the "Third Package" of liberalization, it introduced common licensing criteria and replaced national ownership and control restrictions with the concept of a "community air carrier" where EU airlines must be majority owned or controlled by EU member states. In a practical sense, the Third Package meant that a Dutch operator like KLM could, if it desired, operate flights between the UK and Italy without restrictions. Europe was becoming a single market, increasingly favoring consumers over national carriers with the European Union being the catalyst.

Meanwhile, the first of Airbus's new-generation twin-aisles entered service. Lufthansa began passenger service of the A340-200 in March 1993, and Air France quickly followed suit with the A340-300. Airbus took specialty fitted A340s with an additional five center fuel tanks on a demonstration tour. The airplane broke the nonstop flight record when it flew 10,409 nm—a record that would last until 2005. The A340's range attracted the interest of a growing number of airlines. Cathay Pacific converted six of its A330 orders to A340-300s with a 7150-nm range. Cathay's managing director Rod Edington explained "as carriers like Swissair and Alitalia use MD-11s nonstop from places like Zurich and Rome [to Hong Kong], it's become clear to us that we have to get back to a non-stop philosophy" [10].

While Airbus made twin-aisle inroads, Boeing accelerated the 777 development program under the leadership of 777 General Manager Alan Mulally. Some 240 design teams, with up to 40 members each, addressed nearly 1500 design issues with individual aircraft components. Adding to the challenge was Boeing's groundbreaking initiative for the 777 to become the first jetliner designed entirely by computer using a three-dimensional computer-aided software system called CATIA. This enabled its engineers, in simulation, to "assemble" a virtual aircraft, check for interference, and verify that the thousands of parts fit properly—thus reducing costly rework. Initially, Boeing was not convinced of CATIA's abilities, so the company built a physical mock-up of the nose section to verify its results. The test was so successful that additional mock-ups were canceled [11]. Some 10,000 people were involved in the development program. At its peak, more than half were engineers. The original 747 factory in Everett, Washington, equivalent to 45 football fields, was doubled at a cost of $1.5 billion [12]. The total development bill would reach $5 billion. Suppliers would invest another $2 billion [13]. Other estimates were much higher. Even as Boeing overshot its development budget, it executed a nearly flawless development program. Assembly of the first aircraft started in early 1993, and it first flew in June 1994. It then entered a rigorous 11-month flight testing program—the most extensive since the supersonic Concorde program. This was necessary for it to reach the coveted goal of receiving Federal Aviation Administration (FAA) approval for 180-min ETOPS at service entry [14]. The 777 was awarded simultaneous airworthiness certification by the FAA and European Joint Aviation Authorities (JAA) in April 1995. It completed the development program on time, despite all the new technology and the ambitious, ETOPS-driven reliability targets. How did it manage? Asked years later why the 777 development program was on time, former President Phil Condit cited the imperative to limit the 777 to three major changes: ETOPS, digital design, and the composite fuselage. "Arguably there was a fourth major change—working together," said former Boeing executive Carolyn Corvi. "The entire ecosystem was involved, including customers and suppliers. It worked extremely well."[1] One important but underappreciated factor in the 777's success was Boeing's decision to use the same supply chain

[1]Carolyn Corvi (former VP & General Manager Airplane Programs and Supplier Management – Boeing), interview with author, 13 Feb. 2017.

architecture as the 767. It chose not to introduce new technology and a new supply chain architecture at the same time. This is a lesson that would be lost on the next generation of Boeing leadership with the 787 program.

In parallel with the 777, Boeing responded to the A320 single-aisle challenge in mid-1993 when its board of directors authorized the launch of a next-generation 737, dubbed the 737-X. The aircraft would feature a larger wing, updated avionics, and fuel-efficient CFM56-7 aeroengines. The larger wings enhanced its range to 2900 nm, to give it transcontinental US capability. Boeing forecasted demand for 4100 737-size transports between 1997 and 2010—roughly 300 per year [15]. This was a conservative perspective given the fact that more than 400 737s, A320s, and MD-80s were delivered in 1991 and 1992. US low-cost carrier (LCC) Southwest Airlines launched the new model in late 1993 with an order for 63 737-700s. Southwest's large order underscored the growing influence of low-cost carriers in the US market 15 years after its deregulation. Its fleet reached 178 737s in early 1994 after the acquisition of Morris Air, another low-cost carrier. Its route structure gradually expanded from its Texas base, and in 1993 it started its first service to the Eastern United States at Baltimore Washington International Airport.

While LCCs gained market share, the global airline industry plummeted into the red for much of the early 1990s. Losses exceeded $20 billion for the 1990–1995 timeframe; 1992 was the worst year, with airlines losing $7.9 billion—a –4% net margin. Orders plummeted from 854 in 1990 to 329 in 1994. The boom–bust pattern of the airline industry continued, and aircraft OEMs were whipsawed.

Although Boeing and Airbus had large order backlogs to carry them through the recession, McDonnell Douglas didn't, and was forced into draconian production cuts as deliveries crashed from 171 in 1991 to just 40 in 1994. To survive, it slashed its cost structure by $1.5 billion and its breakeven production rate on the MD-80 from 40 to just 16 [16]. It managed to eke out small profits during the recession, but its market share and financial resources dwindled. Aerospace analysts began to question its long-term viability as a jetliner manufacturer. Still, McDonnell Douglas pressed ahead with its newest member of the MD-80 family—the MD-90—and made its first delivery to Delta Air Lines in early 1995. Its breakeven for the new variant was a modest 100 units [17]. It also launched a 106-seat version with new engines and avionics, the MD-95, and continued modest production in China with its Trunkliner program. It leaned heavily on its aftermarket from nearly 2000 DC-9s and MD-80s in operation to make it through the recession.

A NEW SUPER-JUMBO?

In 1993, Boeing started to review options for a very large commercial transport (VLCT) and enticed the four Airbus partners—Aérospatiale,

British Aerospace, Construcciones Aeronáuticas SA (CASA), and Deutsche Aerospace AG (DASA)—to participate in a trade study to determine the feasibility of forming a partnership for the new aircraft. Each company joined the study group as an individual partner and not representing the Airbus management. This was a threat to the 23-year-old consortium.

In parallel, Airbus initiated its own study of a super-jumbo. In 1994, chairman Jean Pierson announced that it was studying a four-engine, double-decker aircraft, codenamed A3XX, which would seat at least 600 people, compared with fewer than 500 in the biggest Boeing 747. Airbus intended to build the A3XX if the VLCT initiative fell through. It estimated a development cost of $8 billion for the new aircraft. The huge investment requirement, which was conservative compared to other estimates as high as $15 billion, meant that the VLCT would be the least risky option. Airbus commercial director Charles Masefield opined, "Our first choice is the VLCT, but if that doesn't happen, we will have to react. Like Boeing we are double hedging" [18].

The following July, Boeing announced the suspension of the VLCT initiative. Only British Airways and Singapore Airlines publicly expressed interest in the aircraft, and the business case was dubious. Bowing to the inevitable, McDonnell Douglas terminated its own super-jumbo program, the MD-12, in October 1995. This was an abrupt about face from that year's Paris Air Show, where it talked up the prospects of the aircraft and its future in the jetliner business. This exacerbated the doubts about its long-term future.

With Boeing and McDonnell Douglas withdrawing from the super-jumbo stakes, Airbus would press ahead with the A3XX alone.

The year 1995 began lethargically with frequent cancellations and delivery postponements. In 1994, US airlines alone cancelled 325 orders. McDonnell Douglas CEO Harry Stonecipher complained of aircraft production overcapacity—a comment clearly aimed at "newcomer" Airbus [19]. Despite the gloom, Airbus chief Jean Pierson affirmed his desire to achieve parity with Boeing. "We have developed a 30% market share, which is big enough to challenge the dominant competitor. We are much more than a niche player, but we are somewhat fragile....That's why our goal must be to acquire a 50% market share" [21].

In May 1995, Boeing delivered the first 777 to United Airlines. After earning the coveted ETOPS 180-min. rating from the FAA, its first commercial flight took place on 7 June from London Heathrow to Dulles International. A few weeks later, there was more good news for the program when four Asian carriers—All Nippon Airways, Cathay Pacific, Thai Airways, and Korean Air—made 31 commitments for a new stretch version of the aircraft, the 777-300. With a fuselage extension of 10 ft, its capacity was 368 passengers with a range of up to 5700 nm. Boeing touted a 33% fuel burn advantage and 40% lower maintenance costs compared to its 747-1/200s [20]. This not only would open new routes like Tokyo–San Francisco and Bangkok–Sydney, but also would work well on 2- to 3-hour routes in capacity-constrained Asian airports. This commitment underscored the growing influence of Asian airlines. It also raised the question of what the long-haul, twin-engine 777 would mean for the four-engine A340 and Boeing's own 747s.

AND THEN THERE WERE TWO

The year 1995 would prove to be the nadir of one of the great industry recessions of the jetliner era with just 379 deliveries—half of the 1991 delivery total; Douglas garnered just 50 of these (see Fig. 3.4). Yet there were ample signs that the industry recession was passing. Global gross domestic product (GDP) growth recovered to 3% in 1994 and 1995, and air travel demand expanded 8% and 6%, respectively. Airline profitability followed. Global airlines, reeling from five straight years of losses exceeding $20 billion, shifted to the black in 1995 with an aggregate $4.5-billion profit.

In November, Singapore Airlines placed a massive order with Boeing for 34 777-200s and 43 options worth up to $12.7 billion. It was a devastating loss for the A330, because Singapore Airlines already ordered A340s and Airbus was counting on cockpit commonality to sway its decision. Ultimately, the 777's size advantage and the plans for a 777-300 stretch won the day. Another positive development that month was British Airways taking delivery of its first 777 powered by new GE90 aeroengines. Boeing's order book for its new aircraft reached 230, in contrast to 266 for the A330/340 [22]. Just as good times were returning, Boeing's labor relations took a hit when 32,000 machinists in Washington, Oregon, and Kansas went on strike for 10 weeks. Apart from improved pay and benefits, the unions wanted assurance that Japanese subcontractors—very prominent on the 777—wouldn't steal their jobs [23].

The global economy hit full stride in 1996 with 3.3% GDP growth, and airlines boasted another profitable year. There were more than 1000 orders that year—the most since 1989—with Boeing's share being 60%. Boeing launched the 757-300, a stretch of its popular 757-200, with an order from German charter carrier Condor. It also contemplated a 747-5/600 stretch but did not make a formal launch decision. Airbus continued its study of the A3XX super-jumbo, with a study by Aérospatiale estimating $280 billion in sales for the program. It stuck with its $8 billion development estimate—a figure that Boeing disputed as much too low based on its VLCT study; it contended that $12–15 billion was a more reasonable estimate [24].

Boeing rode the strong order momentum while concluding agreements with American Airlines, Delta Air Lines, and Continental Airlines to purchase jetliners exclusively from Boeing for 20 years. This could cut off Airbus and McDonnell Douglas from three of the largest US airlines. Offsetting this bad news for Airbus was USAir's agreement to purchase up to 400

Type	OEM	Model	1990	1991	1992	1993	1994	1995	1996	1997	1998	1999	2000	TOTAL
Single-Aisle	Airbus	A320	58	119	111	71	64	56	72	127	168	222	241	**1309**
	Boeing	707	4	14	5	0	1							**24**
		717										12	32	**44**
		737-3/4/500	174	215	218	152	121	89	76	132	116	42	4	**1339**
		737-6/7/8/900								3	166	253	269	**691**
		757	77	88	99	71	69	43	42	46	50	67	45	**697**
	McDonnell Douglas*	MD-80	139	140	84	43	23	18	12	16	8	26		**509**
		MD-90						13	25	26	34	13	5	**116**
Twin-Aisle	Airbus	A300	19	25	23	22	23	17	14	6	13	8	8	**178**
		A310	18	20	24	22	2	2	2	2	1			**93**
		A330				1	9	30	10	14	23	44	43	**174**
		A340				22	25	19	28	33	24	20	19	**190**
	Boeing	747-1/2/300	8	2										**10**
		747-400	62	62	61	56	40	25	26	39	53	47	25	**496**
		767	60	62	63	51	40	36	42	41	47	44	44	**530**
		777						13	32	59	74	83	55	**316**
	McDonnell Douglas*	MD-11	3	31	42	36	17	18	15	12	12	8	4	**198**
		TOTAL	**622**	**778**	**730**	**547**	**434**	**379**	**396**	**556**	**789**	**889**	**794**	**6120**

Source: Teal Group
* McDonnell Douglas acquired by Boeing in 1997
Excludes Soviet and Russian aircraft

Fig. 3.4 Jetliner deliveries: 1990–2000.

A320s with a firm order for 120 valued at $5.3 billion. USAir cited growing competition from low-cost carriers and fleet rationalization as motivations for the mega-order [25]. The news was distressing for Douglas because it lost another key sales target.

MEGA-MERGER

On 2 Dec. 1996, Boeing and Douglas Aircraft announced a landmark "strategic collaboration" agreement, where Douglas would be a subcontractor to Boeing for manufacturing and engineering services. It would also sign a noncompete agreement, which would prevent collaboration with Airbus on the A3XX program. On 15 Dec., the plan changed. Boeing would buy McDonnell Douglas for $13.3 billion, creating the world's largest integrated aerospace company. The new company, which would keep the Boeing name, would have $48 billion in revenue and 200,000 employees. Boeing CEO Phil Condit called the deal "an historic moment in aviation and aerospace" [26]. Condit would remain CEO, and McDonnell Douglas CEO Harry Stonecipher would become the new president and chief operating officer. This was much more than a jetliner merger. Boeing placed great value on Douglas's military portfolio, including the F/A-18 Hornet fighter, the C-17 Globemaster transport, and the AH-64 Apache attack helicopter. Airbus leader Jean Pierson described it as "a merger to kill Airbus" and warned that it would create a giant with the explicit support of the US government for "monopolizing the civil aeronautics construction sector" [27]. The US Federal Trade Commission approved the merger in July 1997. The European Commission (EC), exerting newfound influence regarding a US merger, took longer and ultimately approved the merger if Boeing cancelled its exclusivity agreements with US airlines. Why didn't the European Commission put up more of a fight? According to Rob Spingarn of CreditSuisse, "Boeing-Douglas was much more about defense than civil aircraft. Douglas had no [jetliner] backlog at the time. Douglas didn't make Boeing a larger commercial player, which is why it got through the EC. This merger was about the military business."[2]

The month after the merger announcement, Boeing signaled a new product strategy when it cancelled the 747-500/600 study program and decided to focus on 767-400 and updates to its 777 variants. The fragmentation of air routes—particularly across the Atlantic—reduced the influence of the 747. In 1984, 62% of the flights across the North Atlantic were operated by 747s. Twenty years later, this figure would dwindle to 16% [28]. Boeing was betting that in a world of efficient twin-engine ETOPs aircraft that passengers would prefer nonstop service in a smaller aircraft to a connecting flight via a super-jumbo in a new 747 or A3XX.

---

[2]Rob Spingarn (Managing Director – CreditSuisse), interview with author, 5 Jan. 2017.

The Changing Nature of Airlines

As the ranks of the large jetliner OEMs dwindled to two, major changes unfolded with the world's airlines. In North America, five low-cost carriers were born in the 1990s following in the footsteps of Herb Kelleher's Southwest Airlines, taking advantage of lower barriers to entry. Air Tran was founded in 1992, Frontier Airlines in 1994, Westjet in 1996, and Allegiant in 1998. The fifth LCC, JetBlue, was started by David Neeleman (Fig. 3.5) in 1999. Neeleman was a serial LCC entrepreneur. He not only founded low-cost charter airline Morris Air in the 1980s, but also co-founded WestJet and would go on to start Azul Brazilian Airlines a decade later. Although not copying all aspects of the Southwest business model, these airlines emulated many of the same techniques that helped make the airline so successful. As the North American airlines adjusted to the new state of the industry, Europe would see similar upheaval, also due to deregulation and the rise of low-cost carriers.

The changing landscape of the European airline industry can first be traced to the signing of the Maastricht Treaty in February 1992. In addition to establishing the European Union and its common currency, the treaty also encouraged greater cooperation among its member states on various issues to improve life for its citizens. Later that year, the 12 European Union members agreed to create a single aviation market, remove government interference, and deregulate the industry by 1997. This deregulation, combined with the increasing popularity of open skies agreements, lowered the barriers for entry and increased competition among the existing airlines and possible new entrants. A major step was taken in 1992 when the Netherlands signed the first open skies agreement with the United States, despite objections from

Credits: James Howes, World Travel & Tourism Council

Fig. 3.5 Low-cost airline pioneers: Herb Kelleher, David Neeleman, and Michael O'Leary.

EU authorities. The agreement gave both countries unrestricted landing rights on each other's soil. This was a major shift from the traditional approach of tightly regulating flights between countries.

Like the US market after deregulation, many of these new entrants, buoyed by easier market entry, positioned themselves as low-cost options to the current legacy carriers. Two significant players in this low-cost carrier market were Ryanair and easyJet. Ireland-based Ryanair was founded in the 1980s, but motivated by the new rules of the game and a visit to Southwest Airlines by CEO Michael O'Leary, it realized it needed to shift to an LCC business model. It decided to go public in 1997, enabling it to purchase 45 new 737-800s in 1998. Another LCC, easyJet, was founded in 1995 by Stelios Haji-Ioannou in the United Kingdom and ascended rapidly through organic growth and acquisitions. Ryanair and easyJet took the low-cost model to new levels of efficiency. Both leveraged online Internet booking to reduce employees and improve productivity. Ryanair flew to underutilized airports to reduce landing fees.

In the Middle East, new carriers emerged in the Gulf region in the late 1980s and 1990s. As more and more Western carriers, utilizing the longer-range aircraft like the 747-400, started to bypass the Middle East on connecting routes, Middle Eastern carriers began planning their growth. Emirates Air was launched by Sheikh Mohammed bin Rashid Al Maktoum and was conceived as a response to Gulf Air's reduction in service to Dubai. Emirates created new routes to London Gatwick, Frankfurt, Istanbul, and Hong Kong and purchased eight A310-300s and nine 777-200s during the decade. Gulf Air became the first airline from the region with service to Australia, also adding routes to Singapore, Rome, and Johannesburg along with six A330-200s to its fleet. State-owned Qatar Airways, founded in 1993, grew steadily as the overall region saw spectacular growth. By the end of the decade, traffic demand among Middle East carriers nearly doubled, jumping from 49 billion revenue passenger kilometers (RPKs) in 1990 to 96 billion in 2000 [29].

The Asia-Pacific region saw tremendous growth in the 1990s as an ever-growing middle class had more disposable income and air travel demand increased. Contrary to their US counterparts, Asian airlines fared well during the early 1990s due to growing demand. The member nations of the Orient Airlines Association turned a $2.2 billion profit from 1990 to 1992 while many US airlines were struggling for survival. However, in Japan, All Nippon and Japan Airlines both dealt with losses in 1993 due to the Japanese recession and high operating costs before returning to profitability in 1994. The Australian government sold Australian Airlines to Qantas, giving the merged carrier complete domestic and international coverage. This era also saw the birth of new low-cost carriers as both Malaysian airline Air Asia (1993) and Indonesia-based Lion Air (1999) were founded. Overall, air travel demand in the region grew every year, with RPKs more than doubling from 346 billion in 1990 to 735 billion in 2000 [30].

By 2000, both Asia and the Middle East had doubled their 1990 levels of RPKs while airlines in North America and Europe took advantage of growing consumer confidence and low oil prices to boast record profits. Collectively, the airline industry recorded $35 billion in profits between 1995 and 1999. Fuel prices, which once contributed to the demise of some airlines, became a boon to the survivors. Staying relatively flat throughout the rest of the 1990s, fuel prices dropped to $0.31 per gallon by February 1999, enabling the incredible economic comeback.

As passenger transport demand grew, a new source of jetliner demand mushroomed in the 1990s: air cargo. Carrying cargo in an aircraft's belly for supplemental revenue was commonplace in the airline business since its inception. All-cargo airlines like Flying Tiger and Cargolux had established successful niche operations. And the introduction of large twin-aisles like the 747 expanded air cargo's business potential for airlines. Two things changed in the 1990s. First, air cargo demand exploded as transnational corporations created dispersed, global supply chains and international trade grew. Computer manufacturers, for example, sourced motherboards, keyboards, displays, and chips from different countries—typically in Asia—which then needed to be transported to the final assembly site. Completed computers were then shipped to customers in other regions. The high value-to-weight ratio of these items meant that air cargo was the preferred mode of transport. Air cargo demand also skyrocketed for low technology perishable items, like flowers and fish.

The second change was on the supply side of the air cargo market as a new type of supplier—air cargo integrators—emerged to offer time-definite delivery services. Integrators typically owned assets along the entire delivery chain—trucks, warehouses, and aircraft—that were knitted together by sophisticated information technology (IT) systems. Thus, they could offer faster transportation, typically anywhere in the world within 24–48 hours, and could deliver within a specific timeframe. This is exactly what many manufacturers and retailers desired as they pursued faster inventory turnover and fewer nonproductive assets. Financial institutions and professional services firms also were major customers. Four companies controlled 90% of the integrated cargo market in the 1990s: FedEx, United Parcel Service (UPS), DHL, and TNT. Global delivery networks were required to provide time-definite services. FedEx established global "hub and spoke" operations, with major hubs in Memphis, Paris, and Subic Bay, Philippines along with many secondary hubs and local logistics and ground transportation operations in approximately 200 countries. DHL had presence in even more countries.

Because of these trends, the global air cargo fleet doubled in the 1990s, reaching more than 1700 dedicated cargo jetliners by 2000. Most of this buildup was from modifying existing mature aircraft to become all-cargo aircraft. This provided extended life for many older aircraft, like the 727, which were retiring from passenger service.

Comparing rosters of the world's largest airlines during the bookends of the 1990s highlights the changing nature of airlines during the decade. By 2000, two air cargo integrators (FedEx and UPS) and low-cost carrier Southwest Airlines were among the 10 largest global fleets. What didn't change was the domination of US airlines, which had 11 of the 15 largest fleets. This would shift in the future.

THE LATE 1990S BOOM

The late 1990s delivered the greatest upcycle to date in air transport history. Over the four-year period of 1997–2000, aircraft OEMs would deliver more than 3000 jetliners and receive orders for another 5300. The global fleet expanded by nearly one-quarter, from 12,800 to 15,800. Several trends drove this expansion. The global economy was on fire, fueled by growing international trade and euphoria surrounding technology and the Internet. Global GDP growth reached 4.3% in 2000—unprecedented in recent history.

Still, there was an important question facing the industry: *What would Boeing do with Douglas?* Answers emerged in late 1997 when Boeing announced it would terminate the MD-80 and MD-90 programs when orders were filled in mid-1999. This wasn't surprising because they were direct competitors to the 737. It would continue production of the MD-11 and would also keep the MD-95, which it would rename the Boeing 717. Boeing was attracted to the 717's low development cost, modern cockpit, and passenger appeal; the aircraft fit neatly into its flagship 737 family with 106 seats in a two-class configuration. Boeing would certify the aircraft in 1999 and deliver it to launch customer Air Tran, a US low-cost carrier. Another decision was to continue operations of Douglas's sprawling Long Beach final assembly plant and its 10,500 employees. It would have a few years to figure out its long-term future.

One challenge with the integration was merging two distinctly different cultures. Boeing was focused on technology and products; financial performance drove Douglas. "We brought a focus on financial performance and cash flow that was totally lacking at Boeing," recalled a former Douglas executive. "Douglas was very focused on assets and margins. Return on net assets (RONA) was the biggest part of the Douglas bonus system. This led to some resentment of longtime Boeing executives who were not financially savvy."[3] Beyond Harry Stonecipher, McDonnell Douglas executives Mike Sears and Mike Cave made the transition into the new organization and began to exert influence. Boeing began to focus more on financial performance. One casualty of the shift was Commercial Airplane Group President Ron Woodard, a talented

[3]Jeff Johnston (former General Manager – McDonnell Douglas), interview with author, 1 Feb. 2016.

salesman who prioritized market share over margins. His goal of achieving 50% share of all jet transport orders against an aggressive Airbus led to larger discounts and subpar profitability, despite the bullish market environment. Aircraft pricing discounts increased from 10% to 18–20%, sometimes reaching 30%. Moreover, a closely held study found that Airbus had a 12–15% cost advantage over Boeing in production costs and tooling [31].

As a result, Boeing, a $56-billion behemoth and market leader, managed to lose money in 1997. Woodard was replaced by Alan Mulally. Well-known industry analyst Paul Nisbet attributed the shakeup to Harry Stonecipher. "What we're seeing is Stonecipher beginning to assert himself and finally feeling comfortable in the changes that should be made," he said [32]. Large layoffs ensued—nearly 20,000 in 1997 and 1998 despite Boeing's large order book—about 8% of the company's workforce. Redundancies, the termination of the MD-80/MD-90, and the desire to improve financial results contributed to the magnitude of the reductions. In 1998, Boeing reversed course and decided to end MD-11 production. This left the 717 as the only legacy Douglas jetliner in its portfolio.

Aircraft OEMs introduced several new aircraft in the late 1990s. Airbus launched the long-range A340-500 and higher capacity A340-600 in 1998. The A340-500 would carry 313 passengers (in a three-class configuration) 8650 nm, which made it the world's longest-range aircraft at the time. Air Canada was supposed to be the launch customer for the A340-500, but filed for bankruptcy in January 2003. Instead, early deliveries went to Emirates, allowing the carrier to launch nonstop service from Dubai to New York—its first route in the Americas. The A340-600 carried 379 passengers and was targeted at replacing early-generation 747s. The variant required a new wing and new powerplants. After discussions with GE about being the exclusive aeroengine supplier broke down, Airbus selected Rolls-Royce Trent 500s for the aircraft. Airbus would spend $2.9 billion developing the variants, which would enter service in 2002 [33]. Airbus addressed the other end of its product spectrum with the 107-seat A318 program in 1999 after it garnered orders from Trans World Airlines, Air France, and lessor International Lease Finance Corporation. This positioned Airbus to fend off the 717 and hopefully capture a slice of the burgeoning regional aircraft segment. There were over 600 orders for smaller regional jets in 1997–1998 alone. Ironically, the A318 had its origins in the AE31X program, a proposed joint venture among Airbus Industrie (39%), China's AVIC (46%), and Singapore Technologies Aerospace (15%) to build a new 100-seat aircraft. The initiative, which began in 1997 during a visit by French President Jacques Chirac to China, would never move forward, but it did pique Airbus's interest in smaller aircraft [34].

Boeing also worked the upper and lower end of its product ranges. Its 737 Next Generation (737NG) development programs were well-executed and generally held their schedules. The first 737-700 (a 128-seat variant replacing the 737-300) was delivered to launch customer Southwest Airlines in early 1998 and received rave reviews for 99% dispatch reliability in early operation [35]. The 737-800, a larger replacement for the 737-400, was delivered to European charter operator Hapag Lloyd a few months later. And Scandinavian Airlines System took the smaller 108-seat 737-600 in September 1998. At the same time, Boeing was developing a high-capacity variant, the 737-900, which would ultimately enter service with Alaska Airlines in 2001. Boeing executed an admirable phase-in for the latest generation of the world's best-selling jetliner; it would deliver 166 737NGs in 1998 and quickly ramp up to 278 and 280 in 1999 and 2000, respectively. Boeing's pricing strategy was value-oriented. The list price of the 737-600/700 was just $1 million more than their replacements; the 737-800 was $4 million more than the 737-400 due to the stretch. Boeing leveraged system commonality, increased factory automation, and lessons learned from the 777 to hold the line on pricing [36]. List prices for 737NGs ranged from $35 to $60 million, although substantial discounts were commonplace—particularly with the competition for market share with Airbus.

On the upper end, Boeing decided that it needed longer-range 777s to counter the A340 and to address airline needs for longer-range aircraft. A new aeroengine would be required to extend the 777s range without compromising capacity, and in 1999 Boeing entered an exclusive agreement with General Electric to power its new aircraft with enormous GE90-115B engines, rated at up to 115,000 lb of thrust. Two variants, the 777-300ER (extended range) and 777-200LR (longer range), would enter service in 2004 and 2006, respectively. The latter could carry 317 passengers 8550 nm, on par with the A340-500 but with two fewer engines.

EUROPEAN RESTRUCTURING

Airbus steadily gained jetliner market share in the late 1990s. In 1999, Airbus pulled in 476 orders worth $20.5 billion—its second-best year ever. More significantly, by its calculations it achieved a 55% share of announced orders—the first time ever that it bested Boeing.[4] As it achieved parity, Airbus upgraded its organizational structure, shifting from a marketing entity selling aircraft produced by its four shareholders to a traditional, integrated aerospace company. Airbus CEO Jean Pierson had long advocated for an integrated company, which he viewed as essential for Airbus to improve its productivity versus Boeing. In a 1996 interview, he acknowledged the current structure was riddled with inefficiencies and duplications, which Airbus would need to eliminate if it were to pursue the A3XX [37]. Although the need was apparent, the political will for an integrated company was lacking.

[4]"Company History—Interactive Timeline," Airbus Industrie, http://www.aircraft.airbus.com/company/history/the-interactive-timeline [retrieved 3 Sept. 2017].

Consolidation of US aerospace and defense prime contractors changed the context. Lockheed started a wave of consolidation in 1995 when it acquired Martin Marietta. Boeing's acquisition of McDonnell Douglas also created a defense behemoth. In contrast, European companies were fragmented. In 1997, British Aerospace (BAe) executive John Weston argued that there was broad commonality of view among industrialists that there was a need to create one major consolidated (European) company. He noted that Europe, with defense spending of $125 billion a year, was supporting three times the number of contractors on less than half the budget of the United States [38]. As early as 1995, British Aerospace and DaimlerChrysler Aerospace had contemplated a merger as a first step in creating an integrated Airbus, but the French Socialist government did not want to privatize Aérospatiale. Then the ice broke. In 1998, Aérospatiale merged with French defense supplier Matra, diluting the government's shareholding. Later that year, British Aerospace and DASA announced they would merge, but British Aerospace quickly reversed course and purchased Marconi Electric Systems instead. Losing its favored partner, DASA and Spanish Airbus partner CASA merged in June 1999. Four months later, DASA agreed to merge with Aérospatiale-Matra to create the European Aeronautic Defence and Space Company (EADS). The new company went live in 2000. EADS and BAE transferred ownership of their Airbus factories to the new Airbus SAS in return for 80% and 20% shares in the new company, respectively.

The A380: The First Moonshot

With Airbus now a single corporate entity, it focused on its next major challenge: the super-jumbo A3XX. Why? The first reason was market growth. Air travel demand accelerated in the late 1990s, and Airbus believed there was a need for 1600 aircraft larger than the 747-400 over the next 20 years. It could base a business case on capturing at least 650 of these aircraft. The A380 could create its own market demand, just like the 747 in the late 1960s and 1970s [39]. The second reason was to create a complete product family, from single-aisle to super-jumbo. Airbus was all too aware of the fate of McDonnell Douglas and Lockheed, failed jetliner OEMs with narrow product ranges. A third reason was that the mega-project would be a way to unify the company in its new structure. With few new development programs on the horizon, it was a way to keep the engineering team engaged. There was another crucial reason for launching the A3XX: *profit pools*. Boeing enjoyed a monopoly with the 747, and Airbus believed that it reaped a $30- to $40-million profit per airplane, allowing them to cross-subsidize other products. Even if the A3XX wasn't profitable, Airbus reckoned, it could level the financial playing field [40]. Finally, one cannot discount the role of ego in the decision—the prestige of having the largest jetliner.

The internal Airbus goal was to have 50 firm orders before program launch. It picked up orders from Qantas, Air France, Singapore Airlines, Emirates, and lessor International Lease Finance Corporation (ILFC). In late 2000, Virgin ordered six, and the threshold was achieved. The Airbus supervisory board approved the new program on 19 Dec. 2000. The new aircraft would be called the A380 (Fig. 3.6). Airbus CEO Noël Forgeard stressed the importance of the decision. "With the launch of the A380 we are now closing the final large gap in our product spectrum. We are now able to offer aircraft in all the categories from single-aisle via wide-body to mega-liner and could therefore fulfill the wishes our customers may have" [41]. Jürgen Thomas, who led the A3XX trade studies and would become known as "the father of the A380," stepped aside to let the younger Charles Champion run the program.

The new double-decker, four-engine aircraft, with a maximum takeoff weight of 1,234,600 lb, could carry 544 passengers (three classes) 8200 nm. The 747-400, the next largest aircraft, had a maximum takeoff weight of 910,000 lb. Composites were used for 25% of its aerostructures, twice that of previous aircraft. In addition to advanced technology systems and avionics, two engine choices were offered: the Rolls-Royce Trent 900 and the GP-7200, a joint venture between General Electric and Pratt & Whitney. First flight was planned for 2004 with entry into service in 2006.

Airbus also unveiled an ambitious supply chain strategy for the new aircraft. EADS and BAE Systems facilities in France, Germany, the United Kingdom, and Spain would produce most of the aerostructures subassemblies, which would be shipped by sea to Bordeaux. There, outsized trucks would carry them to a final assembly facility in Toulouse. Airbus estimated that $10.7 billion would be required for development and facility expansion. It would defer this expense with risk sharing partners such as GKN Westland, Saab,

Credit: Aero Icarus

Fig. 3.6 The A380 entered service in 2007.

and AIDC (Taiwan). In addition, and controversially, about one-third of the development tab would come from low-interest loans from the four "Airbus parent countries," which Boeing and the US government protested, but Airbus argued was strictly in compliance with the 1992 US–EU accord that allowed governments to fund up to 33% of commercially viable aircraft programs in the form of reimbursable loans [42].

In the United States, Wall Street analysts reacted negatively to the news and lampooned the OEM's prediction that it would achieve a 20% internal rate of return and break even by 2009. JSA Research analyst Paul Nisbet anticipated a Boeing response that would defer a program breakeven "for decades." He also questioned the investment assumptions, projecting that EADS would need to raise $8 billion for facilities and in aggregate could face a $20 billion cash drain for research and development (R&D), investment, and the inevitable losses on early aircraft deliveries [43].

Airbus then went to work on the most ambitious European aerospace development program ever. It received good news when, in January 2001, FedEx launched a freighter version of the aircraft with firm orders for 10 A380Fs. FedEx made this decision to keep up with surging air cargo demand, particularly in Asia. The world's most influential cargo airline and integrated logistics supplier shared Airbus's vision. With orders from six airlines and now FedEx, the future appeared bright.

A Second Moonshot: The 787

As Airbus was betting on larger capacity, Boeing took a very different route. In March 2001, Alan Mulally announced the cancelation of the 747-X program. Customers, he said, had provided a "clear direction … and the 747-400 will satisfy a majority of their large airplane needs."[5] At the same time, he unveiled a radical new jetliner concept with a canard, delta wings, and rear-mounted engines with a cruising speed of Mach 0.95 and a range of 9000 nm. Boeing dubbed the aircraft the Sonic Cruiser. American Airlines indicated that it would allow one extra daily flight between New York and Europe, but warned that "cost is the big issue." Mulally was optimistic that the aircraft could be launched within two years with an entry into service by 2008. Critics worried about the large drag penalty as the aircraft approached sonic speeds [44]. Boeing moved to wind tunnel tests to refine its design and settled on an aircraft made primarily of composites and titanium. Virgin Atlantic expressed interest in the new aircraft in mid-2001.

As Boeing and airline partners evaluated the futuristic design, the world and the aviation industry changed fundamentally on 11 Sept. 2001, when terrorists hijacked four commercial airliners, launching separate attacks on

[5]Boeing announcement, 29 March 2001.

New York City's Twin Towers and the Pentagon in Washington, DC. Another attack against the nation's capital was averted because the airline's passengers attempted to regain control of the plane before crashing near Shanksville, Pennsylvania. In all, 2996 people were killed and there was more than $10 billion in damage to the affected infrastructure. The aftermath on air transport was immediate and devastating. US airspace was temporarily closed, and passenger confidence in the air transportation system shattered. Passenger traffic in the United States, already reeling from a recession, fell 5.9% in 2001 and another 1.4% in 2002. US airline revenue declined $23 billion between 2000 and 2002, and for the first time since World War II, capacity declined two years in a row. The global air transportation system didn't fare much better, with traffic declining 2.7% in 2001. In aggregate, global airlines lost $13 billion in 2001 and $11.3 billion in 2002 [45].

Despite the bleak environment, Boeing continued its Sonic Cruiser trade study. The critical question: What is the incremental value of a long-range transport that can go 20% faster? Alan Mulally characterized the Sonic Cruiser as "an appealing concept and unbelievable technical challenge" whereas Airbus, predictably, dismissed its attractiveness and proclaimed that the next 250-seat long-range transport would be "cheap and green" [46]. An October 2002 Boeing meeting with its airline advisory board was telling: not one airline gave a high rating to the 0.95-Mach, high-speed design; they were interested in fuel efficiency [47]. In December, Boeing declared that the Sonic Cruiser would not be built. "Everyone felt it was dead 15 minutes after 9/11," opined one aeroengine official [48].

Instead, Boeing would focus on using the technologies from the trade study to make a more conventional aircraft with excellent operating economics. The following month, it announced the 7E7, a super-efficient airplane roughly the size of a 767 that used many of the technologies from the Sonic Cruiser concept. These included a carbon-fiber composite aircraft structure, heavy use of electronics to replace hydraulic and pneumatic systems, efficient new high-bypass ratio aeroengines, and advanced avionics. Initially, short- and long-range variants were contemplated, the latter seating 200–220 passengers with a range of 7800–8000 nm. As many as 40 airlines provided input to Boeing on its new program [49].

While Boeing evaluated 7E7 concepts, the industry entered a deep recession. Large jetliner deliveries plummeted by 30%, from 823 in 2001 to 571 in 2003. The impact for Boeing, which was heavily dependent on US customers, was pronounced with 737 deliveries falling 40%. Large layoffs of employees in Seattle ensued. In contrast, Airbus maintained relatively steady production rates due to a different customer mix and the high social costs in Europe of laying off employees. As a result, a critical milestone was reached in 2003 when Airbus, for the first time, surpassed Boeing in deliveries, with 297 compared to Boeing's 274. There was more bad news in Boeing's defense business when an ethics scandal erupted and halted a US Air Force deal to acquire 100 767 tankers for $18 billion. CFO Michael Sears and Vice President Darleen Druyun resigned. Something had to change, and in December 2003, Boeing CEO Phil Condit resigned. He left a mixed legacy. As *Bloomberg Businessweek* noted,

> The decision to steer the company more deeply into defense contracting represented a historic shift. The most crucial step in that process was the 1997 acquisition of McDonnell Douglas. Not only did that deal make defense contracting a much bigger part of Boeing's mix, but the smaller acquiree also had an outsized effect on the Boeing culture. Boeing, the jewel of its hometown of Seattle, had always prided itself on treating employees—from designers to line workers—as family. But the values most esteemed at McDonnell were the ability to schmooze with Washington power brokers and win the contract. Building planes sometimes seemed to take second place [50].

Former McDonnell Douglas CEO Harry Stonecipher (Fig. 3.7) came out of retirement to become Boeing's new CEO. To many, the ascent of Stonecipher represented the de facto takeover of the company by McDonnell Douglas. The blunt new leader was critical of Boeing's financial status, which he compared to GE and Coca Cola. "We have returns that can't even see the bottom of Coca Cola," he stated. "That is unacceptable." Responding to the perception that he was only interested in making money, he responded, "You're right, I am" [51].

Credit: Boeing

Fig. 3.7 Boeing CEO Harry Stonecipher.

Five months into Stonecipher's tenure, Boeing finalized the 787's product strategy. The baseline aircraft, the 7E7-8, would carry 217 passengers in a three-class arrangement with a range of up to 8500 nm. A stretch 7E7-9 would carry 257 passengers 8300 nm. And the short-range 7E7-3 would carry 289 passengers 3500 nm. Its first flight was planned for 2007 with entry into service the following year. Boeing also defined its supply chain strategy, which would feature unprecedented outsourcing of aerostructures and aircraft systems to limit the asset requirements and nonrecurring engineering (NRE) cost of the new program.

The 7E7's ultra-long range coupled with modest passenger capacity broke a longstanding jetliner paradigm that longer range meant greater capacity. It would offer range comparable to the A380 with the capacity of a 767. Boeing's bet was that passengers wanted point-to-point service, and a relatively small, efficient, and long-range aircraft would create hundreds of new city pairs and fragment trans-Pacific routes much like the 767 did to the Atlantic. In contrast, Airbus felt that airlines would require a 550-seat aircraft for dense trunk routes between hub airports.

The 7E7 bagged its launch order with All Nippon Airways for 50 787s in April 2004. In the same month, Boeing rolled out the first of its new series of 777s when it delivered a 777-300ER to ILFC and its customer Air France. After a bleak start to the decade, Boeing logged 272 orders in 2004—slightly less than deliveries that year—including 126 more orders and commitments for the 7E7. In 2005 it would revert to its traditional aircraft naming convention when it changed the name of the 7E7 to the 787.

In November 2005, Boeing announced another development program: the 747-800. The 747 stretch would offer General Electric GEnx turbofans, upgraded wings, and common avionics with the 787. Most of the interest was from cargo operators; Cargolux and Nippon Cargo Airlines were the launch customers for the freighter version of the aircraft; Lufthansa would be the lead customer for the passenger version. With a length of 250 ft, this would be Boeing's largest aircraft, and was its competitive response to the A380.

MOONSHOT NUMBER 3: THE A350 XWB

How would Airbus respond to the 787, which appeared to threaten the A330? Initially, it felt no response was needed, but airlines urged Airbus to provide a competitor. It initially proposed the A330-200 Lite, a derivative of the A330 featuring improved aerodynamics, composite wings, and the General Electric GEnx engine like those on the 787. The company planned to announce this version at the 2004 Farnborough Airshow, but did not proceed [52].

Aeropolitics bubbled to the surface as Airbus contemplated its new aircraft. In July 2004, Boeing CEO Harry Stonecipher accused Airbus of breaching a 1992 bilateral EU–US agreement governing large civil aircraft government subsidies. The agreement allowed up to 33% of a program's cost to be met through reimbursable government loans if they were repaid within 17 years with interest and royalties.

Later that year, Airbus CEO Noël Forgeard confirmed that a re-engined A330 was not adequate, and that Airbus was considering a new design. By the end of 2004, the boards of EADS and BAE Systems, then the shareholders of Airbus, gave permission for Airbus to offer a new aircraft, called the A350. It incorporated a new wing, engines, and horizontal stabilizer with heavy use of composites. It would share the A330's metallic fuselage. Thus, it would be a "pseudo white-sheet aircraft." At the 2005 Paris Air Show, Qatar Airways announced an order for 60 A350s, and the industrial launch followed in October with an estimated development cost of €3.5 billion ($4.4 billion). The 250- to 300-seat twin-engine aircraft would have two variants: an A350-800 capable of flying 8800 nm with typical passenger capacity of 253 in a three-class configuration, and the 300-seat A350-900 with a range of 7500 nm. It was designed to be a direct competitor to the 787-9 and 777-200ER [53].

The launch of the A350, coupled with the pending entry of the A380, escalated aeropolitical tensions. Airbus was expecting government loans to cover up to one-third of the A350's development costs. The United States went on the offensive and filed a case against the European Union (EU) for illegally subsidizing Airbus. The EU responded the following day by filing a complaint against the United States for illegal support to Boeing. Airbus contended that Boeing benefited from R&D contracts with NASA and the US Department of Defense. It also contended that Boeing received subsidies from US states and foreign governments. Japanese suppliers, for example, received significant government support for 787 production facilities.

Another monkey wrench appeared when two major lessors urged Airbus to create a white-sheet design for the A350. At a March 2006 International Society of Transport Aircraft Trading (ISTAT) conference, ILFC CEO Steven Udvar-Hazy (Fig. 3.8) told a surprised audience that Airbus needed to decide whether the A350 would be a marketing response to the 787 or a backbone of their wide-body, mid-sized product line [54]. It was a stunning rebuke to the program, because he was possibly the world's most influential jetliner customer, leader of the second largest aircraft lessor, and an existing customer for 16 A350s. "If Airbus sticks with its current design," he predicted, "it will wind up with as little as 25 percent market share against the 787." Udvar-Hazy added that the investment for a white-sheet would be in the $8–10 billion range, and that Airbus needed to make its decision by that year's Farnborough Air Show—just four months away. Moreover, his remarks were endorsed by Henry Hubschman, president of GECAS, the world's largest aircraft lessor [55]. Airbus initially downplayed the critiques, but it was difficult to ignore the input of the industry's two largest lessors—particularly with those lessors accounting for more than 40% of jetliner orders in the mid-2000s.

Credit: Smithsonian's National Air and Space Museum

Fig. 3.8 Steven F. Udvar-Hazy.

Some leading airlines weren't pleased with the A350 either. Qatar Airways CEO urged Airbus to develop a more competitive design with better economics and a cabin that could accommodate lie-flat seats. In December 2005, Qantas and Cathay Pacific eschewed the A350 in favor of the 787 and 777. Emirates wasn't impressed. And Singapore Airlines CEO Chew Choon Seng, after reviewing the 787 and A350, opined that Airbus should have created a white-sheet design.

There was another problem: The proposed GEnx engines, developed for the 787, were optimized for the smaller 787, not the A350.[6] The writing was on the wall: Airbus needed to go back to the drawing board, even as it dealt with a swelling development bill for the A380.

At the July 2006 Farnborough Airshow, Airbus unveiled a redesigned aircraft dubbed the A350 XWB (Xtra-Wide-Body). The new design featured a wider fuselage cross section—about 5 in. wider than a 787—which could accommodate nine-abreast (economy class) in a high-density seating layout. Like the 787, the new fuselage would be made from carbon fiber composites, allowing higher cabin pressure and humidity, with lower maintenance costs. Days after the announcement, Singapore Airlines agreed to order 20 A350 XWBs with options for another 20.

Still, Airbus's leadership was not aligned internally on the new program, and weeks after the announcement the team met for dinner at a converted French abbey to reflect. CEO Christian Streiff, who had taken over from

[6]Robert Lange (Senior Vice President – Airbus), interview with author, 9 June 2016.

Forgeard in 2006, asked the gathered executives to raise their hands if they wanted to move forward on the A350 XWB. Only a handful did, including sales chief John Leahy [56]. Airbus engineers persisted on the design and came up with a cost-effective way to build the composite fuselage using composite panels. By December, the leadership team changed its opinion, and the Airbus board of directors approved the industrial launch for three variants of the new aircraft. The A350-800 would seat 270 (in three classes) with a range of 8500 nm; the larger A350-900 (314 seats) and A350-1000 (350 seats) would cover 8000 nm. Airbus would fund the new development largely from cash flow, with the first delivery of the A350-900 by mid-2013—a two-year delay to the original schedule. It would drop the General Electric GEnx aeroengines in favor of a new Rolls-Royce Trent variant, the Trent XWB, which was optimized to its requirements. Existing A350 contracts were renegotiated due to price increases compared to the original design. In making this decision, Airbus effectively hedged on its bet that the future belonged to massive jetliners like the A380.

Global Airline Restructuring

While Airbus and Boeing pursued their "moonshots," the early 2000s double whammy of weak air travel demand and increased fuel prices created extreme financial challenges for airlines, who were forced to cut capacity, reduce fleet sizes, cancel leases, pursue labor concessions, and lay off staff. The collateral damage was not limited to just US airlines: In the months following the 9/11 attacks, financially weak carriers Sabena and Swissair were pushed into collapse. (Swissair would later re-emerge as Swiss International Air Lines.)

Traditional cost-cutting measures would prove to be insufficient for many carriers, particularly in the United States. Eight US airlines filed for Chapter 11 bankruptcy protection between 2002 and 2006, allowing them to renegotiate most contracts and restructure everything from labor agreements to aircraft leases. These bankrupt airlines included legacy airlines such as United, Delta, Northwest, and US Airways. Some carriers avoided entering bankruptcy, such as American Airlines, who instead received sufficient labor concessions, and Southwest Airlines, who remained profitable through its low-cost carrier business model.

While legacy brands were disappearing in the United States, consolidation was also taking place among European airlines, but in a different manner. In 2004, flag carriers Air France and KLM Royal Dutch Airlines mutually agreed to merge, but kept separate operations under their respective brands while managed by a single Franco-Dutch holding company called Air France-KLM. The reasons for this strategy were three-fold: (1) governments could preserve their flag carrier's identity and would be less likely to oppose the merger; (2) airline holding companies could invest in a diversified set of airlines; and (3) holding companies could improve the financial performance of small, struggling national carriers by optimizing the combined network and seat capacity. Lufthansa would embark on a similar strategy by acquiring Swiss International Airlines, Brussels Airlines, Germanwings, and Austrian from 2005. Not to be left out, Iberia signed a preliminary agreement to merge in 2009. Two years later, they would form the International Consolidated Airlines Group and cease to trade as separate companies.

Aside from consolidation, another phenomenon was transforming the dynamics of air travel in Europe: the rapid emergence and growth of low-cost carriers. Although many of these airlines began operations well before the 2000s, the 1997 deregulation of the market by the European Union reduced entry and growth barriers, thereby opening the flood gates for LCCs, just as had happened in the United States in the 1980s.

During this time, incumbent LCCs like Ryanair and easyJet experienced tremendous growth. For example, Ryanair flew fewer than 10 million passengers in 1998, but its traffic ballooned to approximately 70 million passengers by 2010, making the airline the leading low-cost carrier in Europe [57]. At the same time, new LCCs started operations, such as Jet2 in 2002, Wizz Air in 2003, and Vueling in 2004.

The Asia-Pacific region became another major source of demand growth in the 2000s as the middle class continued to expand, particularly among large populations like those in China and India. As disposable incomes grew with the growing middle class, a multitude of new low-cost carriers began service, emulating similar phenomena in North America and Europe. Such new carriers included Tigerair (2003), Thai Air Asia (2003), Jetstar Asia Airways (2004), Nok Air (2004), Jeju Air (2005), SpiceJet (2005), IndiGo (2006), Air Asia X (2007), and Air Busan (2007). As a result, revenue passenger kilometers, a measure of air travel demand, grew over 6% annually in the Asia Pacific region from 2000 to 2007.

In China, the Civil Aviation Administration of China (CAAC) merged the country's 10 largest airlines into three airlines that were based in the largest cities: Air China in Beijing, China Eastern Airlines in Shanghai, and China Southern Airlines in Guangzhou. Like Pan Am was in the United States, these three airlines would serve as the primary international flag carriers. Furthermore, with these mergers, the CAAC aimed to create a hub and spoke system similar to other developed countries. To support this growth, Chinese carriers quickly filled OEM order books for new aircraft. For example, Chinese airlines accounted for only ~7% of Boeing orders in the 1990–1999 period, but this figure doubled in the following decade.

The Chinese national government also quickly recognized the importance of building an aviation infrastructure and became the driving force behind new development projects. The 2008 Olympics, hosted in Beijing, provided another motivation to showcase world class facilities to visiting crowds.

Although the largest cities received the most focus at first with projects like Capital Airport in Beijing and Pudong Airport in Shanghai, the government also began to shift attention to regional airports to build the hub and spoke network. Dozens of new airports were planned.

Although demand for air travel was rising in Asia, regulatory barriers remained in place that prevented even faster growth. For example, most of China's airspace was under military control, which resulted in air traffic delays and inefficient routings. Asia-Pacific countries signed only a handful of open skies agreements among themselves. One liberalization advocate was the Association of Southeast Asian Nations (ASEAN), which promulgated a vision of a single Southeast Asian aviation market, much like the European common aviation area.

In the Middle East, the new breed of carriers continued to gain clout. To start the new millennium, Dubai-based Emirates placed an order for 25 777s, 8 A340s, 3 A330s, and 22 A380s. Abu Dhabi–based Etihad Airways, founded in 2003, also placed large orders for 777s and A380s. Finally, in neighboring Doha, Qatar Airways recorded its first profit in 2004 and followed an expansion plan similar to those of its rivals in Dubai and Abu Dhabi.

The passenger traffic beginning or ending in these cities accounted for a very small portion of the overall traffic these emerging Middle East airlines carried. Instead, these carriers relied on connecting traffic to destinations beyond the Middle East. Thus, it was not long before Dubai, Abu Dhabi, and Doha were dubbed the new "center of the world" due to their natural geographic connection point between North America, Europe, Africa, and Asia. Turkish Airlines, watching the positive traffic growth in the neighboring Middle East, would also grow its Istanbul hub as a competitor to its UAE and Qatar-based rivals, also flying the "aviation silk road."

The growth of these carriers arguably had the greatest impact on legacy European network airlines, who historically relied on international connecting traffic from North America to Asia, the Middle East, and on the Kangaroo Route to/from Australia. The advantage of the Middle Eastern carriers, according to Boston Consulting Group analysis, was the lower labor unit costs and lack of corporate taxes, which created an ~20% unit cost advantage over their European and North American counterparts. As a result, Middle Eastern carriers had a cost advantage over European carriers on more than half the passenger volume between Europe and Asia [58].

MOONSHOT DELAYS

The robust growth in the airline industry following the 2001–2002 recession provided an important boost to Airbus and Boeing. Deliveries increased from 571 in 2003 to 654 in 2005 and 811 in 2006. There were several notable orders. Airbus scored an order for 120 A319s from European LCC easyJet in

2002, providing a counterbalance to its competitor Ryanair, which flew 737s. By 2003, Airbus's order book swelled to €129 billion ($145 billion), and A380 firm orders reached 129 [59].

In parallel with the gushing business volume, Airbus and Boeing worked on their moonshots. From 2001 to 2005, the perception was that Airbus executed its A380 development program on schedule. It battled the inevitable weight problems of a new aircraft development and successfully tested its new high technology systems. The first A380 rolled out in January 2005 in a ceremony reminiscent of a Broadway production, and took to the skies three months later—a few months behind the original schedule.

THEN ALL HELL BROKE LOOSE. In June 2005, Airbus announced a delay in the program's schedule attributed to the aircraft's complex wiring systems. Different computer systems were used by Airbus's production facilities, and as a result the 330 miles of wire harnesses on the aircraft were incompatible. The problem first emerged in summer 2004, when the first major assemblies arrived in Toulouse. When the electrical cables were installed, the production teams learned that the cables were too short. They reported the problems to leadership, but it would wait nearly one year to publicly admit the issue. In June 2006, Airbus announced a second delay, adding six to seven months onto the schedule. Confidence in the leadership was shattered, and the share price of parent company EADS fell by 26%. The departure of EADS CEO Noël Forgeard, Airbus CEO Gustav Humbert, and A380 program manager Charles Champion followed. Christian Streiff became the new Airbus leader and conducted a thorough review of the situation. In October 2006, he announced a third delay to the program, pushing the first delivery to October 2007 with just 13 deliveries in 2008. What happened?

The New York Times summed up the situation when it labeled the fiasco as a "story of hubris, haste, inattention, and obfuscation." It wrote, "Personal and cultural rivalries, at the very top level and below, got in the way of efficiency and openness. Computers in Toulouse and Hamburg proved incompatible. An overambitious production timetable for the superjumbo jet, with its super-size technical challenges, discouraged dissent that would cause postponements." It estimated the cost of the delay of at least $6.6 billion. Mario Heinen, the new A380 leader, told analysts, "We ended up with a vicious circle where there was apparently no way out" [60]. Airbus ultimately delivered the first A380 to Singapore Airlines in October 2007. The true extent of the cost of this delay would not be understood for many years.

While Airbus struggled with its first moonshot, Boeing endured a leadership crisis as the 787 development program entered full gear. In May 2005, Harry Stonecipher resigned when an internal investigation revealed an inappropriate relationship with another Boeing executive. Boeing hired an outsider, Jim McNerney, as his replacement. The CEO of 3M, not only was McNerney a

Credit: I, Yasobara, CC BY 2.5

Fig. 3.9 The 787 rolled out in 2007.

highly regarded, results-oriented leader, but he also understood aerospace thanks to a stint as the CEO of GE Aircraft Engines. He was considered by many to be the ideal hire for Boeing. The leading internal candidate, Commercial Airplane Group CEO Alan Mulally, was passed over for the top role. He would leave the company within a year to become CEO of Ford Motor Company.

Boeing also dealt with its own development issues on the 787. Borrowing a page from Airbus, it created a dispersed network of contractors and shipped major assemblies to Everett in a specially modified 747, called the Dreamlifter—Boeing's answer to Airbus's Super Guppy. It began assembling its first aircraft in mid-2007 in anticipation of entry into service a year later. It rolled out its first 787 in a splashy ceremony in July 2007 attended by 15,000 and viewed by as many as 100 million people on television and the Internet (Fig. 3.9). It boasted 677 orders from 47 airlines worth $110 billion at list prices, which qualified it as the most successful commercial airplane launch in history.[7]

Jetliner orders caught fire in the mid-2000s. In 2005, net orders exceeded 2000 for the first time. This was followed by 1996 orders in 2006 and 2495 in 2007 (Fig. 3.10). Receiving 6500 orders in three years was by far the best three-year period in history, easily exceeding the previous record of 4100 orders from 1998 to 2000. Boeing Commercial Airplane chief Scott Carson called it "the most robust upturn in the history of the industry, following what was probably the most robust downturn in the days following September 11" [61].

ANOTHER AIRLINE CRISIS

The end of the first decade of the new millennium ended much like it began: with a crisis that posed steep challenges for the economy. As a result

[7]Boeing Press Release, "Boeing Celebrates the Premiere of the 787 Dreamliner," 8 July 2007.

Type	OEM	Model	2001	2002	2003	2004	2005	2006	2007	2008	2009	2010	TOTAL
Single-Aisle	Airbus	A320	257	236	233	233	289	339	367	386	402	401	**3143**
	Boeing	717	49	20	12	12	13	5					**111**
		737-6/7/8/900	281	212	166	197	207	290	324	284	367	370	**2698**
		757	45	29	14	11	2						**101**
Twin-Aisle	Airbus	A330	35	42	31	47	56	62	68	72	76	87	**576**
		A340	22	16	33	28	24	24	11	10	10	4	**182**
		A380							1	12	10	18	**41**
	Boeing	747-400	31	27	19	15	13	14	16	14	8	0	**157**
		767	40	35	24	9	10	12	12	10	13	12	**177**
		777	61	47	39	36	40	65	83	61	88	74	**594**
		MD-11	2										**2**
		TOTAL	**823**	**664**	**571**	**588**	**654**	**811**	**882**	**849**	**974**	**966**	**7782**

Source: Teal Group
Excludes Soviet and Russian aircraft

Fig. 3.10 Jetliner deliveries: 2001–2010.

of the subprime mortgage crisis that began in 2007, banks that were heavily invested in mortgage-backed securities began to face liquidity crises, leading to bankruptcies and government bailouts. The wider global economy then faced a credit crunch along with increased unemployment.

By the beginning of 2008, air travel demand began to decline because of the impending crisis, and airlines scrambled to reduce capacity in pace with falling demand. Introduction of new flights was postponed, aircraft deliveries were deferred, and layoffs were instituted to reduce costs. Even spending in maintenance, which normally accounts for 5–10% of airline operating costs, was reduced by deferring maintenance and burning through inventory rather than purchasing new spare parts. Despite these actions, the world's airlines would still lose a collective $26 billion in 2008.

As a result, the industry quickly became ripe for consolidation. In the United States, the West Coast low-cost carrier America West Airlines kicked off the merger frenzy by acquiring the larger US Airways in a "reverse merger." Delta and Northwest then agreed to merge in 2008. United would buy Continental, and Southwest would purchase AirTran two years later.

Deliveries of the A380 ramped up in 2008. Emirates became its second operator in August, and Qantas followed in October with flights between Melbourne and Los Angeles. Air France became an operator in 2009.

As the A380 eased into service, in 2007. Boeing announced a six-month delay to the 787's entry into service due to supply chain, software, and documentation issues. Program manager Mike Bair was replaced. Three more delays were revealed in 2008 with the first flight pushed back to late 2009. The 787 was following the same path as the A380, but for different reasons. Some airlines began to question Boeing's ability to execute on the program.

As the first decade of the new millennium ended, the major question was whose vision was right—Airbus or Boeing? Would passengers prefer the A380 connecting mega-cities and major hubs or the smaller 787 flying point-to-point?

REFERENCES

[1] Aboulafia, R., *The 2016 World Civil & Military Aircraft Briefing*, The Teal Group, Arlington, VA, 2016, p. 522.

[2] O' Lone, R., "Boeing Plans 777 As First in New Air Transport Family," *Aviation Week & Space Technology*, 22 Oct. 1990, p. 18.

[3] Hughes, D., "ICAO Members Set Noise Guidelines for Restricting Chapter 2 Aircraft," *Aviation Week & Space Technology*, 5 Nov. 1990, pp. 38–39.

[4] Norris, G., "Update Fails to Save SIA MD-11s," *Flight International*, 7–13 Aug. 1991, p. 4.

[5] "MD-12 Divides Douglas," *Flight International*, 7–13 Nov. 1991, p. 5.

[6] Velocci Jr., A., "Taiwan Aerospace Waffling on Taking Stake in MD-12," *Aviation Week & Space Technology*, 25 May 1992, p. 26.

[7] Ibid.

[8] Smith, B., "Quick Changes in MD-12 Plans Damage Douglas' Credibility." *Aviation Week & Space Technology*, 20 July 1992, p. 27.

[9] Ibid.

[10] Mecham, M., "Cathay Shifts to A340s to Regain European Routes," *Aviation Week & Space Technology*, 13–20 Dec. 1993, p. 32.
[11] Norris, G., and Wagner, M., *Boeing 777*, Motorbooks International, St. Paul, MN, 1996, pp. 20–21.
[12] Bauer, E., *Boeing: The First Century & Beyond*, TABA, Issaquah, WA, 2006, p. 300.
[13] Norris and Wagner, 1996, p. 7.
[14] Andersen, L., and Ekstran, C. L., "Boeing's 777 Will Be Tops When It Comes to ETOPS," *Seattle Times*, 16 Aug. 1993, http://community.seattletimes.nwsource.com/archive/?date=19930816&slug=1716209 [Accessed 26 August 2017].
[15] Lenorovitz, J., "Boeing 737-X to Be Offered," *Aviation Week & Space Technology*, 5 July 1993, p. 36.
[16] Ibid.
[17] Ibid.
[18] Bowen, D., "Airbus Will Reveal Plan for Super-Jumbo: Aircraft Would Seat at Least 600 People and Cost Dollars 8bn to Develop," *The Independent*, 3 June 1994, https://www.independent.co.uk/news/business/airbus-will-reveal-plan-for-super-jumbo-aircraft-would-seat-at-least-600-people-and-cost-dollars-8bn-1420367.html [accessed 5 April 2017].
[19] Sparaco, P., *Airbus: The True Story*, Editions Privat, Toulouse, France, 2006, p. 251.
[20] Sparaco, P., "Airbus Chief Sets New Course," *Aviation Week & Space Technology*, 23 Feb. 1995, p. 52.
[21] Shifrin C., and Sparaco, P., "Boeing to Build 777-300X for Asia/Pacific Carriers," *Aviation Week & Space Technology*, 19 June 1995, p. 31.
[22] Mecham, M., "Year's Biggest Order Goes to 777," *Aviation Week & Space Technology*, 20 Nov. 1995, pp. 28–29.
[23] Kemp, K., *Flight of the Titans*, Virgin Books, London, 2006, pp. 134–135.
[24] Shifrin, C., and Sparaco, P., "Boeing, Airbus Differ on Large Aircraft Issues," *Aviation Week & Space Technology*, 9 Sept. 1996, p. 25.
[25] James, B., "Potential for 400 Jets and $18 Billion Sale: Airbus Gets Huge Order from USAir," *The New York Times*, 7 Nov. 1996, https://www.nytimes.com/1996/11/07/news/potential-for-400-jets-and-18-billion-sale-airbus-gets-huge-order-from.html [accessed 5 April 2017].
[26] Ibid.
[27] Sparaco, *Airbus: The True Story*, p. 281.
[28] Kemp, *Flight of the Titans*, pp. 180–181.
[29] Strickland, J., "From Modest Beginnings: The Growth of Civil Aviation in the Middle East," *Journal of Middle Eastern Politics and Policy*, 17 May 2015, http://jmepp.hkspublications.org/2015/05/17/from-modest-beginnings-the-growth-of-civil-aviation-in-the-middle-east [accessed 5 April 2017].
[30] La Croix, S., and Wolff, D. J., *The Asia-Pacific Airline Industry: Economic Boom and Political Conflict*, East-West Center, Honolulu, Hawaii, 1995, pp. 4–6.
[31] Newhouse, J., *Boeing vs. Airbus*, Vintage Books, New York, 2008, pp. 125–126.
[32] Proctor, P., and Velocci Jr., A., "Woodward Ousted in Boeing Shakeup," *Aviation Week & Space Technology*, 7 Sept. 1998, p. 82.
[33] Aboulafia, R., *The 2017 World Civil & Military Aircraft Briefing*, The Teal Group, Arlington, VA, 2017, p. 409.
[34] Sparaco, P., "Airbus, Asians Plan Regional Twinjet," *Aviation Week & Space Technology*, 26 May 1997, p. 24.
[35] Philips, E., "Southwest Reports 737-700s Demonstrating High Reliability," *Aviation Week & Space Technology*, 3 March 1998, p. 55.
[36] Proctor, P., "Lower Costs Drive Next-Generation 737," *Aviation Week & Space Technology*, 16 Dec. 1996, pp. 68–69.

[37] Tagliabue, J., "Airbus Tries to Fly in a New Formation; Consortium's Chief Hopes a Revamping Could Aid Its Challenge to Boeing," *The New York Times*, 2 May 1996. https://www.nytimes.com/1996/05/02/business/international-business-airbus-tries-fly-new-formation-consortium-s-chief-hopes.html [accessed 5 April 2017].

[38] Rothman, A., and Landberg, R., "Europe Defense Firms Feel Pressure t [Accessed 5 April 2017]. *Seattle Times*, 15 June 1997, http://community.seattletimes.nwsource.com/archive/?date=19970615&slug=2544541 [accessed 11 April 2017]

[39] Kemp, *Flight of the Titans*, p. 181.

[40] Ibid., p. 163.

[41] Ibid., p. 182.

[42] Sparaco, P., "Europe Embarks on $11 Billion Gamble," *Aviation Week & Space Technology*, 1 Jan. 2001, pp. 22–23.

[43] Velocci, Jr., A., "Wall Street on A380: Show Us the Money," *Aviation Week & Space Technology*, 1 Jan. 2001, p. 28.

[44] Dornheim, M., "Its Boeing's Time for Something New," *Aviation Week & Space Technology*, 2 April 2001, p. 32.

[45] International Air Transport Association, "The Impact of September 11 2001 on Aviation," 5 Sept. 2011, pp. 3-4.

[46] Sparaco, P., "Sonic Cruiser Team Grows As Boeing Refines Concept," *Aviation Week & Space Technology*, 5 Aug. 2002, p. 38.

[47] Norris, G., Thomas, G., Wagner, M., and Forbes Smith, C., *Boeing 787 Dreamliner—Flying Redefined*, Aerospace Technical, Perth, Australia, 2005, p. 45.

[48] Mecham, M., "250-Seater Will Draw on Sonic Cruiser Engine Technology," *Aviation Week & Space Technology*, 20 Jan. 2003, p. 23.

[49] Norris, Thomas, Wagner, and Forbes Smith, p. 46.

[50] Holmes, S., "Boeing: What Really Happened," *Bloomberg Businessweek*, 15 Dec. 2003, https://www.bloomberg.com/news/articles/2003-12-14/boeing-what-really-happened [accessed 22 August 2016].

[51] Newhouse, pp. 141–142.

[52] Gunston, B., *Airbus: The Complete Story*, Haynes, Yeovil, Somerset, United Kindom, 2009, p. 253.

[53] Ibid.

[54] Ott, J., and Lott, S., "Lessor Gods," *Aviation Week & Space Technology*, 3 April 2006, p. 22.

[55] Gates, D., "Airplane Kingpins Tell Airbus: Overhaul A350," *Seattle Times*, 29 March 2006, http://old.seattletimes.com/html/boeingaerospace/2002896362_boeing29.html [accessed 30 September 2017].

[56] Hepher, T., "Insight—Flying Back on Course: The Inside Story of the New Airbus A350 Jet," Reuters, 21 Dec. 2014.

[57] "Budget Airlines: In the Cheap Seats," *The Economist*, 27 Jan. 2011, https://www.economist.com/node/18010533 [accessed 30 April 2017].

[58] Love, R., "The Rise of Middle Eastern Carriers: Meeting the New Challenges of the Airline Industry," Boston Consulting Group, September 2006, pp. 5–6. https://www.bcg.com/documents/file86775.pdf [accessed 10 August 2017].

[59] Sparaco, P., and Taverna, M., "Perched for Recovery," *Aviation Week & Space Technology*, 10 Nov. 2003, p. 26.

[60] Clark, N., "The Airbus Saga: Crossed Wires and a Multibillion-Euro Delay," *The New York Times* (from the *International Herald Tribune*), 6 Dec. 2006, www.nytimes.com/2006/12/11/business/worldbusiness/11iht-airbus.3860198.html, [Accessed 15 August 2017].

[61] "Prosperity Greets New Boeing Civil Aircraft Chief," *Aviation Week & Space Technology*, 2 Oct. 2006, p. 19.

Chapter 4

The Regional Revolution

While the jetliner segment consolidated to an Airbus–Boeing duopoly, a revolution took place in smaller aircraft—the "regional" aircraft segment. Although jets were operated for the clear majority of passenger transport routes, there was one segment where turboprops still ruled: short-haul flights less than 500 nm. This would change with the onset of a new class of aircraft called regional jets in the 1990s.

The 1980s: Turboprop Dominance

In the early 1980s, the regional aircraft fleet was composed of a fragmented group of small turboprop aircraft. In 1980, more than 200 regional airlines with an average seating capacity of 16 seats flew an average stage length of just 129 miles in the United States [1]. Popular aircraft models included the Fairchild Metro and BAe Jetstream, both 19-seat turboprops, as well as the larger Saab 340 and Shorts SD360. The Saab 340 was a sleek and fast 34-seat turboprop. In contrast, the Shorts model resembled a flying shoebox, a plodding aircraft with an odd rectangular fuselage (see Fig. 4.1). It was affectionately known as the "shed." Regional air travel was gaining in popularity, with growth aided by code sharing arrangements, where regional airlines (then known as commuters) aligned their schedules with major airline hub operations. American Airlines went a step further and acquired most of its commuter feeder operations. A decade of US airline deregulation also contributed [2]. To cope with growth, operators purchased larger, higher capacity aircraft including the ATR42 (42–52 seats), the Fokker 50 (58 seats), and the de Havilland Dash 8 (37–40 seats). The average seating capacity of the US regional fleet increased 50% between 1980 and 1990 with the number of passengers tripling to reach 41.5 million [3].

The situation in Europe was very different; regional airlines operated 70% of their flights to or from the large (category 1) airports. Concern about

Credit: Pedro Aragão

Fig. 4.1 Short Brothers 360: the "shed."

congestion caused by these aircraft led some to speculate that they were destined to be replaced by high-speed rail. European regional operators responded by increasing the size of their aircraft. The average aircraft capacity of European Regional Airlines Association members was 38 seats in 1990—roughly twice the size of US operators [4].

Europe was the birthplace of arguably the world's first jet aircraft dedicated to short-haul operations, the British Aerospace BAe 146 (Fig. 4.2). Originally launched by Hawker Siddeley in 1973 with aid from the British government, the 146 provided targeted 1970s-era turboprops with range, speed, and a comfortable cabin. The high-wing four-engine design enabled

Credit: Wikimedia

Fig. 4.2 British Aerospace BAe 146.

service of smaller airports with short-field operations, such as London City Airport. Europe was a prime target for the new aircraft. With its capacity of 71 passengers, its five-abreast seating matched Boeing 747 standards of comfort in terms of seat width. After the program was mothballed in 1974, British Aerospace (successor to Hawker Siddeley) restarted production in the late 1970s. British operator Dan-Air took delivery of the first 146 in 1983; the first revenue flight was between London Gatwick and Berne Airport [5]. Orders from US carriers Air Wisconsin and Pacific Southwest Airlines would follow.

In contrast to the large jetliner segment, where US original equipment manufacturers (OEMs) dominated, European manufacturers led the way in regional and commuter aircraft. In the Netherlands, Fokker produced the popular F27 and F50 turboprops as well as the F28 and F100 jets. Fokker had a long history in aviation; it was founded in Germany in 1912—*four* years before Boeing—before moving to its new domicile in 1919. Sweden was home to Saab, a sophisticated producer of fighter aircraft as well as the Saab 340. In the United Kingdom, Belfast-based Short Brothers was even older than Fokker and produced the Short 330 and 360 turboprops. British Aerospace made the Jetstream 31 and 41 models as well as the BAe 146. Dornier produced the Do 228 turboprop in Oberpfaffenhofen, Germany—possibly the catchiest name ever for a production venue. And Avions de Transport Regional (ATR), a French–German joint venture, produced the popular ATR-42 turboprop and in 1989 introduced a larger ATR-72 variant with 68–78 seats.

There were other notable manufacturers in the Americas. Texas-based Fairchild Aircraft made the Metro, and to the north, Kansas-based Beechcraft boasted the popular 1900. Perhaps the most unlikely regional aircraft manufacturer was Brazil's Empresa Brasileira de Aeronáutica (Embraer), a state-owned company founded in 1969. In 1973 it rolled out the 15- to 21-seat EMB 110 Bandeirante, a surprisingly popular debut model that would eventually reach 494 units. It followed with the 30-seat EMB 120 in 1985.

Overall, the regional aircraft segment in the late 1980s was fragmented and dominated by European producers, who accounted for 55% of the units and a much higher share of the value of sub-100-seat aircraft in 1989. And except for the BAe 146, the fleet was composed mostly of turboprops with a clear tendency for larger cabins, despite the advent of high-bypass-ratio turbofans, which had the potential to dramatically improve the economics of jets. Shorts developed a design for a 48-seat regional jet using new-generation aeroengines, the FJ-X, but it remained on the drawing board.

All of this would change in the following decade due to an unlikely revolution centered across the Atlantic Ocean in, of all places, Quebec, Canada.

CANADIAN REVOLUTION

Laurent Beaudoin (Fig. 4.3) did not set out to become the CEO of a major aerospace company. Born in Laurier Station, Quebec, he began his career as

Credit: Bombardier Inc. and its subsidiaries

Fig. 4.3 Bombardier's Laurent Beaudoin.

an accountant before joining Bombardier in 1963 as comptroller. The son-in-law of founder Joseph-Armand Bombardier, he rose quickly and was named president just three years later at age 27. Laurent made his mark in first building a globally recognized snowmobile business (Ski-Doo) and then diversifying into rail transportation, where the company won major contracts in Montreal, New York, and the "Chunnel"—the tunnel under the English Channel. He displayed a penchant for acquiring companies with great assets at bargain prices, fixing them, and positioning them for growth. Opportunism was his acquisition strategy. With solid recreational vehicle and transportation businesses, Laurent wanted a new industry, a third industry pillar, to broaden Bombardier's portfolio. What he needed now was an opportunity—a company in need of rescue. Enter Canadair.

Canadair was one of Canada's oldest and most successful aerospace firms. Established in 1911 as a subsidiary of British firm Vickers & Sons, US defense contractor General Dynamics acquired the firm in 1947. It focused on building F-86 Sabre Jets in the 1950s, and in the 1960s developed the CL-41 Tutor Jet (a trainer) and the CL-215 Water Bomber. Like many aerospace firms, it was devastated by the end of the Vietnam War and the recession of the early 1970s. Employment plummeted from 9200 to 2000, and General Dynamics made plans to close its main production facility in Montreal. Not wanting to see a large piece of the Canadian aerospace industry go under, the Canadian government purchased the firm in 1976 with the intention of fixing it up and selling it to the private sector [6].

The key project to restore the firm's fortunes would come from an unlikely source: aviation entrepreneur extraordinaire Bill Lear. Lear was not only a pioneer in avionics and consumer products like the eight-track tape, but also created (nearly single-handedly) the business jet sector through his eponymous company Lear Jet. Lear sold his stake in the firm to the Gates Rubber Company in 1967 and turned his focus to creating a new business jet with a much larger cabin than the Lear models of his former firm. He named his new design the Learstar 600. Canadair bought an option on the Learstar 600 design for $375,000 in April 1975, and its engineers studied the concept while its sales staff tested market demand. Lear settled on a royalty payment agreement with Canadair and agreed to stay on as a consultant until the prototype was finished [7].[1]

With a concept in hand, President Fred Kearns initiated the new business jet initiative; the aircraft would be called the Challenger. It would be one of the first business jets designed with a supercritical wing, and importantly, it boasted a widened fuselage that allowed a "walk-about cabin." Jean Chretien, Canada's minister of industry, signed off on the investment. Development and certification took longer than expected, and by 1981 just 33 aircraft had been delivered versus the projected 113 in the business plan. The cost of the development program reached Cdn$1.2 billion ($1 billion)—twice the original budget [8]. The severe recession of the early 1980s didn't help matters. The financial shortfalls continued, and the privatization of Canadair became an issue in Canada's 1984 election. The political mood in Canada, like much of the Western world, swung decidedly to the right, and the victorious Progressive Conservative party under Prime Minister Brian Mulroney decided to sell Canadair. In preparation for the sale, the federal government assumed Canadair's Cdn$1.2 billion ($1 billion) in debt. At least five bids were submitted, and Bombardier was declared the winner, paying Cdn$120 million ($100 million), with the Canadian government receiving Cdn$100 million (~$83 million) in special shares in Bombardier "to encourage increased research and development, exports and Canadian content" [9]. Laurent Beaudoin had his third pillar.

In parallel with these acquisitions, a radical idea emerged to transform the Challenger "from a racehorse into a workhorse" and create a new type of regional aircraft. The idea is attributed to Eric McConachie, a highly talented consultant working for Bombardier, who wanted to stretch the Challenger and expand seating capacity from 12 to 40+ seats and turn it into a passenger jet [10]. With hub-and-spoke networks growing in popularity following US airline deregulation, McConachie judged that airlines would benefit from an aircraft with greater speed and range than existing turboprop aircraft. This would enable regional airlines to fly faster and open longer routes to improve hub passenger

[1]Lear died in 1978 and had little influence on the ensuing design of the Challenger.

volume for their customers, the major airlines. The prime market would be airlines wishing to provide service on lightly traveled routes from 200 to 850 miles [11]. In a market survey conducted in late 1986, 81% of the 78 operators were interested in the regional jet concept. Bombardier concluded there was a market for 900 regional jets (including competition) within 10 years; sales projections were for more than 400 units [12]. One of the potential competitors would be a regional jet proposed by Shorts (the FJX). British Aerospace also responded, promoting a "deseated" version of the BAe 146.

Bombardier launched the Canadair Regional Jet (CRJ) program in April 1989 with orders for 56 aircraft. West Germany's DLT (later renamed Lufthansa CityLine) would be the launch customer (Fig. 4.4). Powered by two 9400-lb-thrust General Electric CF34-3 engines, the CRJ would carry 50 passengers. Although the CRJ's operating costs per flight were significantly higher than competing turboprops, its speed advantage meant that its cost per passenger seat mile was only marginally higher than turboprops, and it could fly more flights per day. Passengers also preferred the lower cabin noise that jets offered compared to turboprops. Bombardier estimated that with an equal load factor to a turboprop, the CRJ could deliver an additional $2 million in revenue per year [13]. Its initial price was set at $15 million, about 50% more than a turboprop of a comparable size. More good news arrived when the Canadian and Quebec governments confirmed they would provide Cdn$86 million ($72 million) in reimbursable contributions for the program [14].

Two months after the launch of the CRJ, Bombardier bought Short Brothers—another privatization courtesy of the United Kingdom's Thatcher government—for just $48 million while the British taxpayers assumed its $1.2 billion in liabilities. As part of the agreement, Short Brothers agreed to

Credit: Arpingstone

Fig. 4.4 Lufthansa CityLine introduced the CRJ in 1992.

terminate its FJX project for a new regional jet [15]. This eliminated a potential CRJ competitor and expanded Bombardier's wing manufacturing capability, because Shorts built the wings for the Fokker 100. The following year it added to its business jet portfolio when it purchased bankrupt Gates Learjet. And in 1992, it picked up Toronto-based de Havilland Canada Aircraft from Boeing for $1, ending the latter's disastrous four-year ownership of the company. This acquisition turned out to be very significant for the CRJ's success because it addressed two major capability gaps: an experienced airline salesforce and customer support capability oriented to regional operators. At the time, Bombardier's expertise centered on business jet operators. "By combining their resources, skills and experience and building on their united strengths," Laurent Beaudoin told cheering employees at de Havilland's Toronto plant, "Bombardier and de Havilland can look forward to achieving a strategic position in the aerospace industry" [16]. To facilitate integration of its acquisition, it created the Bombardier Regional Aircraft Division based in Toronto.

In just eight years, Laurent Beaudoin and Bombardier swept up four aerospace manufacturers at bargain-basement prices to become a significant regional aircraft and business jet manufacturer.

EMBRAER: THE UNLIKELY COMPETITOR

With Bombardier making progress in pioneering the regional jet, which manufacturers in the crowded regional aircraft segment would respond with their own designs? Brazil's Empresa Brasileira de Aeronáutica (Embraer) was one of the least likely candidates.

Brazil had a long and proud history in aeronautical design dating back to aviation pioneer Alberto Santos-Dumont in the early 1900s. World War II demonstrated the importance of air power to the government, and in 1946 it organized a commission under Marshal Casimiro Montenegro to set up a world-class aeronautics school. Searching for ideas, Montenegro visited leading US engineering schools and eventually hired Massachusetts Institute of Technology (MIT) professor Richard Harbert Smith to develop a plan for the new school, which was dubbed "The Smith Plan." Montenegro would hire experts and renowned foreign professors for the Instituto Tecnológico de Aeronáutica (ITA), with the majority coming from MIT. ITA, based in Sao Jose dos Campos, would soon become known as one of the best and most prestigious engineering schools in Brazil.

It is not surprising that in 1969, Brazil selected Sao Jose dos Campos as the location for Embraer, its new government-owned aircraft manufacturer. A military government had run Brazil since a 1964 coup, and threw its support behind the startup in part due to the promise of a prototype for an unpressurized 19-seat turboprop aircraft called the Bandeirante—a design created by French engineer Max Holste. The development of the prototype was not without its challenges: One week before the Bandeirante was to be shown to the military, the nose of the aircraft fell off [17].

The government took a 51% share of the fledgling company, and Brazil ordered 80 EMB 110 Bandeirantes for the military with generous payment terms. To raise private funds for the other 49%, the government passed a law to allow individuals to convert 1% of their income taxes to Embraer shares [18]. Air Force Major Ozires Silva, Embraer's first CEO, led a development team that worked tirelessly on the program, and by 1972 the Bandeirante was certified. It proved to be popular inside and outside Brazil, and over 500 Bandeirantes would be produced by 1990. Embraer followed the EMB 110's success with mostly licensed military production before the 30-seat EMB 120 was introduced in 1985. The larger turboprop proved to be popular with US commuter airlines. By the late 1980s, Embraer had morphed from a quixotic dream to an up-and-coming turboprop OEM in a crowded field of manufacturers. But where to go next?

Embraer decided to develop not one but *two* new aircraft. The first, the CBA-123, was a pressurized 19-seat Bandeirante replacement. The aircraft would be a joint venture with Argentina's Fabrica Militar de Aviones. The program captured the zeitgeist of the late 1980s as economic integration between Brazil and Argentina via the Mercosur trading bloc was increasing. Hence *CBA* stood for *Cooperação Brazil-Argentina*. The aircraft was full of advanced technology, including two Garrett TPF351 turboprop engines mounted on pylons on the rear fuselage in a pusher configuration, wings with supercritical airfoils, and digital avionics and engine controls.

The second aircraft would be a direct response to the CRJ, the EMB-145 (Fig. 4.5). The 45- to 48-seat regional jet would feature a stretched EMB

Credit: Wikimedia

Fig. 4.5 Embraer EMB-145.

120 fuselage and a list price of $11 million. Embraer had been actively considering a jet as early as 1986 or 1987. Per former Embraer executive Horacio Forjaz, "We were visiting EMB 120 operator DLT [forerunner to Lufthansa CityLine], and an executive there had an article hanging on the wall arguing that regional jets were coming. We saw the trend of regional turbofans replacing turboprops, and advancements in technology pointed this way as well."[2] To reduce development costs, the EMB-145 would have as much as 75% commonality with the EMB-120. The jet engine configuration would use the "over wing" approach—wing-mounted engines like turboprop designs [19]. Following in the footsteps of Bombardier and the Canadair Regional Jet (CRJ), the EMB-145 would become known as the "ERJ."

Brazil's economy and Embraer enjoyed robust growth in the late 1980s. Then the wheels fell off. Brazil's inflation rate skyrocketed, budget pressures mounted, and progress payments from sponsoring government did not arrive as the country coped with a political crisis when President Collor de Mello was accused of corruption. The news got worse. Iraq invaded Kuwait in August 1990, and global oil prices skyrocketed. The CBA-123's advanced technology and associated $5 million price tag suddenly became a liability. It suffered accumulated losses of $280 million before its inaugural flight in 1990 [20]. Unable to fund two development programs (let alone one), Embraer cancelled the CBA-123 and decided to focus on the regional jet.

Embraer faced more bad news. The ERJ 145 needed a new swept-wing design to meet its speed targets, abandoning the straight-wing EMB 120 design. The engines also needed repositioning. In 1991 the company decided on a rear fuselage–mounted twin-engine design like the Canadair Regional Jet. By the early 1990s, Embraer had accumulated $1.2 billion in debt, which was exacerbated by the need to take out short-term loans at high interest rates [21]. Employment plunged from 12,000 in 1989 to 6000 by 1991 [22]. The ERJ project was temporarily shelved. New leadership was required, and Ozires Silva, the company's original CEO, rejoined the firm after a six-year stint with another firm. He placed Satoshi Yokota in charge of the program as Embraer evaluated its options for survival. The team was motivated. "The employees remaining after the layoffs were the hardest-working and the aviation fanatics," per Yokota, "This was our design and they wanted to build it—they didn't want to build cars or televisions."[3]

Needing capital, Embraer then began searching for risk-sharing partners. The early results weren't encouraging. The best-known aerospace companies were not keen to participate, because they wanted good guarantees and low risk. Then Embraer had a stroke of luck. The McDonnell Douglas MD-12 program had just been cancelled and Spanish supplier Gamesa lost a

[2]Horacio Forjaz (former Vice President—Embraer), interview with author, 24 Sept. 2015.
[3]Satoshi Yokota (former Vice President—Embraer), telephone interview with author, 11 May 2016.

significant contract. At the same time, the Basque government in Spain was very interested in building up the local aerospace industry. Yokota recalled what followed:

> We asked them if they wanted to build the wings for the ERJ and they quickly accepted. Within a matter of weeks, we met with the Basque government and concluded the agreement. They lacked capability for a project of magnitude so we helped them put together a team of local employees and Embraer experts and retirees. Gamesa contributed some $70 million to the program and built a new factory. They weren't as worried about risks or ROI as they wanted to become a major supplier.[4]

After Gamesa was on board, other risk-sharing agreements were easier to conclude. Major partners included Sonaca (fuselage and wings), Enaer (empennage components), Allison (GMA 3007A engine), Parker Aerospace (flight controls and fuel systems), and C&D Technologies (interiors). These partners proved to be essential because the aircraft redesigns resulted in a three-year delay compared to the original schedule.

Despite the difficulties, Embraer held letters of intent (but few firm orders) from 14 airlines in the early 1990s. These airlines saw economic advantages of the ERJ 145 relative to the CRJ200, even though it would arrive four years later. It had lower fixed costs per flight, and the list price was $17.6 million compared to approximately $21 million for its Canadian competition [23]. European airline Regional was the first airline to purchase the aircraft.

As if all this activity wasn't enough, Embraer was privatized on 7 Dec. 1994. A 55% controlling stake in Embraer was sold for $89 million to a group of investors; the Brazilian government retained a 20% ownership and a veto power through golden shares. Another 10% was given to Embraer employees, and the rest was sold to the public. The purchasers assumed $290 million of debt and immediately injected $35 million into the company [24].

Exit Fokker

While Bombardier and Embraer pursued their new regional jet initiatives, in the Netherlands Fokker pursued its own regional aircraft strategy. Like Embraer, it pursued simultaneous development of two new aircraft. The first program was the F50 turboprop, an update to the F27 Friendship that had been in production since 1958. The straight-wing F50 entered service in 1987 and was a light seller. The second program, discussed in Chapter 2, was the Fokker 100. A major development cost overrun on the F100 program brought the company to its knees. Near death, it was bailed out by the Dutch government in 1987. One of the conditions for the bailout was to bring in strategic partners;

[4] Ibid.

Credit: Arpingstone

Fig. 4.6 Fokker F70.

British Aerospace and DASA were touted as strong candidates. Fokker even discussed setting up a second assembly line with Lockheed [25].

Despite the financial duress, Fokker looked to round out its product family. Should it grow the F100 to a larger size or shrink it to compete in the growing regional jet sector? Fokker made the fateful decision to "go small" with the F70 (Fig. 4.6). It called the F100/F70 combination the "JetLine" range. On paper, the F70 was the lower risk option from the standpoint of development costs. It would use the same engines, wings, and systems as the F100, but with a shorter fuselage. Additionally, to meet US "scope clause" restrictions, it would be limited to 70 seats. Scope clauses were agreements between major airlines and pilots' unions that limited the number and/or size of aircraft that a company could contract out to lower-paying regional airlines.

Fokker had initially offered an 80-seat version with new technology engines, but per chief engineer Rudi den Hertog, "airlines wanted the aircraft as soon as possible…they didn't want to wait for new engines" [26]. Fokker therefore stuck with the Tay 620 aeroengines used on the F100. The combination of older technology engines and a shortened fuselage equated to a relatively heavy aircraft (22,673 kg) with suboptimal fuel economy. And the F70 had a $23 million price tag. A middling value proposition coupled with an early 1990s recession led to disastrous financial performance.

In 1993 and 1994, the company cut nearly 4000 jobs and transferred in-house production of structures and components to a separate company. This wasn't enough. In mid-1994 the company reached a $1.16 billion agreement with the Dutch government and Daimler Benz Aerospace (DASA) where the latter ended up with a 51% share of Fokker [27].

Two years later, the curtain fell on Fokker when DASA ended its financial support of the company after major losses. Excess production capacity, significant competition, and the guilder's unfavorable exchange rate versus the US dollar all played a role in this ignominious ending. The impact on the Dutch economy was sudden and dramatic. More than 5000 employees were furloughed—the largest corporate layoff in Dutch history—and vendors and suppliers throughout the country were also hit. "This is a tragedy," lamented Dutch Prime Minister Wim Kak [28]. After 77 years as a major aircraft manufacturer, Fokker was gone. The contest for regional jet supremacy now had one less competitor. This was good news for Bombardier and Embraer because the F70 was considered a key regional jet competitor.

Regional Jet Demand Growth

The 1990s started out slowly for Bombardier as the industry dealt with a deep global recession. No new orders for regional jets were booked at the 1992 Farnborough Air Show, and the company held orders for just 36 aircraft—enough to carry the company to mid-1994. This was a long way from the program's breakeven of 400 aircraft. Nonetheless, CEO Laurent Beaudoin remained confident that economic recovery would yield growth [29]. He was prescient, as in 1995 regional jet deliveries exceeded 100 for the first time (Fig. 4.7). In Europe, home of BAe 146 and the CRJ launch customer Lufthansa CityLine, the number of regional passengers increased 12.9% from 1995 to 1996 to reach 50 million. By 1997, 26% of the European fleet of 859 regional aircraft were jets, up from 14% in 1992. The average flight length was 422 km (267 nm) [30]. Regional jet pioneer Lufthansa CityLine changed the very nature of its operations. It handed over its Fokker 50 turboprop operations to affiliate Contact Air and returned its Dash 8s to de Havilland so that it could focus entirely on jet operations. By 1996 its all-jet fleet had reached 28 CRJs and 15 Avro RJ85s [31].

Likewise, CRJ customer Comair restructured its operations in North America. Comair, like many US regional airlines, was an affiliate to a major airline (Delta Air Lines). In this capacity, it delivered passengers to Delta's major hubs from surrounding communities, known as regional feeds, and also increased its frequency of service on mainline routes during times when demand did not warrant use of large aircraft. The advent of CRJs enabled Comair to expand its zone of operations considerably and build up its Cincinnati, Ohio, hub. From 1996 to 1997, its passenger enplanements grew 15% as revenues expanded by more than 20%. The average flight was 330 miles—a combination of new routes and turboprop replacements. For part-owner Delta, Comair's CRJs dramatically expanded the number of feed passengers more than smaller turboprops did—"more than Delta ever thought was achievable," said Comair marketing executive Charles Curran. This growth resulted in Cincinnati, once a sleepy second-tier airport, becoming Delta's second largest hub (Fig. 4.8) [32]. Other major North American

OEM	Model	1990	1991	1992	1993	1994	1995	1996	1997	1998	1999	2000	TOTAL
British Aerospace	BAe146	25	26	13	15	28	22	26	21	10	23	14	**223**
Fokker	F70/100	35	36	58	53	30	41	17	7	0	0	0	**277**
Fairchild	328-JET	0	0	0	0	0	0	0	0	0	15	33	**48**
Bombardier	CRJ1/200	0	0	3	22	26	41	51	61	77	83	99	**463**
Embraer	ERJ 135/140/145	0	0	0	0	0	0	4	32	60	96	156	**348**
	TOTAL	**60**	**62**	**74**	**90**	**84**	**104**	**98**	**121**	**147**	**217**	**302**	**1359**

Source: Teal Group

Fig. 4.7 Regional jet deliveries: 1990–2000.

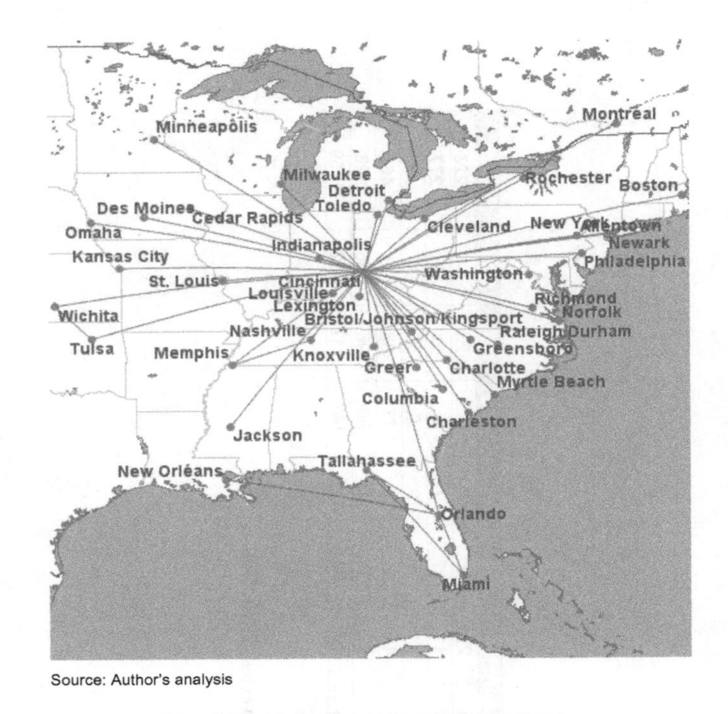

Source: Author's analysis

Fig. 4.8 Comair route map: mid-1990s.

CRJ deals followed. Orders from SkyWest and Air Canada in 1993 doubled Canadair's backlog and raised the possibility that Canadair's production rates would need to increase. Delta affiliate Skywest used CRJs to turn Salt Lake City into a major Delta hub by adding distant locations such as Vancouver, San Francisco, and Albuquerque to its route network while using turboprops for shorter flights. Orders from Atlantic Coast Airlines and Atlantic Southeast Airlines for CRJs followed a few years later.

The US regional airline business model depended on pilot salaries that were dramatically lower than those of major airlines. In 2001, for example, the starting salary for a captain at Comair was $50,800 whereas starting copilots earned just $16,000—the 40-hour-week equivalent of $7.50 an hour [33]. This led to the creation of *scope clauses*, a contract between an airline and a pilot union that limited the number and/or size of aircraft that the airline could contract out to its lower-cost regional feeder. The goal was to protect union jobs at the major airline from being outsourced to regional airlines. The 2005 Northwest Airlink scope clause, for example, set a maximum size limit of 69 seats and a numerical limit of 54 regional jets with 50–69 seats. This meant that it could not buy 70-seat regional jets. Other scope clauses set the maximum size even lower. Bombardier created the 44-seat CRJ440 by replacing two rows of the CRJ200's seats with a closet to comply with a Northwest Airline 50-seat scope clause. Scope clauses meant that, in some instances, regional airlines over-ordered smaller aircraft to compensate for their inability to operate larger regional jets.

In Europe, the concept of regional airlines as short-haul feeder services, popular in North America since the 1980s, took hold. British Airways was the first European carrier to conclude a franchise "express" feeder arrangement when it signed up CityFlyer Express to feed its Gatwick hub in 1993. Other European carriers followed its lead in establishing franchise agreements, including Lufthansa, Air France, Sabena, Air Europa, Alitalia, and Iberia [34]. This development added to regional jet demand.

While Bombardier snagged early regional jet orders and built a solid position, Embraer gained its own momentum. After few small initial orders, Continental Express placed a firm order for 25 ERJs valued at $375 million (list price) and options for another 175. ERJ program manager Satoshi Yokota recalled telling Continental Express President David Seigel, "We appreciate the offer but don't see you taking all of the options." Seigel retorted, "If you quit smoking we'll take the options." Yokota did give up the habit, and Continental Express ended up taking the full quota of aircraft.[5] Another major domino fell when American Eagle, an affiliate of American Airlines, ordered 42 ERJs at the 1997 Paris Air Show—Embraer's largest-ever order. The company appeared to be on its way, but not without a legal challenge from its Canadian competitor.

REGIONAL JET AEROPOLITICS

In 1996, following the large Continental Express order, Bombardier accused Embraer of receiving illegal subsidies to reduce the list price of its ERJs from $15 million to effectively $12.5 million. The CRJ, in contrast, had a list price of $18 million [35]. Both companies appealed to their respective governments to file formal complaints to the World Trade Organization (WTO). At the center of the dispute was Brazil's Proex program, which offset the country's "risk premium" in making loans to foreign customers. Because of Brazil's history of hyperinflation, its banks and government could borrow only at a substantially higher cost of capital (as much as 4–6%) than counterparts in Europe and North America. Brazil effectively provided a direct subsidy to foreign customers, which pushed borrowing rates below US Treasury Bills. In the case of the Continental Express order, its borrowing rate was 2.5% below US Treasuries; for American Eagle's ERJ order, it was 2.5%. Brazilian President Fernando Henrique Cardoso and Canadian Prime Minister Jean Cretien met in early 1997 and agreed to appoint independent mediators to resolve the dispute [36].

---

[5]Ibid.

Mediation was not successful, and a WTO adjudication panel found that Brazil's subsidies were indeed illegal. It agreed to reduce its Proex interest rate subsidy to 2.5%. The Technology Partnerships Canada program, which supported Bombardier, was also judged to be out of bounds, forcing Canada to amend the program.

Canada returned to the WTO in 1999, charging Brazil with failing to comply with the earlier decisions. The WTO asked for a further reduction in interest subsidies, but Embraer refused to cancel existing contracts. In 2000, the Canadian government obtained authorization from the WTO to impose $1.4 billion in retaliatory tariffs on Brazilian exports. Rather than moving forward with the sanctions, Canada opted to match the interest provided by the then-updated Proex program to help finance Air Wisconsin to purchase 75 CRJs. In 2002, the WTO sided with Brazil that these were also illegal subsidies and awarded $248 million in retaliatory tariffs [37]. In the end, although the level of countermeasures was lopsided, public perception was that both countries were subsidizing their local player.

THE FEE-FOR-SERVICE DREAM HOUSE

While the aeropolitics raged, regional jet demand blossomed. Passengers preferred jets to turboprops, and regional jets offered major airlines the opportunity to strengthen their hub-and-spoke networks due to their lower operational costs. Regional jets had 30% higher unit costs per seat mile than larger jets like the 737 and A320, but their costs *per trip* were lower and their breakeven load factor was ~50% [38]. This meant that for thinly traveled routes, the revenue from the small number of passengers could cover the trip costs of a regional jet when a larger jet would be uneconomical. Regional jets could also offer increased frequencies on thicker routes operated by larger jets [39]. Airlines concluded that regional jets could expand their hub fortresses, allowing them to serve new cities, increase route frequencies, or steal traffic from the hub catchment areas of competitors. This led to fee-for-service commercial arrangements where major airlines paid regionals a fixed amount per segment, regardless of the number of passengers. Regional operators accrued impressive profits. Industry analyst Richard Aboulafia referred to it as "a dream house."[6]

A study of US regional jet operations between 1996 and 2000 concluded that:

- 32% of introductions were for new services.
- 23% were frequency supplements (with larger jetliners).
- 23% were capacity reduction.
- 22% were turboprop replacements.

---

[6]Richard Aboulafia (Vice President—Teal Group), telephone interview with author, 25 Feb. 2017.

The motive for regional jet introduction varied with route characteristics. On very small spoke routes, more than 40% of RJ introductions replaced turboprop service. In contrast, on large spoke routes, almost half of RJ introductions were for the purpose of supplementing frequency on the route. Finally, on point-to-point routes, 56% of introductions were new service [40].

British Aerospace Responds

While new competitors emerged in the Western Hemisphere, British Aerospace steadily increased its footprint with the BAe 146. It offered two variants: the 85-seat 146-200 and the 100-seat 146-300. Both were powered by four Avro Lycoming ALF 502 turbofans, which produced short-field capability and relatively quiet operations, but high fuel and maintenance costs relative to two-engine competitors. Production rates averaged 20–25 per year in the late 1980s, and in 1989 reached 36. The BAe 146 proved to be popular in Australia, were Ansett Australia and Australian Airlines were operators, and in the United States, where customers included Pacific Southwest, Air Wisconsin, and Aspen Airways.

British Aerospace updated the 146 family in the early 1990s with the RJ70 (70 seats), RJ85 (112 seats), and RJ100 (128 seats). All three featured an improved cabin and more efficient Lycoming LF507 engines. Deliveries began in 1993.

Falling demand for regional aircraft in the mid-1990s made it clear that there were still too many regional aircraft manufacturers. In January 1996, ATR partners Alenia and Aerospatiale as well as British Aerospace subsidiaries Avro International Aerospace and Jetstream International merged into Aero International Regional (AIR). The objective of this initiative, called *Minibus*, was to reduce competition and secure Europe's position in regional aircraft production. The combined revenue was $1.4 billion with aggregate production of more than 100 aircraft in 1996, including Jetstreams, ATRs, and the BAe RJ family [41]. One of the future projects would be a twinjet 70-seat regional jet.

70-Seaters: The Next Wave

With the momentum for 50-seat regional jets in full force and a substantial lead on the competition, Bombardier began to look for its next challenge. It foresaw the need for larger regional jets to compete with the BAe 146 and forthcoming Fokker 70. In 1995, it began work on a new 70-seat regional jet dubbed the CRJ700 (see Fig. 4.9). Rather than shrinking a larger airframe like its European competitors, it would stretch the CRJ's fuselage and maximize system commonality. The aircraft would seat 66–78 passengers and would feature a new wing with leading-edge slats and GE CF34-8 engines with 12,670 lb of thrust each. Its trip cost would be 15% higher than the 50-seat

Credit: RHL Images from England

Fig. 4.9 Lufthansa CityLine CRJ700.

CRJ, but its seat-mile costs would be 20% lower. The list price was set at $23 million. French regional Brit Air became the launch customer in early 1997 [42].

Embraer began talking about its own 70-seat aircraft in 1996 and 1997. In 1998, it completed a detailed market feasibility assessment to understand customer needs and operations. "We used a new approach with the customer," recalls former Embraer CEO Mauricio Botelho. "We established committees with customers to discuss in very deep ways what the aircraft should be. In the beginning, we had 5–10 customers; at the end of the research, we had 30–40 customers."[7] The dialogue uncovered a critical opening for Embraer to differentiate itself from Bombardier: a wider fuselage. In Australia, for example, Embraer learned that BAe 146s were more popular than CRJs, and its wider fuselage was an important reason.[8] What emerged from Embraer's research was a new clean-sheet aircraft with a wider, roomier fuselage than the CRJ and a "double bubble" layout designed to create more cabin and baggage space. It would take a family approach for a 70-seat E170 (see Fig. 4.10) and larger 108-seat E190. The family would be dubbed E-Jets.

Embraer launched the E-Jet program at the 1999 Paris Air Show with a landmark $1.7-billion order from European carrier Crossair for 30 E170s and 30 E190s, with options for another 100. Including the options, the deal was worth $4.9 billion—the largest order ever for regional aircraft by any manufacturer. Embraer's euphoria was matched by Fairchild Dornier's disappointment as Crossair dropped its orders for the 728 jet (it was also the launch customer for

[7]Mauricio Botelho (former CEO—Embraer), telephone interview with author, 24 May 2016.
[8]Forjaz interview, 25 Sept. 1995.

Credit: Antti Havukainen

Fig. 4.10 Embraer E170.

this aircraft) for the E170/190. The cost of the program would be $850 million [43]. In contrast to the ERJ, the new aircraft would feature underwing engines: CF34-8 for the E170 and CF34-10 for the E190. Honeywell would again be the primary avionics supplier with its Primus Epic suite. The Crossair order was a game-changer for Embraer. "At the time Crossair was the most influential regional operator in the world," Forjaz explained, "This changed the whole perception of Embraer, and after this order, suppliers were anxious to join the team."[9] Embraer changed its supply chain strategy for the new program. Instead of the hundreds of suppliers for the ERJ, it would have just 38 suppliers, of which 16 were risk-sharing partners [44]. The details of this ambitious supply chain strategy are addressed in Chapter 8.

In 2000, Bombardier responded with another stretch—the 86-seat CRJ 900. The $135-million development program would use most of the same suppliers as the CRJ 700. "This is not earth-shattering in terms of development," said Bombardier regional aircraft president Steve Ridolfi, "We've kept the same partners and are keeping it simple. The CRJ 700 was the big mid-platform change transitioning from a small to a big jet; the CRJ 900 is an incremental jet" [45].

As Bombardier opened the door on the CRJ 900, it shut the door on an even larger aircraft—a 108- to 115-seat white-sheet aircraft dubbed the BRJ-X. Bombardier had studied the concept since 1998, and it would have featured five-abreast seating like the DC-9. The program's $1 billion development costs and restrictive pilot scope clauses at US airlines, which prevented regional

[9]Ibid.

carriers from flying 70+-seat aircraft, were cited by Bombardier officials as the reasons for terminating the program [46]. There was another reason cited by former Bombardier engineering chief John Holding: "We were concerned that Airbus and Boeing would kill us."[10] The idea of a larger aircraft, however, didn't die and would return in the future.

More Attrition

German turboprop manufacturer Dornier had its own ambitions in the fledgling regional jet market. Rather than targeting the 50-seat segment, it would go small and pursue an underserved niche: 30-seat regional jets. Like Embraer's approach with the EMB 120 Brasilia, it would add turbofans to its Dornier 228 design. But it needed development funds, and its financial position was precarious. It sold 80% of the company to US manufacturer Fairchild Aircraft in June 1995. Fairchild would end the 228 program, focus on its own 19-seat Metro, and fund the 30-seat jet program. The new majority owner guaranteed to keep 1200 jobs at the firm's Oberpfaffenhofen, Germany plant until 1999 and would produce its new regional jet there [47]. The 328 resembled a shortened BAe 146—a stout, sturdy design. The company didn't stop there. At the 1998 ILA Berlin Aerospace Show, it announced a family of regional jets—the 528 (50 seats), 728 (70 seats), and 928 (90 seats). It also initiated a slightly larger version of the 328—the 428. An $850-million investment was planned, and provisional orders for 60 728s—the first aircraft to be built—were placed by both Lufthansa CityLine and Crossair [48].

Fairchild Dornier was attempting to break out of the pack and become the strongest rival to Bombardier and Embraer. It looked like it was on its way, but poor sales of the 328, which first flew in 1998, weakened the company's balance sheet. Competition from the new 37-seat EMB 135 didn't help either, as the 328 lost opportunities like Continental Express because it lacked an aircraft family. Orders for the 728 in the vital North American market didn't materialize as forecasted due to scope clause restrictions, and it lost one of its key launch customers, Crossair, to Embraer. Contributing to its sluggish demand was the sense that Fairchild Dornier overdesigned the 928 and 728 to compete with Airbus and Boeing, which made it too heavy and not competitive.[11] The aircraft manufacturer also negotiated with Aviation Industry of China (AVIC) to build a 528 production facility in China to improve competitiveness [49]. Then the 9/11 terrorist attacks and a global recession struck. By early 2002 the company's financial situation was precarious. Orders

[10]John Holding, (former Executive Vice President—Engineering & Product Development—Bombardier Aerospace) interview with author, 29 July 2016.

[11]Ben Boehm (former Vice President—Commercial Aircraft Programs—Bombardier Aerospace), interview with author, 8 Jan. 2016.

OEM	Model	2001	2002	2003	2004	2005	2006	2007	TOTAL
British Aerospace	BAe146	10	2	4	0	0	0	0	**16**
Fairchild	328-JET	29	8	9	9	1	0	0	**56**
Bombardier	CRJ1/200	109	135	155	108	52	22	10	**591**
Bombardier	CRJ7/900	23	45	66	67	78	76	49	**404**
Embraer	ERJ 135/140/145	164	131	104	102	58	26	44	**629**
Embraer	E170/175/190	0	0	0	46	73	92	123	**334**
	TOTAL	**335**	**321**	**338**	**332**	**262**	**216**	**226**	**2030**

Source: Teal Group

Fig. 4.11 Regional jet deliveries: 2001–2007.

that were supposed to provide cash flow to fund its ambitious development program failed to materialize or were deferred. In March 2002, it held talks with Boeing but the negotiations stalled over differences in the value of the business and Boeing's strategic priorities. In April 2002, despite holding $11.7 billion in orders (50% firm), Fairchild Dornier ran out of cash and filed for insolvency [50]. The 728, weeks away from its first test flight, never flew. The 328 production line was shuttered. Just 83 had been delivered. Bombardier considered purchasing the company but walked away in June 2002, sealing its fate. Thousands of jobs disappeared. It was a shocking and ignominious final chapter for a proud company.

British Aerospace would also face the gallows. Its membership in the AIR partnership proved to be a diversion, and in 1997 a proposed 70-seat twinjet was shelved. Alenia was the only partner of the four-company consortium willing to support the project; even British Aerospace declined further investment [51]. Just 78 BAe 146s were delivered from 1997 to 2001; during the same period Bombardier and Embraer delivered more than 950 regional jets (Figs. 4.7 and 4.11). The company did continue investment in an update to its RJ family—the RJX—which would replace the LF507 with four modern Honeywell AS977 turbofans. It also studied, but rejected, a design alternative to re-engine the 146 with two CFM56 engines. In 1999, British Aerospace and Marconi Electronic Systems merged to create BAE Systems. The new firm was very focused on the defense industry, and in 2001 it decided to exit the regional aircraft segment. It was hemorrhaging money at a rate that threatened corporate survival, and in the post–9/11 environment, leadership decided that the four-engine RJX was no longer practical. "In spite of every effort to drive our costs down, the price the market currently expects to pay for aircraft of this type means we would incur a significant loss for each one sold," BAE Systems executive Nick Godwin told *Aviation Week & Space Technology* [52]. BAE honored its remaining orders, and then production of the Avro RJ ended with the final four aircraft being delivered in late 2003 (Fig. 4.12). Another important European regional aircraft manufacturer, a pioneer in short-haul jets, was gone. BAE exited the industry having produced 166 Avro RJs and 239 BAe 146s. It left an important legacy. "You could say that the BAe 146 was the first regional jet," said former Embraer

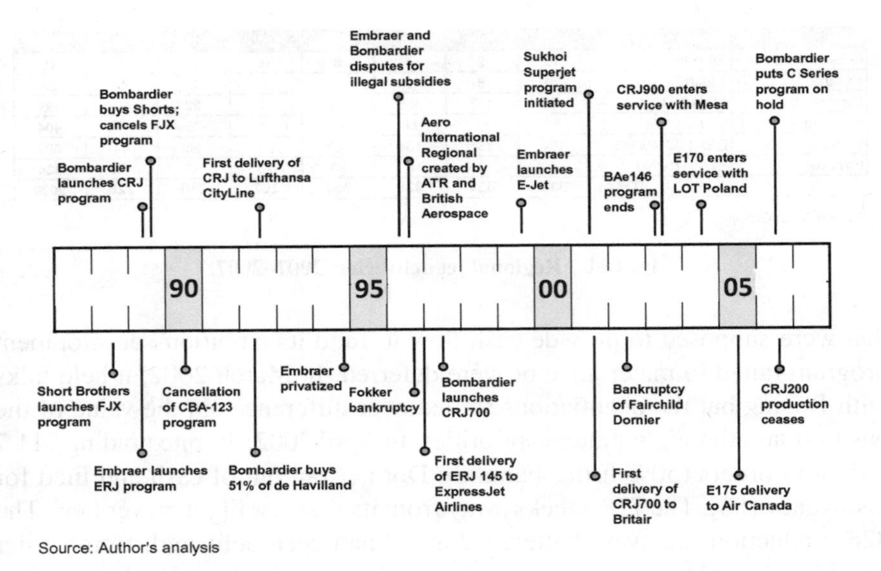

Fig. 4.12 Regional jet milestones: 1988–2007.

CEO Mauricio Botelho. "It was a good aircraft but it was killed by operating costs of four engines."[12]

As Fairchild Dornier and BAE Systems exited the market, a new competitor threw its hat into the ring. In 2002, China's 10th Five-Year Plan targeted commercial aircraft production as a key national objective. So, in September of that year the AVIC I Commercial Aircraft Company (ACAC) was founded with the objective of making China a player in the jetliner business. Its entry would be a 70- to 95-seat regional jet, the ARJ21. Closely resembling the MD-80 and MD-90, the ARJ21 was produced in the same Chinese factories that had produced parts for McDonnell Douglas's Trunkliner program in the 1980s. China claimed that the ARJ21 was a completely indigenous design, although Ukraine's Antonov Design Bureau was responsible for the wing. Western suppliers would also participate, including General Electric with its 17,000-lb CF34-10A turbofans. Entry into service was planned for 2007, with its first flight in 2005. Deemed to be "disastrously heavier" than its nearest competitor (the E170), the ARJ21 was dismissed by a well-known aerospace analyst as "a random collection of imported technologies and design features flying together in loose formation" [53]. Nonetheless, the ARJ accumulated 71 orders from Chinese carriers and lessors in 2003–2004. It would soon fall years behind its original development plan; by 2006 the initial flight was still several years away.

[12]Botelho interview, 24 May 2016.

REGIONAL JET TRANSITION

Bombardier's sales quickly transitioned to its larger models as the regional jet market matured. In 2001, it delivered 23 CRJ 700s with Air France carrier Brit Air the launch customer. The CRJ 900 entered the market in 2003. By 2005, it was producing 78 of both models, which helped to offset the decline of 50-seaters. Important orders included Delta Connection Carriers ASA and Comair, Mesa, and Horizon.

Embraer delivered 46 E-Jets in 2004, its inaugural year. The first 70-seat E170s were delivered in March 2004 to LOT Polish Airlines, followed by US Airways subsidiary MidAtlantic Airways and Alitalia. Surprisingly Air Canada—headquartered in Bombardier's backyard—took deliveries of the first E175s. E-Jet demand quickly ramped up and reached 92 by 2006.

By the mid-2000s, the wave of 50-seat deliveries quickly faded. CRJ 1/200 production, which peaked at 155 units in 2003, ended by 2008. In the same year, Embraer delivered just six ERJs. In 2011, Embraer delivered its last ERJ. What happened? Industry analyst Richard Aboulafia offers a concise explanation:

> The wave of bankruptcies in the 2000s exposed the fallacy of pursuit of market share at all cost by US majors. There's nothing like the gallows to focus one's mind, and they [the majors] fundamentally changed the contract structure with regionals. The fee for service model was a dream house, and it went away as majors grabbed profits. Demand for feeders fallowing this bubble was the same as before it all began. The legacy of the entire episode may be today's 90-seat jets.[13]

It turned out that 50-seat regional jets were a 15-year phenomenon, artificially inflated by scope clauses, cheap fuel, and herd mentality among US major airlines to expand feeder networks. Combined, Bombardier and Embraer would deliver about 1900 units over the 1992–2011 timeframe. But the impact of regional jets was real. The number of passengers carried by US regional carriers increased by an order of magnitude between 1980 and 2005 to reach 152 million (see Fig. 4.13); during the same timeframe, the average passenger trip increased from 129 to 442 miles, while average seating capacity tripled. Major regional carriers emerged. By 2010, US regionals American Eagle, SkyWest, and ExpressJet all operated more than 200 regional jets, while a further six operated more than 100. There were important operators in Europe as well including Lufthansa CityLine, KLM Cityhopper, and Régional.

There is another side of this legacy, because the attrition of regional jet manufacturers was severe. This included Fokker, British Aerospace, and Fairchild Dornier. "Regional jets are the story of losses on an epic scale…a rainbow coalition of bankruptcy," said Richard Aboulafia. "This is the only

[13]Richard Aboulafia, (Vice President—Teal Group), interview with author, 6 May 2015.

Statistic	1980	1985	1990	1995	2000	2005
Passengers Emplaned (millions)	14.7	27.0	41.5	58.3	82.5	152.6
Average Passenger Trip Length (miles)	129	174	194	217	296	442
Average Seating Capacity	16	26	24	30	37	50
Revenue Passenger Miles (billions)	1.9	4.7	8.0	12.6	24.4	67.4

Source: 2015 Regional Airline Association Yearbook

Fig. 4.13 US regional airline statistics.

aerospace sector where companies consistently lose money. Embraer and Bombardier were the only people to buck that trend."[14]

BOMBARDIER'S 747

With the regional jet segment maturing and the E190 gaining momentum, Bombardier decided to dust off its BRJ-X concept from the late 1990s. Pierre Beaudoin, Laurent Beaudoin's son and then president of Bombardier Aerospace, decided to bring in an outsider to lead the initiative.

Gary Scott was the needed new blood to lead a new larger aircraft program at Bombardier. A long-time Boeing veteran, he was program manager for the 737 and 757 in the early 1990s and later the Vice President of Commercial Aircraft Business Strategy. He was also rumored to be the next CEO at CAE, a Canadian simulator and training services supplier. Scott opined:

> "When Bombardier asked me to join, there were three things going on in my mind: First, I knew there was a market for this aircraft based on experience. Second, I knew that physics was on our side, as the segment was served primarily by out-of-production aircraft. Finally, I knew that for Bombardier the C Series would be the equivalent of what the 747 was for Boeing. It was bet the company."[15]

Addressing the C Series opportunity would require a new mindset. "We realized we were in a regional jet mindset," said former Bombardier executive Ben Boehm. "We put this away and started thinking single-aisle. Much of our effort in 2004 was to change the mindset throughout the organization…rather than straddling 90–110 seats—both worlds—we needed to think 100 seats as the minimum. We focused on new routes we could open up."[16]

---

[14]Aboulafia telephone interview with author, 25 Feb. 2017.

[15]Gary Scott (former President—Commercial Aircraft, Bombardier Aerospace), telephone interview with author, 21 Sept. 2015.

[16]Boehm interview, 8 Jan. 2016.

The C Series would be 110–130 seats with a 3000-mile range. The development program, Bombardier's largest to date, would target the perceived gap between E-Jets and larger 737 and A320 single-aisles. Scott brought in a new approach based on his Boeing experience, which emphasized the importance of program management. There were two triangles: the first was engineering, operations, and supply chain; the second was sales, marketing, and customer support. In between these two triangles was program management. Scott leaned heavily on Rob Dewar, the C Series's Vice President of Integrated Product Development, to execute the initiative.

Bombardier faced several headwinds in getting the program on track. The first was the magnitude of the task and potential competitive responses, including price discounting from Boeing and Airbus to defend their turf. Second, two of Bombardier's most important potential customers—Delta and Northwest—were in bankruptcy. And it was having trouble finding an engine supplier. CFM International and International Aero Engines—suppliers to the 737 and A320—both declined to participate, citing insufficient market demand. Only Quebec's Pratt & Whitney Canada seemed interested, but it had never built an engine in the thrust range required for the C Series and would take many years to develop the new engine [54].

"By the middle of 2005 we were getting more airlines interested in the concept and they opened their books to us for the evaluations," said Ben Boehm. "After we crunched the numbers, we realized that we needed more, and that the commonality of incumbents would overwhelm the performance advantages of the C Series." Gary Scott added, "The initial C Series was good, maybe 10% better performance, but it wasn't good enough. Airlines weren't overwhelmed by it. We finally decided it would be a mistake."[17] So the C Series was put on hold. After two years of research and development and an expenditure of $100 million, Bombardier needed a better value proposition. Pierre Beaudoin asked Gary Scott to retain 50 of the C Series's 350 employees to work on a new business plan [55].

In 2005, 13 years after Bombardier first introduced the CRJ, the future in the regional jet segment appeared to be set. Bombardier would continue to focus on its CRJ 700 and CRJ 900 models as it contemplated yet another fuselage stretch. Embraer would continue to ride the popularity of its well-positioned E-Jet program. The regional jet segment, which was once messy, competitive, and highly political, had become a well-defined duopoly. Embraer and Bombardier controlled the market, and even Brazil and Canada were at peace with the subsidy disputes. Little did the regional aircraft manufacturers know that a major disruptive innovation in aeroengine technology was in the works that would shatter this outlook.

---

[17]Scott interview with author.

REFERENCES

[1] US Regional Airline Association, *2015 Annual Report*, Regional Airline Association, Washington, DC, 2015, p. 15.
[2] "US Airline Commuter Growth to Parallel Majors in 1990s," *Aviation Week & Space Technology*, 20 Nov. 1989, p. 83.
[3] US Regional Airline Association, p. 15.
[4] "European Regionals Fight for Access," *Flight International*, 8–14 Aug. 1990, p. 46.
[5] "Dan-Air's New BAe 146," *Flight International*, 4 June 1983, p. 1635.
[6] MacDonald, L., *The Bombardier Story*, John Wiley and Sons, Toronto, 2001, p. 133.
[7] Rashke, R., *Stormy Genius: The Life of Aviation's Maverick Bill Lear*, Houghton Mifflin, Boston, 1985, pp. 331–332.
[8] MacDonald, pp. 133–134.
[9] Salpukas, A., "Canadair to Be Sold to Bombardier, Inc.," *The New York Times*, 19 Aug. 1986, https://www.nytimes.com/1986/08/19/business/company-news-canadair-to-be-sold-to-bombardier-inc.html [accessed 7 March 2018].
[10] MacDonald, p. 160.
[11] "Canadair Surveys Need for Stretched Challenger Jet," *Aviation Week & Space Technology*, 1 Dec. 1986, p. 61.
[12] MacDonald, pp. 165–166.
[13] "Canadair Challenges RJ Competitors," *Flight International*, 19–25 Sept. 1990, p. 34.
[14] MacDonald, pp. 167–168.
[15] "Bombardier of Canada Wins Competition to Buy Short Brothers," *Aviation Week & Space Technology*, 12 June 1989, p. 63.
[16] Farnsworth, C., "Bombardier Agrees to Buy De Havilland from Boeing," *The New York Times*, 23 Jan. 1992, https://www.nytimes.com/1992/01/23/business/company-news-bombardier-agrees-to-buy-de-havilland-from-boeing.html [retrieved 21 Feb. 2017].
[17] Rodengen, J., *The History of Embraer*, Write Stuff Enterprises, Ft. Lauderdale, FL, p. 42.
[18] Rodengen, pp. 39–43.
[19] Fotos, C., "Embraer to Launch Two Civil Aircraft Programs," *Aviation Week & Space Technology*, 4 Sept. 1989, p. 65.
[20] Rodengen, p. 102.
[21] Ibid., p. 118.
[22] North, D., "Embraer Slashes Positions to Match Reduced Demand," *Aviation Week & Space Technology*, 7 Sept. 1992, p. 127.
[23] Rodengen, p. 122.
[24] Ibid., p. 132.
[25] Mordoff, K., "Fokker to Use Lockheed Personnel; Second Assembly Line Talks Continue," *Aviation Week & Space Technology*, 5 Dec. 1988, p. 89.
[26] Moxon, J., "Reducing the Risk," *Flight International*, 16–22 Feb. 1994, p. 65.
[27] Sparaco, P., "Fokker Initiates New Rescue Plan," *Aviation Week & Space Technology*, 6 March 1995, p. 30.
[28] Sparaco, P., "Curtain Falls on Fokker," *Aviation Week & Space Technology*, 25 June 1996, p. 25.
[29] Hughes, D., "Bombardier Optimistic Despite Slow RJ Sales," *Aviation Week & Space Technology*, 28 Sept. 1992, p. 42.
[30] Sparaco, P., "Europe Regionals Expand in Deregulated Market," *Aviation Week & Space Technology*, 12 May 1997, pp. 61–62.
[31] Jeziorski, A., "CityLine Moving Turboprops to Contact Air," *Flight International*, 13–19 Feb. 1995, p. 8.

[32] Ott, J., "Canadair CRJs Mold New Comair Network," *Aviation Week & Space Technology*, 16 June 1997, p. 118.
[33] Adams, M., "Regional Airline That Could Be Poised to Change Industry," *USA Today*, 3 April 2001, http://usatoday30.usatoday.com/money/biztravel/2001-04-03-comair.htm [retrieved 10 Nov. 2016].
[34] Kingsley-Jones, M., "Sharing the Loads," *Flight International*, 7–13 May 1997, p. 38.
[35] MacDonald, p. 185.
[36] Ibid., pp. 193–194.
[37] Ibid., pp. 195–196.
[38] Proctor, P., "SkyWest Balances Turboprop–Turbofan Fleet," *Aviation Week & Space Technology*, 2 March 1998, p. 49.
[39] Ibid.
[40] Forbes S. J., and Lederman, M., "The Role of Regional Airlines in the U.S. Airline Industry," *Advances in Airline Economics*, Vol. 2, 2007, pp. 193–208.
[41] Sparaco, P., "Air Seeks Wider Role in RJ Market," *Aviation Week & Space Technology*, 26 Feb. 1996, p. 35.
[42] Shifrin, C., "Bombardier CRJ700 Gains Market Momentum," *Aviation Week & Space Technology*, 24 Feb. 1997, p. 30.
[43] McKenna, J., "Record Embraer Order Launches Embraer Program," *Aviation Week & Space Technology,* 21 June 1999, p. 31.
[44] Stewart, D., and Michaels, K., "Winds of Change," *AeroStrategy Newsletter*, March 2007.
[45] "Bombardier Launches CRJ900," *Flight International*, 25–31 July 2000, p. 9.
[46] Moorman, R., "Bombardier Shelves BRJ-X; CRJ700 Wins Type Rating," *Aviation Week & Space Technology*, 4 Dec. 2000, p. 48.
[47] Jeziorski, A., "Fairchild Redesigns 228 as 228 Faces the Axe," *Flight International*, 19–25 June 1996, p. 26.
[48] Sparaco, P., "Fairchild-Dornier Launches All New Regional Jets," *Aviation Week & Space Technology*, 25 May 1998, p. 20.
[49] Flotau, J., "Fairchild Dornier Joins Export Subsidy Battle," *Aviation Week & Space Technology,* 11 June 2001, p. 55.
[50] Michaels, D., "Plane Maker Fairchild Dornier Files for Bankruptcy Protection," *Wall Street Journal*, 3 April 2002, https://www.wsj.com/articles/SB1017739587113659160 [retrieved 14 May 2016].
[51] Sparaco, P., "European Regional Transports in Flux," *Aviation Week & Space Technology*, 2 Feb. 1998, p. 39.
[52] Sparaco, P., "British Withdraw from Regional Jet Market," *Aviation Week & Space Technology*, 3 Dec. 2001, p. 40.
[53] Aboulafia, R., "China's Commercial Aircraft Industry: The Limits of an Autarkic Industrial Policy," Teal Group private newsletter, 2010.
[54] Anselmo, J., Taverna, M., and Wall, R., "Catching the Wave," *Aviation Week & Space Technology*, Vol. 3, 20 June 2005, pp. 29–30.
[55] Fiorino, F., "Shrinking CSeries," *Aviation Week & Space Technology*, 6 Feb. 2006, pp. 39–40.

Chapter 5

MODERN MIRACLES

In December 2014, the influential publication *Bloomberg Business Week* celebrated its anniversary by highlighting the 85 most disruptive ideas that had changed the world since its founding in 1929. The list included innovations like the microchip and DNA sequencing as well as transformational ideas like the Green Revolution. Topping the list as the single most influential idea was the jet engine. It wrote:

> By the 1960s this one invention had shrunk the world. For the first time the entire surface of the planet was reachable—or at least viewable—and its wonders opened up. Food, art, leisure, commerce and relationships were redefined. Life became richer and more hectic. Jet travel created true global citizens while also making it easier to conduct war and spread disease [1].

Jet engines are more than transformative for society. They are technological marvels. Consider Rolls-Royce's latest engine, the Trent XWB, which has a fan case that is wider than the fuselage of a Concorde. Its fan blades ingest more than 1.3 tons of air, the equivalent of a squash court, every second at takeoff. The force on each fan blade at takeoff is almost 90 tons—the equivalent of nine London double-decker buses hanging off each blade. Each one of its 68 high-pressure turbine blades generate 800 hp at takeoff—the same as a Formula 1 racing car. And the temperatures of hot gases surrounding those same turbine blades can exceed 2800°F—hundreds of degrees *above* the melting temperature of its super alloy material.[1] Only internal cooling air from

[1]Ruth Sunderland, "Rolls-Royce's Engine of Growth: Famous Firm Is Banking on Trent XWB to Propel Revival," This Is Money, 18 Dec. 2014, http://www.thisismoney.co.uk/money/markets/article-2879738/Rolls-Royce-s-engine-growth-Famous-firm-banking-Trent-XWB-propel-revival.html [retrieved 22 May 2016].

the cherry-red compressor and coatings prevent the melting of these blades. The Trent XWB is expected to operate as many as 16 hours per day, 7 days a week, and for 5–7 years between overhauls while achieving the highest levels of safety and reliability. It demonstrates why jet engines are modern miracles.

Who were the pioneers of jet engines, or "aeroengines" as they are referred to in this book? And how did this critical aerospace segment evolve?

Jet Engine Origins

Today's jet engines trace their origin to two pioneers, one British and one German. In the late 1920s, a young Royal Air Force officer named Frank Whittle authored a white paper on approaches to faster aircraft, and by 1929 realized that gas turbines could be substituted for pistons. He formed Power Jets Ltd. in 1936, and after years of hardship received a British Air Ministry contract to build a flyable version of his design. Rover Company was brought in to assist with production, and the first flight of Whittle's design took place on 7 April 1941. Whittle's relations with Rover were contentious at best, and seeking a new partner he met Ernest Hives, manager of the Rolls-Royce factory in Derby. In late 1942 Hives met with Rover and traded Rolls-Royce's tank engine factory in Nottingham for the Rover jet factory in Barnoldswick. Rolls-Royce was now in the jet engine business. It continued development of jet engines and delivered its Welland engine to power the Meteor, which began jet operations in 1944.

Meanwhile, in Germany, Hans von Ohain developed his own jet engine design in parallel to Whittle. His first design ran in 1937, and on 27 Aug. 1939 his engine powered the world's first flight of a jet-powered aircraft, the Heinkel He178. Other German designs proved superior to Ohain's approach, and in 1944 the Jumo 4 was fitted to an ME262 aircraft, which reached a speed of 900 km/hr (559 mph).

During the war, Whittle was sent to the United States to support a parallel jet development program. He worked with General Electric (GE), the primary supplier of turbochargers in the United States. Results came quickly, and the first US jet aircraft flew in 1942. General Electric would then continue development of its own jet engine designs. Westinghouse and Allison Engine Company also were involved in early development programs. Pratt & Whitney, a major piston engine supplier during World War II, was also keen to enter the jet business, but was directed by the US government to focus on production of its popular Wasp piston engines.

Not widely appreciated is the fact that Japan also flew a jet during the war. The Ishikawajima Ne-20, based on a German design, powered Japan's first jet flight on 7 Aug. 1945, the day after the US nuclear bombing of Hiroshima.

The context changed considerably following the war. Rolls-Royce saw a clear opportunity to exploit the civil market and developed the Dart, a turboprop engine, and the Avon, its first axial-compressor jet engine. In axial compressors, the gas flows parallel to the axis of rotation and is compressed by rotating blades (airfoils). The Avon proved successful and by 1950 was powering the country's most important military aircraft. It was then matched to a variant of the world's first jetliner, the de Havilland Comet 4. In parallel, Rolls-Royce pursued an important new concept: the *bypass turbojet*. English engineer Alan Arnold Griffith began thinking about bypass as early as the 1930s, and following World War II, Rolls-Royce felt that existing engines like the Avon were advanced enough that it was time to start work on the new concept.

In a bypass approach, some of the compressor air bypasses the hot aeroengine core and is channeled around the outside of the core to the back of the engine, where it surrounds the high-velocity jet exhaust. By not being accelerated by hot section combustion, the colder bypass exits at a lower velocity closer to the aircraft's velocity in relation to the surrounding atmosphere, improving the aeroengine's efficiency. Thus, fuel efficiency is greatly improved, and the engine is also quieter thanks to the tube of colder, slower moving bypass air. Jet engine designers adopted *bypass ratio*, the ratio of mass flow rate that bypasses the engine core versus passing through the core—as an important new engine parameter. This set the stage for the Rolls-Royce Conway (see Fig. 5.1), the world's first "bypass turbojet," which had a bypass ratio of 0.25. Soon, these types of engines would be referred to as turbofans (see Fig. 5.2).

Rolls-Royce also licensed one of its engine designs to Pratt & Whitney, which would become the J42. By 1947, Pratt & Whitney was producing its own design, the axial flow T34. It set an ambitious goal of becoming a

Credit: Nimbus227

Fig. 5.1 Rolls-Royce Conway.

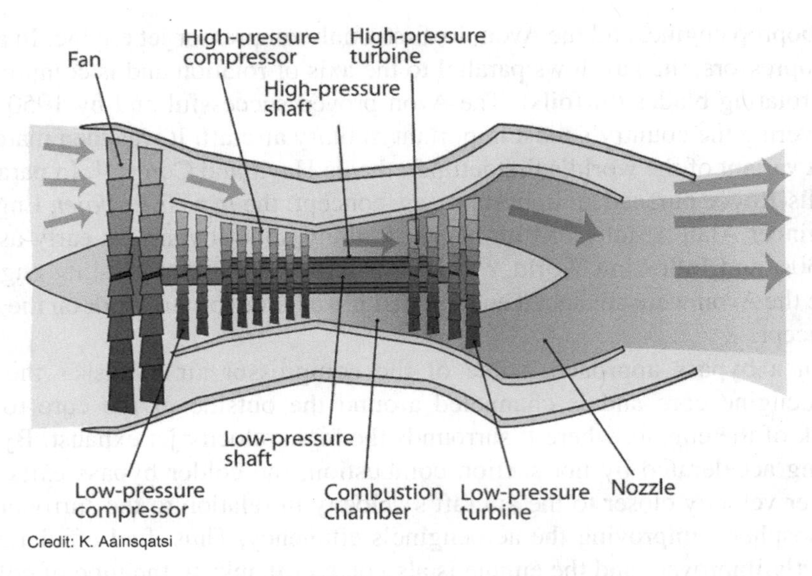

Fig. 5.2 Turbofan engine.

jet engine leader by 1950. The Cold War was heating up, and US suppliers including Pratt & Whitney benefitted. Pratt & Whitney developed the J57 military aeroengine, a novel two-spool design that would power the B-52 bomber and the KC-135 tanker. Designated the JT3C, it was the first US 10,000-lb-thrust-class engine.

Meanwhile, GE focused on military engines. Its axial-flow J47 engine arrived just as the United States ramped up military aircraft development on the F-86 and new bombers. As its order book swelled for these engines, it looked for additional production capacity. It settled on a massive, vacant former Curtiss-Wright aircraft engine factory in Evendale, Ohio. Led by brilliant (and German-born) chief engineer Gerhard Neumann, it developed the J79 engine, which would go on to power several huge military aircraft including the McDonnell F-4 and Lockheed F-104 fighters. More than 17,000 would be produced. The J79 incorporated *variable pitch* stators, a design approach that allowed the static stator blades to be rotated to optimize engine performance at different conditions. This wasn't GE's invention. Due to its relationship with the US military, GE got access to captured German engines during World War II that incorporated variable pitch stators and could patent the invention. GE's exclusive access was one factor that drove Pratt & Whitney to a twin-spool design on its JT3C.[2] GE also created a civil derivative, the CJ805, which was not suited for airline operations. As former GE Aviation CEO Brian Rowe lamented, "We soon learned that the CJ805 was costly to

[2]Paul Adams (former President—Pratt & Whitney), email to author, 26 Dec. 2016.

maintain and not really rugged for airline service." He considered it to be a "light, high-thrust, but potentially fragile thoroughbred" in comparison to the "JT3 workhorse" [2].

PRATT & WHITNEY'S EMERGENCE

Whereas GE's focus was on military engines, Pratt & Whitney was squarely focused on the civil jet market. In 1951, it moved its pistons to its Canadian subsidiary to focus purely on jets. It developed a 10,000-lb-thrust JT3 turbojet engine in 1950. It powered several important military aircraft including the Boeing B-52, Convair F-102, Douglas A3D, Lockheed U-2, McDonnell F-101, and North American F-100. In the late 1950s, Pratt & Whitney followed the lead of the Rolls-Royce Conway and incorporated a fan at the front of its JT3 engine to create its own turbofan. The two-spool design in a turbofan architecture, dubbed the JT3D, was a breakthrough that increased thrust 35% to an eventual maximum of 21,000 lb. Fuel burn dropped 15–22%. Boeing soon selected the JT3D for its 707 program, and the combination of superior performance, reliability, and the right aircraft program propelled the JT3D to become the undisputed leader of the pack. Douglas followed and selected the engine for its DC-8 jetliners. More than 8000 JT3Ds would eventually be built.

While Pratt & Whitney developed the JT3D, General Electric unveiled a unique rear-mounted turbofan engine in 1958. The CJ805 integrated an aft-fan module that was attached to its J79 military turbojet engine. It boasted a 3:1 bypass ratio—significantly higher than the Conway and JT3D—but it lost to the JT3D on a large US KC-135 military tanker program. It did find a home on the Convair 880 and 990 jetliners, which were poor sellers.

The 1960s brought more growth for Pratt & Whitney with its JT8D turbofan. It hit a motherlode of opportunities when, in the 1963–1967 timeframe, it was selected for the 727, DC-9, and 737 programs; all would ultimately become successful. In contrast, Rolls-Royce's two major programs, the BAC-111 and Trident, were not commercially successful. In 1964, the US Air Force opened competition for engines to power a gigantic new transport aircraft. GE took a major gamble and proposed a radically new design: an 8:1 bypass ratio engine with an efficient compressor, advanced turbine cooling technology, and fuel consumption less than half of the best engines in service [3]. In 1965, GE won a $459-million contract, its largest to date, to develop this new engine, designated the TF39. The engine would be mounted to the airframe winner, Lockheed's C-5 Galaxy. The age of high-bypass-ratio turbofans had arrived. Boeing and Pratt & Whitney would make the most of their situation by leveraging their losing designs to create the 747 and JT9D, respectively. The Boeing 747 with JT9Ds would enter service five years later. General Electric made the fateful decision not to compete for the 747 given its commitments to the C-5 and the Boeing 2707 Supersonic Transport programs. The 2707 would eventually be cancelled, and General Electric would spend a decade explaining its decision to withdraw from the 747 competition.

As Pratt & Whitney established early leadership, new jet engine suppliers developed in France. In 1945, the French government created Snecma (la Société nationale d'études et de construction de moteurs d'aviation), which integrated assets from Gnôme & Rhône—a leading piston engine manufacturer—as well as sundry design bureaus, factories, and workshops. Its domestic rival was Hispano Suiza, which received preferential treatment from the British government to utilize its jet engine technology under license. This included production of the Rolls-Royce Tay in 1954. Another French firm, Turboméca, produced the Adour engine for the Jaguar fighter in cooperation with Rolls-Royce. Snecma would separate itself from its rivals in 1959 when it entered a contract to produce Pratt & Whitney's popular JT8D engine under license. In 1968, it would acquire Hispano Suiza's jet engine business to cement its leadership position in France.

Across the English Channel, Rolls-Royce evaluated replacements for the Conway and concluded that a novel three-spool architecture was the best approach to deliver higher efficiencies. In this configuration, three groups of turbines spin three separate concentric shafts to power three sections of the compressor area running at different speeds. This design allows each stage to run at its optimal speed, while also being more compact and rigid. The tradeoff is that it is more complex to manufacture. Rolls leveraged the three-spool approach with the RB211 series and won positions on the A300 in 1967 and Lockheed L-1011 in 1968. It was particularly confident in Lockheed's tri-jet program and discounted the DC-10's potential competitive threat. Financially stretched by high development costs, the situation deteriorated further when in 1970 its new carbon-fiber Hyfil composite fan blades failed during bird ingestion and cross-wind testing. This led to a costly redesign as Rolls-Royce switched to heavier titanium fan blades. In September 1970, Rolls-Royce reported to the government that development costs for the RB211 had risen to £170.3 million—almost twice the original estimate. In addition, the estimated production costs increased 50% and exceeded the £230,000 selling price of each engine [4]. Rolls-Royce went into receivership and was ultimately nationalized by the British government in 1971. The first engine, designated the RB211-22B, would enter service with Eastern Air Lines in April 1972—about one year late. Limited development funds led Rolls-Royce to eschew the DC-10 and walk away from a new European twin-aisle opportunity, the Airbus A300.

These decisions opened the door for GE to bring its high-bypass-ratio technology into the civil market via the CF6 engine; GE won positions on the Douglas DC-10 and the A300 in 1971 and 1973, respectively. The CF6 also secured a position as an option on the 747-200. To secure the engine order in jobs-sensitive Europe, GE proved highly creative in establishing a revenue-sharing arrangement whereby engine makers MTU of West Germany and Snecma of France produced CF6-50 engine parts at a rate that was 25% equivalent to the total engine production costs. The Snecma partnership would forge the beginning of what would become an epic partnership.

Despite its status as a government-owned company, Rolls-Royce pursued the 747 with a redesigned version of its RB211 with 50,000 lb of thrust. Designated the RB211-524, it would be able to power new variants of the L-1011, as well as the Boeing 747. In 1977, British Airways became the first operator to take delivery of this new variant. Rolls-Royce continued to improve the performance of this engine, which won important 747 orders from Cathay Pacific, Qantas, and South African Airways.

Based just outside of Paris, Snecma was best known for its military engines and its joint venture with Rolls-Royce to produce the powerplant for the Concorde. In the late 1960s, it began research into the next generation of high-bypass-ratio turbofans in the 10-ton (20,000-lb) thrust class. It realized it needed a partner to compete. Rolls-Royce, the logical choice, was ruled out due its financial circumstances. Pratt & Whitney, the clear market leader, offered a share of a refanned JT8D [5]. Snecma selected GE Aviation as its partner after meetings between Gerhard Neumann and René Ravaud from Snecma at the 1971 Paris Air Show (Fig. 5.3). One obstacle was for GE to obtain an export license to use technology from its F101 engine program on the new engine. The initial request was rejected, but both the French and GE

Credit: CFM International
*CFM International is a 50/50 joint company between GE and Safran Aircraft Engines.

Fig. 5.3 René Ravaud and Gerhard Neumann: CFM's founding fathers.

continued to push the Nixon Administration for permission, and the go-ahead finally arrived after a 1973 meeting of Presidents Nixon and Pompidou in Reykjavík. In 1974, GE and Snecma would form a 50:50 joint venture, CFM International, and share equally in the revenue of their new engine—the CFM56.

After more than five years of failing to secure an aircraft application, CFM International launched a new CFM56-2 engine to re-engine a fleet of DC-8 aircraft for three customers—United Airlines, Flying Tiger Line, and Delta Air Lines—with the CFM56-2 engine replacing the JT3D. GE and Snecma were just weeks away from cancelling the program had these orders not materialized.[3] Ironically, former US astronaut and Cincinnati-area resident Neil Armstrong, head of United's propulsion evaluation team, helped to seal the CFM56-2 selection by recommending it for the four-engine aircraft.[4] CFM International received more good news when France and the United States awarded CFM major contracts to re-engine their fleets' KC-135 tankers with CFM56-2s; the latter contract was for more than 300 aircraft.

Despite CFM's success, two decades into the jet age the undisputed aeroengine leader was Pratt & Whitney through the success of its JT3D and JT8D models on single-aisle aircraft, and the JT9D on twin-aisles, particularly the 747. By 1980, 79% of all jet engines in service were Pratt & Whitney; Rolls had 12% and the upstart GE garnered the remaining 9%.

CHANGING FORTUNES

With the availability of new engine technology and higher oil prices, Boeing and Douglas both began to examine updated designs for their 737 and DC-9 models. Although CFM's offering was clear, Pratt & Whitney made a once-in-a-generation decision that would haunt it for decades to come: it would address the opportunity through an incremental upgrade to the JT8D rather than pursuing a true "white-sheet" approach. Its new engine, the JT8D-200, would feature a 2.1 bypass ratio and an overall pressure ratio of 21. The corresponding figures for the newer and more fuel efficient CFM56-3 were 6 and 27, respectively. Underpinning this performance differential was the core of the engine, which featured a single high-pressure turbine stage (rather than the usual two), which resulted in a shorter, lighter, and lower maintenance engine. "Pratt & Whitney was drunk on JT8D spare parts revenue," said former GE chief engineer Fred Herzner. "They were vulnerable to a disruptive design like the CFM56."[5] With a positive experience on the KC-135 re-engine program, Boeing chose the CFM56-3 for its new 737-300 aircraft (Fig. 5.4). To cope with relatively low wing clearance, the engine would have a reduced

[3]Fred Herzner (former Chief Engineer—GE Aviation), interview with author, 29 June 2015.
[4]Rick Kennedy (Manager, GE Aviation Media Relations), email to author, 26 Feb. 2018.
[5]Fred Herzner (former Chief Engineer—GE Aviation), interview with author, 12 Dec. 2016.

Credit: Davidelit

Credit: CFM International

Fig. 5.4 737-300 nacelle and the CFM56-3.

profile through gearbox and engine accessory designs and a thin nacelle with an odd, flat bottom. CFM's bet on the 737 was by no means a slam dunk; sales of the 737 to that point had been choppy. According to former GE Aviation CEO Brian Rowe, "In the first review I had with Jack Welch [CEO] on this project, it was clear that our adventure on the 737 was considered quite a joke. No one thought we had much of a chance of convincing anyone to buy this allegedly old-fashioned airplane, new engines or not" [6]. Fortunately, GE's iconic CEO was wrong on this call. US low-cost carrier Southwest Airlines was an important supporter, and Boeing would deliver 651 737-300s from 1984 to 1989.

The success of the 737-300 captured Airbus's attention and renewed its interest in developing a single-aisle. This was the airplane concept that convinced Snecma a decade earlier of the need for a 10-ton engine. Airbus was insistent on offering engine choice, and not surprisingly CFM secured positions with the CFM56-5, a higher-thrust variant of the CFM56-3. What about the other engine? With the JT8D-200 no longer a viable option, Pratt & Whitney joined with Rolls-Royce to form an international consortium to challenge CFM International. They called their company International Aero Engines (IAE) and brought in MTU (Germany), Fiat (Italy), and Japanese Aero Engines (a conglomeration of Kawasaki Heavy Industries, Ishikawajima-Harima Heavy Industries, and Mitsubishi Heavy Industries) as lesser partners. Their answer was the V2500A1, a 25,000-lb-thrust engine with a 5.4 bypass ratio. This was a novel collaboration of the aeroengine "who's who" outside of CFM partners. The responsibility breakdown of the partners was as follows:

- *Japanese Aero Engines:* Fan section and low-pressure compressor
- *Rolls-Royce:* High-pressure compressor
- *Pratt & Whitney:* Combustors and high-pressure turbine
- *MTU:* Low-pressure turbine
- *Fiat:* Accessory gearbox

Engine choice would be a major theme for Boeing's two new mid-range aircraft of the early 1980s, the 757 and 767. The 757 would be powered by the Pratt & Whitney 2037, its highest bypass ratio engine to date (6.0) and the first civilian full authority digital electronics control (FADEC) engine. FADECs were digital computers that controlled all aspects of engine operation and offered numerous advantages, including improved fuel efficiency, engine health monitoring capability, and reduced pilot workload. For its part, Rolls-Royce achieved a breakthrough in gaining the second position on the 757 with its RB211-535C. This was the first to use the "wide chord" fan, which increased efficiency, reduced noise, and provided additional protection against foreign object damage. Although it had inferior fuel efficiency compared to the Pratt & Whitney offering, it had advantages in reliability and noise. A seminal moment for Rolls-Royce in the 1980s was when American Airlines chose it for a massive order of 50 757s. This was its first endorsement by a major US airline since the early 1970s and the L-1011. On the 767, customers were given the choice of Pratt & Whitney JT9D or General Electric CF6 turbofans, marking the first time that Boeing had offered more than one engine option at the launch of a new airliner. Rolls-Royce's RB211-524 would later become a third option on the aircraft.

With new competition, Airbus saw the need to improve the A300. It launched the A300-600 in 1980, which required new, more efficient engines. Enter General Electric, with the CF6-80C, an upgrade to its CF6-80A. Around the same time, Pratt & Whitney decided to replace the JT9D with a new white-sheet engine, the PW4000. The new powerplant, covering 48,000–60,000 lb of thrust, would offer a 7% fuel consumption improvement over the latest JT9D with 60% fewer parts. It would also incorporate technology developed for the PW2000, including a FADEC. At the time, Pratt & Whitney's leadership anticipated that 8000 to 10,000 air transports would be delivered in the next 20 years, of which 60% would be twin-aisle candidates for the PW4000. Target platforms would include the 767, 747, A300-600, and A310. Pratt & Whitney planned to certify the engine in 1986 and phase out the JT9D in the late 1980s [7]. GE sensed the competitive threat for the A300-600 and revised its new engine. This, in turn, delayed introduction of the A300-600 by one year (1985), but yielded a much improved CF6-80C2 engine with 62,000 lb of thrust. This would prove to be a fruitful decision.

Despite all the progress in taking on Pratt & Whitney, GE and Rolls-Royce were facing a behemoth. In 1983, Pratt & Whitney delivered 223 jet engines compared to 54 for GE, 41 for Rolls-Royce, and 33 for CFM International. Moreover, 76% of the installed base belonged to the Pratt & Whitney engine. Facing significant development costs with their CF6-80C2 and RB211-535E4 programs, GE and Rolls-Royce decided to collaborate. Under the agreement, the two companies would participate in the development, production, and marketing of one of the other's jet engines in exchange for 15% of the revenue—a figure that was expected to grow to 25% by 1988 [8].

All engine OEMs were now investing heavily in new-generation high-thrust engines, and with good reason. McDonnell Douglas's MD-11 was initially slated to be powered by a single engine, GE's CF6-80C2. After a protest by Northwest Airlines, a loyal Pratt & Whitney customer, it added the PW4000 as a second option after Pratt & Whitney agreed to pay the certification costs.[6] Boeing's 747-400 would also include these two models, as well as a third option, the RB211-524G from Rolls-Royce. This new variant incorporated wide-chord fan technology and FADEC technology. GE wasn't pleased with Rolls-Royce's decision to compete against the CF6-80C2, and the fledgling partnership between the two firms was ended by agreement after just three years.

The late 1980s brought several interesting aeroengine experiments. GE developed a GE36 unducted fan (UDF) for Boeing's proposed 7J7 aircraft—a 150-seat aircraft with half the fuel burn of the A320. A hybrid of a turbofan and a turboprop, the engine featured GE's military F404 core driving giant 12-foot composite fan blades (see Fig. 5.5). McDonnell Douglas was also interested and flew a modified MD-80 at the 1988 Paris Air Show. The project was shelved due to falling oil prices and customer concerns about noise and the unproven UDF technology. One upside was that General Electric gained valuable knowledge about composite fan blades—insight that it would apply to a future aircraft program.

The second group of experiments were around *geared turbofan* architectures. The geared turbofan design architecture, which is described in Chapter 6, increased fuel efficiency and reduced weight relative to conventional designs. IAE proposed a geared turbofan for the Airbus A340 in 1987, which shared the core of its new V2500 engine. It had a projected

Credit: Andrew Thomas

Fig. 5.5 The GE-136 unducted fan on a MD-80 aircraft.

[6]Adams email, 26 Dec. 2016.

maximum thrust of 28,000–32,000 lb while having only 80% of the V2500's fuel consumption. Although several customers signed preliminary contracts, IAE decided to stop the development a few months later. There was too much technology risk in the geared turbofan architecture, and falling fuel prices hurt the business case. Pratt & Whitney would revisit the geared turbofan years later. This opened the door for a new CFM56 variant, the CFM56-5C, to power the A340.

The end of the 1980s brought another notable change in the aeroengine market—the privatization of Rolls-Royce in 1987. This was part of British Prime Minister Margaret Thatcher's broad initiative to improve economic competitiveness and roll back what she perceived to be creeping socialism in the United Kingdom. More than 50 firms were sold or privatized, including three other firms in aviation and aerospace: British Airways, British Aerospace, and BAA. An 18 April 1987 editorial in *Flight International* captured the prevailing mood. It celebrated it as a "transformed company" that had made significant strides in improving productivity and expanding its product portfolio in civil and military markets. However, it ended on a sanguine note: "It faces two, rich, powerful, ruthless American competitors, neither of which will be disposed to giving an inch to the gentlemanly British" [9].

As the 1980s ended, the playing field consisted of a fading leader with up-and-coming challengers that were gaining strength. In 1989, Pratt & Whitney still led in market share of engines delivered (32%) and held an overwhelming advantage in the number of jet engines in service—more than twice as many as all its competitors combined. Yet the new CFM International joint venture captured 26% of the deliveries that year, followed by Rolls-Royce (14%), General Electric (11%), and International Aero Engines (5%). The aeroengine market was changing, and changing fast.

Trent: More than a River

The new and private Rolls-Royce faced big decisions about what to do with its high-thrust range. It had successfully grown the RB211-524 range to 58,000 lb, but Airbus's requirements for the TA9 (A330) program were much, much more. Boeing was also looking at a large twin, and the MD-11 required more thrust. Moreover, Extended-Range Twin-Engine Operational Performance Standards (ETOPS) were changing the playing field to favor twin-engine aircraft. Airlines were asking for new aircraft that were ETOPS-certified at the time of introduction, which would require a new approach to aeroengine development and certification. The RB211 was long in the tooth; a new solution would be required. Fortunately, the three-spool design of the RB211 was a good basis for a new engine family because it allowed the low-pressure, intermediate-pressure, and high-pressure modules to be scaled, mixed, and matched. At the 1988 Farnborough Air Show, Rolls-Royce announced the launch of the Trent series of engines. This revived the Rolls-Royce tradition of naming its jet engines after British rivers.

The Trent 600 would be aimed at the MD-11 and MD-12, but it was an uphill battle. British Caledonian was to be the launch customer with Rolls-Royce engines (then called RB211), but it was acquired by British Airways and the order was cancelled in 1987. British charter airline Air Europe became the new Trent launch customer when it ordered two MD-11s in late 1988, but it collapsed in March 1991 in the aftermath of the Gulf War. Rolls would never get an MD-11 order, and the Trent 600 program was downgraded to a demonstrator program. To former Director of Engineering and Technology Philip Ruffles, this was a good thing. "We were struggling to overcome several technical problems, and it was taking away from our other development efforts. It was a great sigh of relief when it went away."[7] With the Trent 600 sidelined, Rolls turned its attention to a greater prize: Airbus and its promising new A330. This would be the task of the Trent 700. Rolls-Royce had one big advantage per Ruffles: The new Trent was optimized for the A330 requirements—much more so than the competition.[8] However Airbus faced a major headwind: history. "Airbus hadn't forgiven us for our withdrawal from the A300 program nearly 20 years earlier, and we were also having issues with the V2500," opined former Chief Commercial Officer Charles Cuddington, "Airbus weren't our biggest fans. We felt we were in third place going against Pratt & Whitney and GE for the A330."[9] Rolls-Royce broke through with orders from Cathay Pacific (10 A330s) and Trans World Airlines (20), and the Trent 700 was on its way. It achieved certification in January 1994, and 90-min ETOPS approval was achieved 14 months later; this would be extended to 180 min in 1996. Cathay Pacific began operations with the new engine in March 1995.

BRIAN ROWE'S BIG GAMBLE

Sometime in the late 1980s, GE Aviation chief Brian Rowe (Fig. 5.6) walked into GE CEO Jack Welch's office with a radical idea. He explained that the world was moving to ever-larger twin-engine aircraft that would require massive aeroengines with unprecedented thrust capability. He unfurled a drawing of a radical new engine on Welch's desk and took him through the details of a brand-new engine, loaded with new technology—including massive composite fan blades—that would require at least $1 billion investment. "I want to do this program," he exclaimed, but the iconic CEO—perhaps put off by the risk—wasn't listening. Miffed by his boss's lack of interest, he ordered his assistant to roll up the drawings and began to

---

[7]Philip Ruffles (former Director Engineering & Technology—Rolls-Royce), interview with author, 1 June 2016.

[8]Ibid.

[9]Charles Cuddington (former sales executive—Rolls-Royce), interview with author, 1 June 2016.

Credit: GE

Fig. 5.6 GE's Brian Rowe.

storm out of his office. Fortunately, Welch asked Rowe for a second hearing on the engine concept. This was the start of a new program called the GE90.[10] GE had never done a brand-new engine exclusively focused on the civil market; all its engines were derived from taxpayer-funded military programs. And the technology risk was significant. The aerodynamic requirements of the compressor were very demanding with an overall pressure ratio of 40:1; this meant that the air discharged was compressed to a ratio 40 times the compressor's inlet pressure. And then there were the composite fan blades. "The last composite fan blade we did [in a conventional jet engine] was in 1970, and when we did the bird strike test it turned to dust," opined former GE chief engineer Fred Herzner[11]; in other words, the same result as Rolls-Royce with its ill-fated Hyfil blades in the early 1970s, an experiment that contributed to its bankruptcy.

Brian Rowe had to do more than convince his boss of the merits—he had to convince an aircraft manufacturer. Airbus, committed to the quad-engine A340, was hostile to the idea. Why introduce an aircraft to obsolete its own new model before introduction? Boeing was the obvious target. Its 767 and 747 models were at clear risk from the newer, more efficient Airbus models, and it needed to respond. But would customers embrace a long-range twin-engine aircraft? Fortunately, Boeing now was concluding that an incremental 767 stretch might not be enough for emerging requirements and the competitive threat of the A340; it was pivoting to the need for the 777.

[10]Herzner interview, 29 June 2015.
[11]Ibid.

In 1990, GE announced its intent to build the GE90, and began working closely with Boeing, defining its new engine as Boeing evaluated a new aircraft. “After many meetings,” said Brian Rowe, “I believe I convinced Boeing that they needed to bite the bullet and go ahead with the airplane” [10]. GE would design its new engine not only for the immediate 70,000-lb-thrust requirements of the new aircraft, but also for future requirements well beyond 100,000 lb for much larger aircraft—an extraordinary notion at the time. This meant that the GE90 would be heavy relative to the competition should Boeing move forward with the program, but would be well-positioned for the future.

Pratt & Whitney, which had launched its own new high-thrust program a few years earlier, saw an opportunity for its new PW4000 engine family should Boeing go forward with the new aircraft. It had an important advantage over the competition: Its engine was in operation. The first battle in this three-way competition went to Pratt & Whitney. In October 1990, 777 launch customer United Airlines placed a $4-billion order for 150 PW4000s. The new PW4000 variant included a 112-in. fan and two new turbine sections to increase its output to 84,000 lb of thrust. The transaction also included an additional 265 PW4000s for 747-400s. United also planned to use the engine for 767s. The commonality of the new variant, which shared 75% of the core components of the engines used on the carrier’s 747s and 767s, was a factor in the decision [11,12]. So was United’s longstanding relationship with Pratt & Whitney. (The two had once been part of the same company.) Ironically, that week’s *Aviation Week & Space Technology* featured a large MD-11 on the cover with the news of the 777’s launch receiving only a cover subtitle. Pratt & Whitney also won the next competition with All Nippon Airways (ANA). This wasn’t surprising, because ANA had a longstanding relationship with Pratt & Whitney and tended to be technologically conservative. In early 1996, industry analyst Richard Aboulafia noted that despite the PW4000’s teething problems, Pratt & Whitney was in the best shape in terms of order book and customer base for twin-aisle aircraft—particularly for the 777. In contrast, he thought that “GE could be facing a tough situation,” noting that its GE90 was sized for 75,000 lb of thrust and above, whereas most of the twin-aisle orders required smaller engines [13].

Brian Rowe’s big gamble was zero for two thus far. Had Jack Welch’s skepticism been right all along?

THE ROLLS-ROYCE FORCE OF CONVICTION

After its breakthrough on the A330, there was another large prize for Rolls-Royce to pursue with the Trent: the Boeing 777. The Trent 700’s 97-in. fan would not be big enough, so Rolls proposed a new model with a 110-in. fan diameter, the Trent 800. The obvious launch customer was British Airways (BA). The British government had invested heavily in Rolls-Royce, and surely the British flag carrier—despite being recently privatized—would support its compatriots. BA was also a current Rolls-Royce operator with RR211s on its 747s and 757s.

GE knew it faced long odds in pursuing BA, but was encouraged when Lord King and Colin Marshall introduced GE leadership to important ministry people. As Brian Rowe recounted in his autobiography, *The Power to Fly*, "Rolls-Royce engines were not doing well on BA airplanes.... They had forgotten the basic lesson of take no customer for granted and treat each customer as Number One" [14]. GE had also learned that the newly privatized BA was looking to sell its engine maintenance shop in Wales. The BA 777 competition reached its final stages in summer 1991, and fearing last-minute Rolls-Royce and/or political intervention, GE cut a down payment check for the Wales facility. BA liked the engine plus the opportunity to divest its engine maintenance facility. GE pulled off one of the great upsets of aeroengine competition and captured a much-needed endorsement for its new engine.

In Derby, shock prevailed. If Rolls couldn't capture its own flag carrier, what did that say about the future of the Trent? All Nippon Airways cited BA's lack of confidence in the Trent as contributing to its decision to go with Pratt & Whitney [15]. A small consolation prize arrived in September 1991 when Thai Airways selected the Trent. Rolls had its launch customer, but it was for only six aircraft. There were no orders in 1992 or 1993. By 1994, there was a raging debate within Rolls-Royce regarding whether to cancel the Trent 800. As the storm raged, CEO Ralph Robins (Fig. 5.7) provided steadfast leadership and a beacon of hope that it would all work out. Former sales executive Charles

Credit: Photograph courtesy of ©Rolls-Royce Plc

Fig. 5.7 Rolls-Royce CEO Ralph Robins.

Cuddington recalled one particularly testy meeting with an executive where, after berating Robins to cancel the Trent, they stepped onto an elevator. "We were on the 35th floor heading down and Ralph made his case [to the doubting executive]. By the time we hit the ground floor, the thought was gone. It was the ultimate elevator speech."[12]

As Rolls toiled for orders, it was a groundbreaking redesign of its engine development approach, Project Derwent, launched by Engineering Director Phil Ruffles that shifted the firm from "functional silos," where each engineering discipline operated quasi-independently, to an integrated project team that brought together engineering functions and manufacturing. As a result, the Trent 800 development program was a major success. Ralph Robins called it "the best development program ever," even as his firm spent a whopping $379 million on R&D in 1993 [16].

Singapore Airlines rewarded Robins's conviction with a landmark order for the Trent on its 777s in 1995. Then the largest jetliner order in history, Singapore would buy 77 Trent-powered 777s for $12.7 billion—including 157 Trent 800 engines (including spares) worth $1.87 billion [17]. This was the break that Rolls-Royce was after. Singapore Airlines was a loyal Pratt & Whitney customer and one of the most respected airlines in Asia. Its decision would have knock-on effects. Emirates, an up-and-coming airline in the Middle East, would also select Rolls. The momentum for the Trent 700 and 800 would build rapidly in the ensuing years. By late 1996, Rolls-Royce reached 32% of the orders for A330s and 777s against its much larger US rivals [18]. The sting of losing British Airways to GE in 1991 began to ebb.

Rolls-Royce also made several significant moves to broaden its global footprint. In 1990, it formed a joint venture, BMW Rolls-Royce, with the famous West German car manufacturer to focus on propulsion for jetliners with 75+ seats and corporate jets. Crucially, this gave Rolls a foothold in continental Europe and access to West German development funding. Based in Dahlewitz, West Germany, it also took a 20% stake in the Tay program and 5% in the Trent.

The globalization drive continued with a much larger move—the acquisition of US OEM Allison Engines in late 1994. Allison brought Rolls turboprop and turboshaft capabilities and large fleets of well-known engines including the T-56 (C-130 transport) and Model 250 (light helicopters). In turbofans, its 7000-lb-thrust AE3007 engine would power the Embraer regional jet, which would enter service in the late 1990s. The roots of this engine went back to 1988, when Allison and Rolls-Royce began joint studies of an engine to power the ill-fated Shorts FJX. Rolls-Royce withdrew from the project in 1989, but deep relationships were formed. And like the move with BMW, the Allison

[12]Cuddington interview, 1 June 2016.

acquisition transformed Rolls-Royce into a global company with access to the massive US defense market.

By the late-1990s, Rolls-Royce was a transformed company. Under the leadership of Ralph Robins, it was a leaner and more global aeroengine manufacturer boasting a broad range of turbofans, turboprops, and turboshafts. Against stern odds, Robins backed the Trent project against much larger rivals; he had played *The Sporty Game*, arguably betting the company on this vision, and was succeeding. Shortly after retirement, marketing executive Robert Nuttall asked Robins about the difficulty of the Trent launch decision. His response was characteristically clear-headed: "We had three choices: one, we don't do the Trent and the company fails; two, we try the Trent, it doesn't succeed, and the company fails; or three, we try the Trent and we succeed. It was no decision at all—so here we are!"[13]

A New Aeroengine Market: Regional Jets

As engine suppliers jockeyed for position at the upper thrust range, demand for much smaller engines supporting the "regional revolution" began to take off in the 1990s (Fig. 5.8). General Electric was the first beneficiary with the CF34-3, a 9200-lb engine with more than 20 years of history. It was derived from the TF34, which built up millions of hours powering the A-10 Warthog

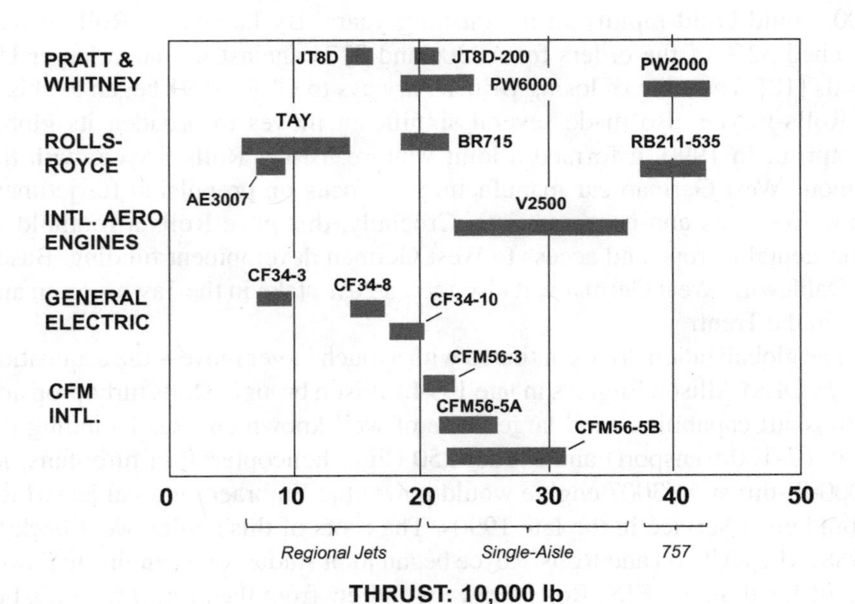

Fig. 5.8 Regional and single-aisle engines: early 2000s.

[13]Robert Nuttall (former marketing executive—Rolls-Royce), interview with author, 1 June 2016.

and S-3 Viking. It also flew on the Bombardier Challenger since the early 1980s. The CRJ was derived from the Challenger, so GE rode into the regional jet market courtesy of Bombardier. The TF34 design for the A-10's close air support mission required that it reach takeoff power about seven times per flight. This would make it well-suited for the high-cycle demands of regional jets.[14] After delivering six CF34s in 1992 demand accelerated, and by 2000, GE delivered 198 CF34s from its Lynn, Massachusetts, production facility.

Meanwhile, the Rolls-Royce acquisition of Allison Engines paid immediate dividends as ERJ sales ramped up. Embraer selected what was then called the GMA-3007 engine in 1990, even though Rolls-Royce pulled out of the project in 1989 to focus on its fledgling Trent project. The 5:1 bypass ratio engine would leverage the core from its military T-406. The name would be changed to AE3007 following its divestiture by General Motors and acquisition by Rolls-Royce. Although the AE3007 started four years behind GE, delivering its first engines in 1996, it soon caught and surpassed GE's production levels. In 2000, it delivered 314 AE3007s. Apparently, Ralph Robins had purchased a winning lottery ticket.

The new wave of regional jets offered just one engine choice. Why? As always, economics played a prominent role. High engine certification costs coupled with much lower unit prices than mainline jets generally dictated a single engine choice. For an aeroengine OEM, selecting the right aircraft model was crucial in this new, growing segment.

When Bombardier stretched its CRJ to the larger CRJ700 and CRJ900 models, it selected the CF34-8, an improved and higher thrust variant (13,800–14,500 lb) of its CF34-3 design. GE upgraded the compressor of the baseline engine's 1970 technology; the overall pressure ratio of the new engine was 28—double the CF34's pressure ratio of 14 and significantly better than the AE3007's 18–20.

With GE following its launch customer Bombardier to larger regional jets, would Rolls-Royce hold serve with Embraer? It was a bridge too far for the older AE3007 technology and the significantly higher thrust requirements for the E-Jets. "Rolls was not able to offer a proven, updated product for the family, and GE was eager to be onboard for the new family," per Embraer executive Horacio Forjaz.[15] With no real competition and a strong desire to grow, the CF34-8 won the E170 and E175 programs. GE would go on to build an even larger variant, the 20,000-lb-thrust CF34-10, for the E190 and E195. By the mid-2000s, the demand for 50-seat regional jets dried up as interest in larger models expanded. The fortunes of GE and Rolls-Royce then reversed once again. In 2006, GE built engines for 190 regional jets; Rolls-Royce—wholly dependent on the 50-seat ERJ—built engines for just 26 aircraft. GE seized control over the regional jet aeroengine market.

---

[14]Kennedy email, 26 Feb. 2018.

[15]Horacio Forjaz (former Executive Vice President—Embraer), interview with author, 24 Sept. 2015.

SINGLE-AISLE ROLLER COASTER

The market for single-aisles got off to a strong start in the 1990s. The CFM56-3 and CFM56-5 engines benefited from strong 737 and A320 demand (Fig. 5.9). By 1991, unit production of these models exceeded 600—an astonishing accomplishment for a seven-year-old company. Pratt & Whitney delivered nearly 1000 JT8D-200s and PW2000s for MD-80s and 757s from 1990 to 1992. And Rolls-Royce gained traction on the 757 with its reliable RB211-535 as it continued steady Tay production for Fokker F70s and F100s.

The bust that followed in the mid-1990s restructured this picture. Fokker's bankruptcy in 1996 was devastating for the Tay, which delivered 532 engines to Fokker from 1990 to 1995. Suddenly this revenue stream disappeared. The news wasn't much better for Pratt & Whitney. The combination of a weak market and an increasingly uncompetitive product led to MD-80 production falling from 139 in 1991 to just 23 three years later. Boeing's acquisition of McDonnell Douglas a few years later and the unsurprising phase-out of the MD-80 sealed the JT8D-200's fate, and it would be out of production by 2000. Even the up-and-coming CFM International suffered as its production rates were halved. During this abyss was one of the more fateful decisions in aeroengine history: In 1994, Boeing selected CFM International and its CFM56-7 as the sole engine for its 737NG. Despite CFM's incumbency, the decision wasn't a slam-dunk because it faced competition from IAE. Boeing required a contribution to the nonrecurring development costs for the new program—a risk that IAE was not prepared to take [19]. CFM faced a difficult engineering task in making a significant performance improvement without

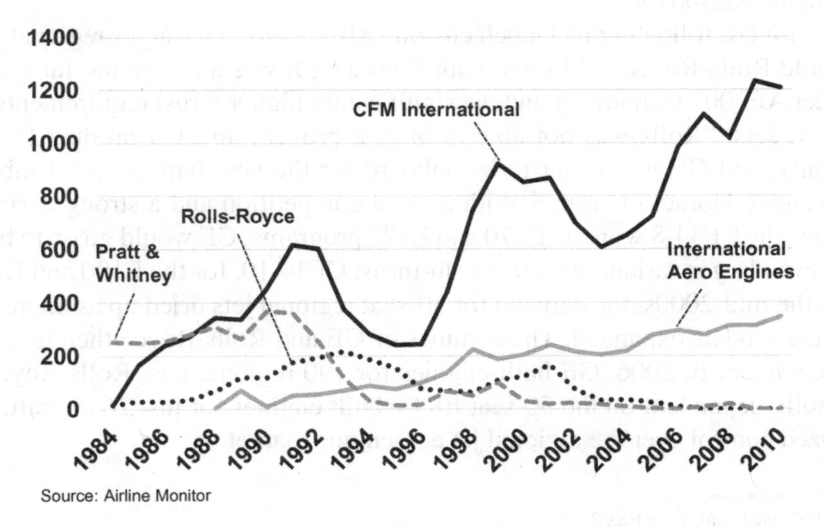

Fig. 5.9 Single-aisle aeroengine deliveries.

increasing the CFM56-3's fan size, because clearance under the 737's wing was limited. CFM addressed these issues by incorporating a wide-chord fan (its first), the CFM56-5B core and low-pressure turbine, and an advanced FADEC. It also benefitted from its unique single-stage high-pressure turbine design, which reduced maintenance costs relative to competing aeroengines with two high-pressure turbine stages.

The dire situation reversed again in the late 1990s (see Fig. 5.8) as global demand for single-aisle aircraft began an unprecedented boom. CFM would capture most of the upside as the 737NG began deliveries in the late 1990s and A320 demand continued to grow. Deliveries for these two aircraft would reach an unprecedented 922 units by 1999.

In contrast, rival IAE continued a slow but steady march of progress on the A320 after a rough introduction. It won a huge $2.5-billion order with British Airways to power up to 188 A320s, and by 1999 unit sales exceeded 200 engines. It trailed the CFM56-5 in unit market share but won kudos for its superior fuel burn. Yet even as IAE gained traction, its partner companies seemed intent to carve up the consortium. Pratt & Whitney won an exclusive position on the Airbus A318, beating out the CFM56 and the BMW Rolls-Royce BR715. Pratt & Whitney's white sheet PW6000 promised 30% better maintenance costs and 40% fewer rotating parts than the competition. Pratt & Whitney's aggressiveness would ultimately lead to major problems when its five-stage high-pressure compressor design failed to meet fuel specification standards. To address the problem, Pratt & Whitney recertified an updated design utilizing a six-stage high-pressure compressor designed by German company MTU Aero Engines. Meanwhile, BMW Rolls-Royce planned for a new turbofan family offering up to 35,000 lb of thrust, which could be launched in 2003 [20].

The boom–bust pattern continued with the arrival of the new millennium. CFM's unit sales plunged to 600—a 50% reduction from 1999—and IAE remained steady because of a strong backlog. As always, the downturn produced program casualties. The venerable 757 went out of production in 2004, and along with it, the RB211-535 and PW2000 programs. With the Boeing 717 termination in 2006 went the BR715 sales for jetliners, although it would continue to sell to the business jet market. Thus, Pratt & Whitney and Rolls were now wholly dependent on their IAE joint venture for single-aisles. Just as these programs were terminated, the market would again begin a recovery that would take demand to new heights. CFM would exceed 1000 engines in 2007.

LARGE-THRUST FIREWORKS

The three-way battle for high-thrust engines continued to rage in the late 1990s as GE and Rolls pursued new applications and market share for their new Trent and GE90 models while Pratt & Whitney looked for PW4000 opportunities. Unexplained compressor surge issues on its engine hindered Pratt & Whitney's sales opportunities. The worst incident was in 2000 when a Sudan Air A300-600R had both PW4000s surge after takeoff and narrowly avoided crashing into the Red Sea before the pilots could get enough thrust to resume climb. Ultimately, Pratt & Whitney determined that the problem resulted from the engine's case construction, which would allow the compressor clearances to open as the engine aged. Pratt & Whitney fixed the problem, but its reputation and PW4000 sales suffered.

The next major prize was two new long-range variants, the 777-200LR and 777-300ER, that Boeing was planning for the mid-2000s to meet burgeoning demand for ultra-long-range routes. This would require even larger engines—with more than 100,000 lb of thrust—for what was considered a niche market of 500 or fewer units. At the same time, development costs for this engine were estimated to be $400–500 million. Given these economics, GE and Pratt & Whitney were in favor of a sole source award. Among potential customers, United Airlines was adamantly opposed to a sole source, whereas Malaysia Airlines and Singapore Airlines were more concerned with optimal performance than fleet commonality. An intense six-month competition ensued. Rolls proposed the Trent 8115, a derivative engine not yet developed based on a demonstrator engine that had achieved 110,000 lb of thrust. Pratt & Whitney bid a new centerline engine based on PW6000 core engine technology—a paper engine [21]. GE offered a 115,000-lb growth variant of its GE90 (Fig. 5.10), the lowest risk option considering that the engine was designed with growth in mind. Boeing decided to make a radical break with the past and move forward with a single engine rather than the three-engine

Credit: GE

Fig. 5.10 GE90-115, the world's largest aeroengine.

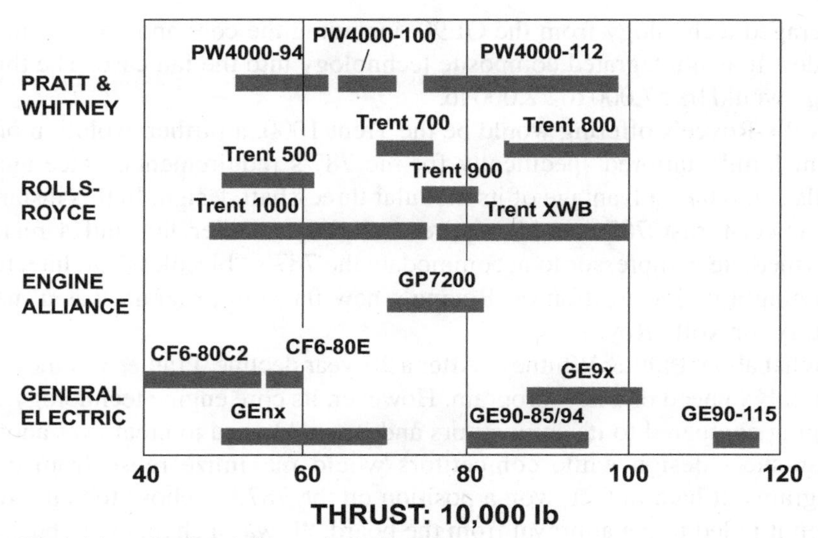

Fig. 5.11 Twin-aisle aeroengines: 2010.

approach of recent history. GE was declared the victor in July 1999. Ironically, GE CEO Jack Welch, who had at one time opposed the GE90 program, was deeply involved in the negotiations and locked up a deal with Boeing CEO Phil Condit early in the competition. Welch's first call after the decision was to the retired Brian Rowe to tell him he was right about the GE90 design strategy 10 years earlier.[16] This was also a signature win for GE Aviation chief Jim McNerney.

The next major aeroengine program was the A380, which would require four engines in the 70,000–80,000-lb-thrust range. Airbus chose to go with two alternatives. The first was Rolls-Royce, who had been pursuing the A3XX program since the mid-1990s. Its offering would be the Trent 900. The other supplier would be the Engine Alliance (EA), a 50/50 joint venture between General Electric and Pratt & Whitney formed in August 1996 to pursue the cancelled 747-5/600 program. Its engine would be the GP7200 (Fig. 5.11).

Boeing's launch of the 787 in 2004 represented a far larger, juicier opportunity that all three competitors pursued with gusto. Boeing initially favored a single engine source, but customer feedback convinced it that it needed to provide engine choice.[17] On 6 April 2004 Boeing announced its decision: GE and Rolls-Royce would power the new aircraft. GE's win wasn't surprising given its growing relationship with Boeing and sole-source positions on the 737 and 777. It proposed a new engine family, GEnx, which

---

[16]Kennedy email, 26 Feb. 2018.
[17]Cuddington interview, 26 Feb. 2018.

leveraged technology from the GE90, including the core and composite fan blades. It also integrated composite technology into the fan case. The thrust range would be 57,000 to 72,000 lb.

Rolls-Royce's offering would be the Trent 1000, a further evolution of its Trent family tailored specifically for the 787's requirements. Once again, Rolls could take advantage of its modular three-shaft design. In this instance, the lower-thrust 787 would require a slightly smaller fan and a resized intermediate compressor to accommodate the 787's "bleedless" architecture. Winning a major position on Boeing's new flagship program was a major victory for Rolls-Royce.

What about Pratt & Whitney? After a 20-year decline, Pratt & Whitney was certainly in need of a new program. However, its core engine technology was lagging compared to its competitors and it would need to create yet another clean-sheet design while competitors would maximize reuse from other programs. It had, in fact, won a position on the 787 but chose to walk away when it failed to get approval from the board. "It was a challenging business case," recalled former Pratt & Whitney President Louis Chênevert, "It would have been a $2 billion commitment."[18] Chênevert was also concerned that committing vast resources to the 787 would limit the company's ability to develop its geared turbofan engine concept. He had no program in hand for the geared turbofan, but feared that he couldn't invest in the 787 and a future GTF program at the same time.[19] Customers and industry observers began to question Pratt & Whitney's commitment to remain in the civil aeroengine business.

The next competition was for the A350. The GEnx was the first to win a position; GE planned to begin deliveries for the A350-800 in 2010. Rolls-Royce garnered the second position with the Trent 1700, a variant of the Trent 1000, which would follow in 2011. Airbus effectively reset the competition when it changed plans and moved to the larger A350 XWB, which would require more thrust. GE feared that the larger variants of the XWB would cannibalize its sole-source position on the 777. It also would need to invest to upgrade the lower-thrust GEnx to meet the XWB's requirements, which were more than 15,000 lb higher than the Mark 1. "Why invest $1 billion to complete with ourselves?" a company spokesman stated [22]. Rolls, in contrast, was enthusiastic about the higher thrust A350 and felt it could offer a new member of the Trent family to match the thrust requirements of the new aircraft. Rolls proposed the Trent XWB and won the competition as de facto sole source engine after GE dropped out of the competition. At the 2007 Paris Air Show, Qatar Airways ordered 80 A350 XWBs and handed Rolls its largest contract in history, worth $5.6 billion at list prices. Thus, Rolls-Royce

[18]Louis Chênevert (former President—Pratt & Whitney), interview with author, 1 Sept. 2015.
[19]Ibid.

completed a 35-year journey, from walking away from the A300, to third choice on the A330, to sole source on Airbus's most important twin-aisle aircraft.

REVOLUTIONARY TECHNOLOGY

The technology in high-thrust engines like the Trent XWB and GEnx highlights the remarkable progress that was achieved in the wake of the industry's first turbofan, the Conway, 65 years prior. Looking at four notable twin-aisle engines—the JT9D-3A, PW4084, GE90-115, and Trent XWB (Fig. 5.12)—highlights the progress.

Thrust grew from 45,000 lb for the JT9D-3A on the quad-engine 747 to 115,000 lb on the GE90-115 for the twin-engine 777. Bypass ratios, just 0.25 for the Conway, nearly doubled from 5.0 on the JT9D to 9.3 on the Trent XWB. Larger, slower turning fans drove improved propulsive efficiency. Facilitating these huge fans were advanced materials—carbon fiber–reinforced plastic for GE and titanium for Pratt & Whitney and Rolls-Royce—to minimize the weight penalty. This massive increase in bypass air comprised more than 80% of the thrust for new generation engines.

What about engine cores? The overall pressure ratio, a measure of the effectiveness of the turbomachinery, reached 52 for the Trent XWB—a 150% increase compared to the JT9D-3A's 21.5. This improvement was facilitated by improved design tools, materials, and manufacturing techniques. The turbine inlet temperature, which measures the hottest part of the engine, reached 2800°F—hundreds of degrees higher than the melting point of the turbine blades. To operate in these extreme environments, new nickel-based superalloys incorporating chromium, cobalt, and rhenium were developed, as were internal blade cooling systems and advanced thermal coatings. Aside from alloy improvements, a major breakthrough was the development of directional solidification (DS) and single crystal (SC) production methods.

	JT9D-3A (1970)	PW4084 (1995)	GE90-115 (2004)	Trent XWB (2015)
Dry Thrust (K lb.)	45.8	74-98	115	84
Fan Diameter (in.)	92	112	128	118
Bypass Ratio	5.2	5.8-6.4	8.4	9.3
Pressure Ratio	21.5	34-42	42	52
Specific Fuel Consumption*	0.65	0.55	0.52	0.48-0.50 (e)

* In lb/lbf/hour; 35,000 ft altitude, 0.8 M, uninstalled

Source: Industry sources, Flight Global, author's estimates

Fig. 5.12 Aeroengine performance improvement: 1970–2015.

These advancements led to a 25% improvement in fuel efficiency compared to the original JT9D. On top of these improvements was a step-function change in reliability ushered in by ETOPS, improved designs, and advanced engine health monitoring systems. Perhaps most impressive are the intervals between engine overhauls, which increased from approximately 5000 hours for the JT9D-3A to greater than 20,000 hours for modern engines. Labeling aeroengines as "modern miracles" is no understatement.

MODULE REUSE: A KEY TO PROFITABILITY

What makes some engine programs profitable and others unprofitable? Clearly, placing the right bets and winning the right programs is essential—particularly with $1- to $2-billion investments required for a white sheet program. But equally important is developing reliable and high-performance hot and cold section modules and then finding ways to reuse or scale them on as many engine programs as possible. This not only reduced nonrecurring engineering expenses, but also reduced the high technical risks inherent in aeroengine designs. It also offered the prospect of lower manufacturing costs.

Module reuse was key to General Electric's entry into the civil aeroengine business. Its CF6 series was based on the core of the TF39 military transport engine. GE's hot section for the hugely successful CFM56 family was derived from the military F101 engine. The same is true for its CF34 regional jet engine, which was a modified military TF34. Only the GE90 was a genuine white-sheet design. Another benefit of reusing military engine modules was that US taxpayers funded much of the module's nonrecurring engineering. It was a financially efficient approach that minimized risk.

Rolls-Royce went one step further in module reuse with its unique three-spool engine design, which made it easier to reuse not only engine cores, but also the fan modules. "Our three-spool design uncouples the initial stages of core compression from the fan, which enables the compressor to spin faster on its own 'intermediate pressure' shaft. This also enables flexible scaling of cores and fan sets between different engine models," explained Rolls-Royce marketing executive Richard Goodhead. He offered an interesting metaphor for their approach. "Think of our design approach as analogous to Lego."[20] An examination of seven Trent models bears this out (Fig. 5.13). The Trent 700 and 800—the first Trents—both had new core modules and fan sets. The Trent 500 utilized the same fan diameter as the Trent 700 and a scaled core from the Trent 800. Three of the seven Trent cores were shared or scaled versions of another engine, saving Rolls-Royce billions in nonrecurring engineering expenses.

In contrast to GE and Rolls-Royce, Pratt & Whitney tended to utilize point solutions for its engine designs. This led to difficulties and delays in

[20]Richard Goodhead (Senior VP-Marketing—Rolls-Royce), interview with author, 14 June 2016.

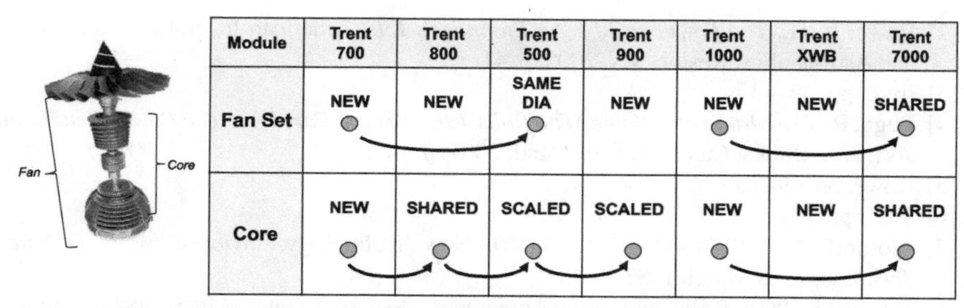

Module	Trent 700	Trent 800	Trent 500	Trent 900	Trent 1000	Trent XWB	Trent 7000
Fan Set	NEW	NEW	SAME DIA	NEW	NEW	NEW	SHARED
Core	NEW	SHARED	SCALED	SCALED	NEW	NEW	SHARED

Sources: Photograph courtesy of ©Rolls-Royce Plc, author's analysis

Fig. 5.13 Rolls-Royce module reuse.

development programs including the PW2000, PW4000, and PW6000 families. This was an issue that it needed to address.

EXIT PRATT & WHITNEY?

With the major engine decisions over the 2004–2007 timeframe for the 787 and A350 XWB, and Bombardier's decision not to enter the single-aisle market, the future appeared to be set for the next decade or so.

In single-aisles, Boeing and Airbus would continue to produce their popular A320 and 737 models and would develop new white-sheet aircraft sometime around 2020. In twin-aisles, Airbus would develop its new A350 XWB, try to turn the A380 into a winner, and ride the popularity of the A330. Boeing would bring the 787 to market circa 2011 to complement its popular 777 while making the most of its aging 747 series.

As for the engine manufacturers, it appeared that GE, its CFM International partner Snecma, and Rolls-Royce were in strong positions and poised for growth. The outlook for Pratt & Whitney was the opposite. Its PW4000 was long in the tooth, the PW6000 had been a disaster, and it walked away from the 787—the largest twin-aisle program in a generation. The only glimmer of hope was its 32.5% of International Aero Engines. *Aviation Week & Space Technology* captured the prevailing sentiment when in 2005 it wrote, "It's distinctly possible that Pratt & Whitney will become a niche player in the large commercial engine field" [23].

Industrial history, however, is sometimes nonlinear. Disruptive innovations and a bit of luck can change the competitive landscape. Few fathomed the impact that a disruptive innovation brewing in East Hartford would have on the jetliner business.

REFERENCES

[1] Vance, A., "The 85 Most Disruptive Ideas in Our History," *Bloomberg Business Week*, 8–14 Dec. 2014, https://www.bloomberg.com/businessweek/85ideas. [retrieved 12 Dec. 2016]

[2] Rowe, B. H., *The Power to Fly: An Engineer's Life*, American Institute of Aeronautics and Astronautics, Reston, VA, 2005, p. 41.
[3] Ibid., pp. 42–43.
[4] Pugh, P., *The Magic of a Name: The Rolls-Royce Story, Part Two: The Power Behind the Jets*, Icon Books, Cambridge, England, 2001, p. 143.
[5] Rowe, pp. 69–70.
[6] Ibid., p. 73.
[7] Mordoff, K., "PW4000 to Use JT9D, New Technology," *Aviation Week & Space Technology*, 28 March 1983, p. 43.
[8] Marcus, S., "GE, Rolls Set Engine Agreement," *New York Times*, 4 Feb. 1984, p. 46.
[9] "Privatization Rolls On," *Flight International*, 28 April 1987, p. 1.
[10] Rowe, p. 128.
[11] "Pratt & Whitney Wins First Contract to Power 777s with PW4000s," *Aviation Week & Space Technology*, 22 Oct. 1990, p. 22.
[12] Rowe, p. 138.
[13] Aboulafia, R., "Engines Face Slow Times," *Aviation Week & Space Technology*, 8 Jan. 1996, p. 95.
[14] Rowe, p. 131.
[15] Prokesch, S., "Rolls vs GE in Sales Dogfight," *New York Times*, 17 Oct. 1991, https://www.nytimes.com/1991/10/17/business/rolls-vs-ge-in-a-sales-dogfight.html. [retrieved 10 September 2015].
[16] Shiffrin, C., "Trent 800 Rolls' Best Start Program," *Aviation Week & Space Technology*, 14 Feb. 1994, pp. 28–29.
[17] "Singapore Airlines to Buy Up to 77 of New Boeing Plane," *Los Angeles Times*, 15 Nov. 1995, http://articles.latimes.com/1995-11-15/business/fi-3360_1_singapore-airline. [retrieved 24 Nov. 2016].
[18] Norris, G., "A Coming of Age," *Flight International*, 11–17 Sept. 1996, pp. 52–53.
[19] "The CFM Story," *Flight International*, 19–25 May 1999, p. 31.
[20] Jeziorski, J., and Norris, G., "Power Struggle," *Flight International*, 23–29 Sept. 1998, pp. 34–35.
[21] Kandebo, S., "GE Wins Exclusive Deal to Power Long Range 777s," *Aviation Week & Space Technology*, 12 July 1999, p. 22.
[22] Norris, G., "Parting of the Ways?" *Aviation Week & Space Technology*, 25 June 2007, p. 36.
[23] Kandebo, S., and Rosenberg, B., "Up in the Air," *Aviation Week & Space Technology*, 14 April 2004, p. 26.

Chapter 6

GEARHEADS

NO EXPERIENCE NECESSARY

Sometime in the late 1980s, Pratt & Whitney engineer Michael McCune (Fig. 6.1) received an unusual offer from the Advanced Projects Group: Would he be interested in working on gearboxes? "But I have no expertise in gears," McCune retorted. "That's alright," the group leader responded, "we are looking for people with *no* experience in gears."[1]

Pratt & Whitney had a long fascination with gears as a means of changing the aeroengine competitive landscape. Despite the dramatic progress aeroengine designers had made in expanding the bypass ratio of modern turbofans, they faced major design tradeoffs as bypass ratios approached double digits. To maximize propulsive efficiency, jet engine designers wanted larger fans turning at moderate velocities and moving large volumes of air. Although counterintuitive, this is more effective than a fan spinning faster and accelerating smaller volumes of air. The dilemma is that in a two-spool configuration, the fan is attached to the same low-pressure shaft as the low-pressure compressor (LPC) and low-pressure turbine (LPT), which are more efficient if they turn faster, not slower. The compromise of sharing the same shaft, Pratt & Whitney believed, was holding back the potential of aeroengine performance. The solution would be to place a reduction gear inside the engine to uncouple the fan from the LPC and LPT spools. Thus, the fan could slow down and grow larger, while the other two spools could rotate significantly faster, shrink, and reduce the number of required stages in each. The engine would be a *geared turbofan* (see Fig. 6.2).

---

[1]Michael McCune (Fan Gear Drive Design Manager—Pratt & Whitney), interview with author, 1 Sept. 2015.

Credit: Michael McCune

Fig. 6.1 Pratt & Whitney's Michael McCune: "father of the gear."

The idea for geared turbofans (GTFs) wasn't new. Garrett (now Honeywell) introduced a gear into its popular TFE731 business jet turbofan in the early 1970s. Its motivation for the integrating gear was to adapt the core of its TSCP700 auxiliary power unit (used on the DC10-10) to a turbofan configuration. Similarly, Lycoming used a gear to add a fan to its T55 turboshaft in 1980 and create the ALF 502 engine used on BAe 146s and

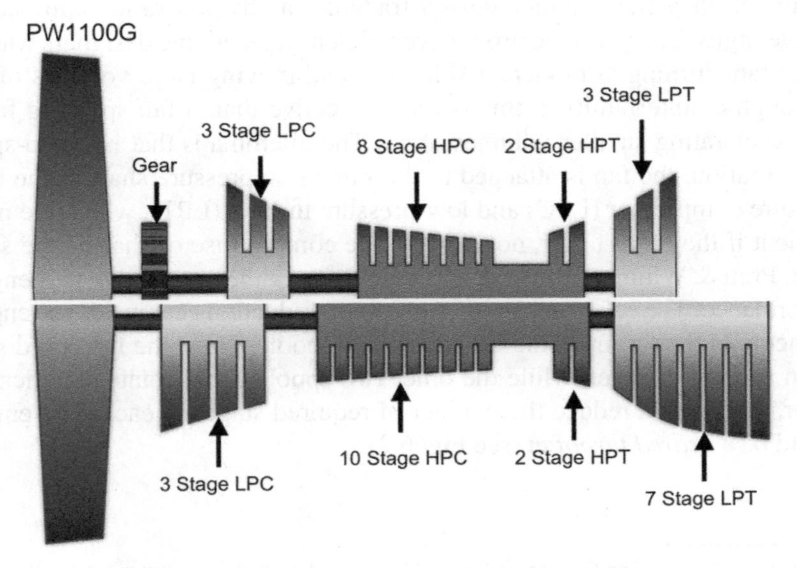

Source: Pratt & Whitney

Fig. 6.2 Geared turbofan architecture.

Challenger 600s. An improved, higher thrust version, the LF507, would be used on the Avro RJ update to the BAe 146. The purpose of the gear in these engines was more to adapt engines and cores to the aircraft's design constraints than to change the nature of gas turbine efficiency. Thousands of GTFs were in operation in the late 1980s, but Pratt & Whitney was contemplating using gears at a fundamentally different scale. The TFE731 (3500 lb) and LF502 (7000 lb) were low-thrust engines. It took four LF502s to power a single regional jet. Pratt & Whitney's vision was to use gears for the full ranges of jetliners—from 20,000-lb thrust to eventually 100,000 lb or more. The shaft-horsepower (shp) input to the gearbox would be more than 30,000 shp compared to the LF502's 6000 shp. Moreover, the existing GTFs operated on low-utilization business jets and a niche regional jet—a far cry from the massive demands of large jetliner businesses. And then there was reliability. The LF502 was a notoriously unreliable engine, with overhaul intervals sometimes less than 1000 hours. Operators would expect a GTF to last 15,000–20,000 hours (or more) between overhauls.

GE and Rolls-Royce both thought about geared architectures, but ruled them out. Rolls-Royce believed its three-spool architecture provided many of the same advantages, but without the risk. GE had suffered a negative experience with a NASA demonstrator program in the 1970s, and its plate was full with the GE90, CFM family, and other development programs. Pratt & Whitney had also concluded at the end of its own open rotor program, which featured a reduction gearbox, that gearbox technology would not work on a large scale. What it needed was a fundamentally different type of gear with unparalleled power input and high reliability. But how to get there?

One of the first tasks for McCune's team was a two-year world tour of gearbox overhaul shops. "We studied every aviation gearbox out there. We collected a list of the field problems and what manufacturers did to fix them. We derived out of the study what the gearbox would have to look like to operators at a large scale and high power." His team learned several valuable lessons, perhaps the most important being that all designs to date used hard-mounted gearboxes. "A gear or bearing has infinite life if you keep it aligned and lubricated. The problem with gas turbines is that at takeoff with hard mounting they don't stay aligned. This results in corner-loaded gear teeth as an aircraft takes off and rotates…so we designed a gearbox that floated." A second problem was with the ALF 502's lubrication system. The challenge wasn't to lubricate the gears, but rather to route the oil out of the gearbox as soon as possible to avoid heat buildup. Pratt & Whitney learned that 70% of heat comes from the "churning" of the oil, not friction of the gearbox teeth. "Oil is your own worst enemy in the gearbox," explained McCune, "on one hand you need it for lubrication, and on the other hand you don't want it to churn too much." Through trial and error, Pratt & Whitney determined the optimal configuration of the oil system and the location of baffles to minimize churning effect. After years of trial and error, Pratt & Whitney arrived at its basic reduction gearbox configuration in 1998. It was a planetary reduction gear system consisting of five outer gears, which Pratt & Whitney called star gears, revolving about a central, or sun gear. The sun gear was connected to the low-pressure shaft and the star gears and carrier to the engine's far fan. A 3:1 reduction was achieved at greater than 99% mechanical efficiency with light weight, high reliability, and low maintenance costs. The company then built a demonstrator at its single-aisle thrust range and ran it in 2001. McCune now had the confidence in the gearbox to approach leadership. "We believe we can make this work," he told them.[2]

Pratt & Whitney had a useful partner in its mission to develop the GTF: Pratt & Whitney Canada (PWC), the world's largest producer of turboprop engines. Turboprops, after all, also utilize gearboxes, albeit at much lower power levels. PWC began talks with customers for a new PW800 GTF and ran an Advanced Technology Fan Integrator demonstrator engine in April 2001. Its own gearbox handled an 11,000-shp load—roughly one-third of what parent Pratt & Whitney would require for single-aisles.

With a gearbox concept in hand, Pratt & Whitney then began design on an actual engine incorporating the gearbox. In 1998, it launched the PW8000 program to target single-aisle aircraft. The engine would share a core with another development program, the PW6000. Just a year later, Pratt & Whitney shut down the PW8000 program. "This was the low point of the GTF's history—all of that effort came to nothing," recalled Pratt & Whitney technical fellow Wes Lord, "one executive told us that Pratt & Whitney wasn't putting another nickel into the GTF, but we were free to try to get funding from NASA."[3] Fortunately, several NASA programs provided a lifeline and, in 2004, Pratt & Whitney proposed to NASA a 29,000 lb -thrust demonstrator engine for its Engine Validation Noise and Emissions Reduction Technology (EVNERT) program. Ironically, the proposal was submitted at the same time as Pratt & Whitney's 787 engine proposal to Boeing.

Next came one of the great decision points in modern aerospace history. As detailed in Chapter 5, Pratt & Whitney walked away from the 787 program while it won a contract on the EVNERT program. It would pivot away from twin-aisle aircraft and bet the company on the GTF and single-aisle aircraft. If the program didn't succeed, Pratt & Whitney would effectively be out of the commercial aeroengine business. It was "all in" on the GTF.

In 2004, Pratt & Whitney began to approach customers about the GTF concept. "The first thing customers would say is what about the poor performance of the ALF 502," McCune recalled. "We realized that we needed to come up with demonstrators to make sure it works." Pratt & Whitney built

[2]Ibid.
[3]Wes Lord (Technical Fellow—Pratt & Whitney), interview with author, 23 Nov. 2016.

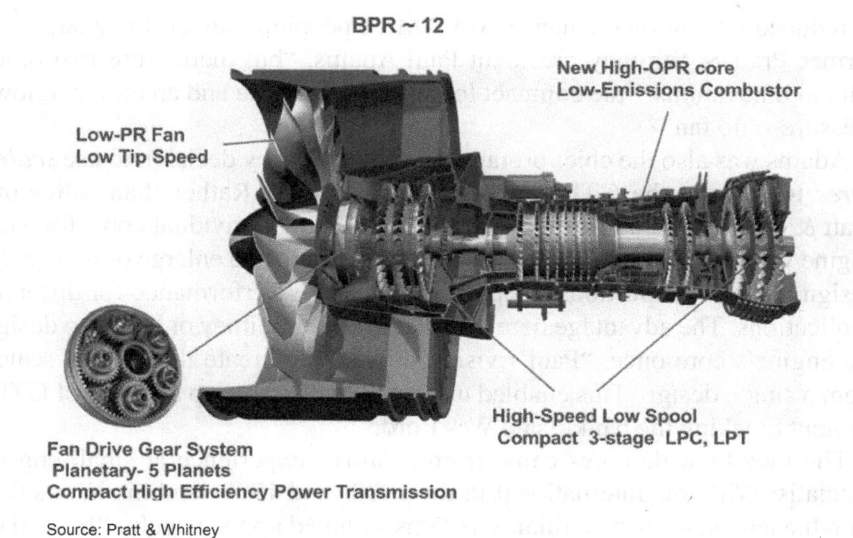

Fig. 6.3 Pratt & Whitney's GTF architecture.

a gearbox test facility in parallel with building engine demonstrators for the EVNERT program. It then brought in customers to demonstrate the gear's durability. In one test, it simulated a continuous 5-*g* turn for 24 hours.[4]

EVNERT funding for building a flight demonstrator engine was cut in 2005 after a NASA budget crisis, and Pratt & Whitney self-funded the development engine and partnered with Goodrich for the nacelle and variable-area nozzle to build a flight-worthy GTF demonstrator. The GTF first ran in 2007 at Pratt & Whitney's Florida test facility and then took to the air on a 747 flying test bed in New York. Airbus expressed interest in demonstrating it on its own A340 testbed.

As it gained momentum, Pratt & Whitney made several critical design decisions. After experimentation, it hit the sweet spot to balance performance with cash operating cost. It would use the 3:1 reduction gear to enable a compact three-stage, low-pressure turbine turning at much higher speeds than conventional turbines. The low-pressure turbine would spin at 12,000–15,000 rpm, while the engine's fan, an efficient low-pressure-ratio design, would rotate at 4000–5000 rpm. Thus, the slow-turning fan maximized propulsive efficiency while the compact fast-turning LPT reduced the engine's size and weight (Fig. 6.3). Conventional aeroengines typically had seven low-pressure turbine stages. There were also two fewer high-pressure compressor stages. By removing these components, Pratt & Whitney reduced the number of airfoils in the engine by more than 2000, which would lead

[4]Ibid.

to reduced engine maintenance cost. "Most people focus on the gear," said former Pratt & Whitney president Paul Adams, "but there were two other critical innovations—the compact low-pressure turbine and an efficient, low-pressure-ratio fan."[5]

Adams was also the chief protagonist of another key decision to use *scaled cores* to address the GTF's potential thrust range. Rather than following Pratt & Whitney's traditional behavior of designing individual cores for each engine variant, it would develop a single core design and enlarge or reduce the design in exact proportions to optimize the GTF's performance for different applications. The advantage here was that Pratt & Whitney only had to design the engine's core once. "Paul's vision drove us to create three cores scaled from a single design. This enabled us to eventually develop a family of GTFs without breaking the bank," said Wes Lord.[6]

The idea to scale cores came from Adams's experience at small-engine specialist Williams International in the 1980s and 1990s and his interaction with the legendary Sam Williams. Adams summed up the benefit: "Instead of point solutions, we did one design, scaled it up and down, and eventually got eight GTF engines."[7]

As Pratt & Whitney was perfecting its new creation, oil prices began an inexorable increase in the early 2000s. The price of oil, which was $20/bbl or less for most of the 1990s, hit $60/bbl in 2006. By 2008, it would average nearly $100/bbl The impact on airline profitability was swift and vicious. Aviation fuel, which accounted for 10% of airline costs in the 1990s, reached 28% by 2007 and 40% by mid-2008, when oil prices peaked at $147/bbl [1]. The need for fuel-savings technology was skyrocketing. But would anyone take a chance on the geared turbofan?

Breakthrough: The Mitsubishi Regional Jet

One of the great ironies of the jetliner business was that Japan, one of the leading locations for advanced manufacturing and the world's second largest economy, was not a producer of mainline jets in the 21st century. A consortium of Japanese companies built 182 YS-11 turboprops between 1962 and 1974, and local firms were leading suppliers of aerostructures, engine components, and aircraft systems, but Japan was notable in its absence from the ranks of countries producing civil jetliners.

Mitsubishi gained confidence not only from sophisticated indigenous programs like the F-2 fighter, but also from designing the wings for the Bombardier Global Express business jet and manufacturing the famous carbon fiber wings for the 787. In 2002, it started studying reentering the

[5]Paul Adams (former President—Pratt & Whitney), interview with author, 19 June 2016.
[6]Lord interview, 23 Nov. 2016.
[7]Adams interview, 19 June 2016.

Credit: Mitsubishi Aircraft Corporation

Fig. 6.4 Mitsubishi Regional Jet.

jetliner business. "We started with a 30–50 seat design—we didn't want to compete with Bombardier and Embraer," recalled Mitsubishi Aircraft executive Yugo Fukuhara. "We took the concept to airlines in 2003–04 and concluded that the market was too small to justify the investment. By 2005–06, we concluded that we needed to go larger to 70–90 seats. But we had to find a way to achieve differentiation."[8] At the 2007 Paris Air Show it detailed the specifications for the new aircraft, the Mitsubishi Regional Jet (Fig. 6.4). It would feature a carbon fiber fuselage and a sleeker fuselage than the CRJ and E-Jet competition with four-abreast seating. There would be two variants between 70 and 96 seats, requiring two aeroengines between 15,600 and 17,600 lb thrust each. GE's entry would be a variant of the ubiquitous CF34, and Rolls-Royce would also compete. Could this be an opportunity for Pratt & Whitney to launch its rapidly maturing GTF? [2].

Pratt & Whitney decided to give it a shot. "At first, Mitsubishi didn't appear to be taking us seriously," recalled Pratt & Whitney Vice President Bob Saia, "but they allowed us to submit a proposal." There were two camps in the Mitsubishi organization. One favored the CF34-10 powering the E190. The other wanted to embrace the new technology as a means of competitive differentiation. Saia added, "We were dead last in a three-way horse race coming into the final turn, and put on a full court press—visits, teleconferences, demonstrations—to make up ground. We asked for a delay

[8]Yugo Fukuhara (Vice President-Sales & Marketing—Mitsubishi Aircraft,), interview with author, 13 July 2016.

in their engine decision in April 2007, which they granted, and this gave us hope. They were listening!"[9]

"Our initial reaction was to ask 'what is a gear doing in the engine?,'" recalled Fukuhara, "Then we dispatched experts to Hartford and conducted a detailed investigation. Our team concluded that the risk of the gear was small, and the advantage was huge. Environmental issues were becoming more important, and it also had a noise advantage. We made a decision to go with the new engine."[10]

In October 2007, Mitsubishi Heavy Industries announced the selection of the Pratt & Whitney GTF for its Mitsubishi Regional Jet (MRJ), which would enter service in 2013. The aerospace industry was stunned. After 20 years, billions in investment, and a fateful decision to walk away from the 787, Pratt & Whitney bagged its first GTF customer. It was back in the jetliner business. Pratt & Whitney president Steve Finger felt this was the start of something big: "This is the game changer. You can incrementally improve the engines you have, but that will fall significantly short in fuel burn and economics. Or you can wait and hope there's going to be something better in 10 or 15 years" [3]. In March 2008, All Nippon Airways placed firm orders for 15 MRJs, with delivery planned for 2013.

Pratt & Whitney established a foothold, but would other aircraft OEMs follow?

Getting Real: The Bombardier C-Series

As Mitsubishi was defining its new aircraft, Bombardier kept the flame burning for its C Series dream. After shelving the C Series in 2006, Bombardier kept a small team in place to monitor technology and evaluate new configurations. With surging fuel prices, there was renewed hope of a relaunch, and in 2007 an internal decision was made to pursue an updated version of the C Series featuring advanced materials and new aeroengines. An early priority was to find partners to help fund the large investment requirement of the new aircraft. China was a clear priority given its huge market for new aircraft and competitive aerostructures suppliers. The Aviation Industry Corporation of China (AVIC) showed interest in making a significant investment in exchange for major aerostructures work. Another priority was customers. Lufthansa was in the market for new aircraft for its Swiss subsidiary and was one of the most enthusiastic target customers. Former Lufthansa executive Nico Buchholz had twin motivations: "We saw it as a superb 100- to 150-seat replacement aircraft…and a chance to push Airbus and Boeing to innovate."[11] And the

[9]Robert Saia (former Vice President-Advanced Programs—Pratt & Whitney), interview with author, 12 Nov. 2015.

[10]Fukuhara interview, 13 July 2016.

[11]Nico Buchholz (Former VP Group Fleet Planning—Lufthansa), interview with author, 29 June 2016.

third priority was engines. "Pratt & Whitney was seeing the demise of the JT8D-200 and was hungry," recalled Bombardier marketing executive Ben Boehm, "Pratt & Whitney wanted to get back *into* the market, and we wanted to get *in* the market. We needed each other."[12] But could Bombardier, which was unaware of Mitsubishi's decision, become comfortable with the gear?

Bombardier President Gary Scott assembled his engineering brain trust, including Rob Dewar, Francois Caza, and John Holding, and descended on East Hartford. Their overarching question: "Why should we *not* be afraid of this engine?" Pratt & Whitney's preparation, demonstrators, and the confidence of its leadership team were impressive—particularly Pratt & Whitney President Louis Chênevert. "He was a major advocate and took us through the entire development history," Scott recalled, "and at the end of the day, we convinced ourselves that this engine would work." Another factor in Pratt & Whitney's favor was Bombardier's strong relationship with Pratt & Whitney Canada, a key supplier of business jet engines that shared Bombardier's hometown of Montreal. As both parties were close to finalizing a deal, Pratt & Whitney admitted that it was also negotiating with Mitsubishi to supply the GTF for another program. "We were pleased that they had another buyer, but not pleased that we could have another competitor," explained Scott.[13] Bombardier called Pratt & Whitney in November 2007 with the news that it had won the competition. The Bombardier board gave the green light to sell the new aircraft the following February, and "version 2.0" was formally launched at the 2008 Farnborough Air Show, with Lufthansa signing a letter of interest for up to 60 C Series.

A second major domino fell for Pratt & Whitney.

The updated C Series was a very different machine compared to the original version. It would have composite wings manufactured by a new resin transfer infusion (RTI) method developed at its Belfast, United Kingdom Shorts production facility. The single-piece wings were not only lighter and more aerodynamic, but also were simpler to build because RTI dramatically reduced the need for fasteners and mechanical parts. It would also feature the industry's first aluminum–lithium fuselage (more on this in Chapter 12), which would be manufactured by Shenyang Aircraft in China (part of AVIC), a risk-sharing partner. Combined with the GTF, the C Series would offer 15–20% fuel savings compared to incumbent aircraft. Two models were developed—a 108- to 133-seat C110 and a 130- to 160-seat C130, later renamed the CS100 and CS300, respectively (see Fig. 6.5). Like the MRJ, they would enter service in 2013. Bombardier estimated a development cost of $3.3 billion—$2.6 billion for research and $700 million in capital investments. Research costs would be shared among Bombardier, its suppliers, and government assistance.

[12]Ben Boehm (former VP-Marketing—Bombardier), interview with author, 8 Jan. 2016.
[13]Gary Scott (former President—Bombardier), interview with author, 21 Sept. 2015.

Credit: Bombardier Inc. and its subsidiaries

Fig. 6.5 Bombardier CS100 and CS300.

Supplier assistance would include a major investment from China's AVIC, which would build the center fuselage. The governments of Canada and the United Kingdom would provide $800 million in repayable investments [4]. The AVIC decision was more significant in another important way—it was the largest and most comprehensive scope of work with design authority ever given to a Chinese aerospace supplier on a western aircraft program. This was done both to position the C Series to secure a strong foothold in the Chinese market and to achieve cost targets.[14]

Lufthansa would convert its letter of intent to a firm purchase order for 30 C Series valued at over $1.5 billion in March 2009. Nico Buchholz, then Lufthansa's Senior Vice President—Corporate Fleet, recalled that there were three motivations for being the launch customer:

> First, we needed to replace our Avro RJs; second, we wanted to bring in new technology; and third, we wanted to send a message to Airbus. I called Airbus after the decision and they were quite upset. I told them if you reengine the A320 we will be a launch customer.[15]

Lufthansa CEO Wolfgang Mayrhuber texted Buchholz after the decision with a short text: "You just started another chapter in aviation."[16]

With the C Series launch and the recent oil price spike, attention shifted to Boeing and Airbus. Would they really wait until 2020 for a new white-sheet aircraft or would they contemplate reengining like the 737? Strangely enough,

[14]Sylvain Lévesque (VP-Corporate Strategy—Bombardier), email to author, 26 Feb. 2018.
[15]Buchholz interview, 6 March 2017.
[16]Ibid.

Bombardier's Scott was rooting for Airbus to reengine the A320. "Once one of the two big players adopted the GTF, the issue of the perceived risk of the new engine on our aircraft would go away," he reasoned. His gut feel, however, was that the two companies would pursue a new aircraft. "When we launched, my sense was that Airbus and Boeing would respond with a white-sheet aircraft. Looking at history, most reengines except for the 737 had not worked. I thought they'd take the new engine and put it on a new aircraft."[17]

Boeing CEO Scott Carson downplayed the competitive threat of Bombardier invading the lower engine of its single-aisle market and the need for a response. But Embraer CEO Frederico Curado wasn't convinced that détente was at hand, and sounded an ominous warning. "The A319/737-700 market is big. I don't think Airbus and Boeing will leave it to Bombardier. If I was them, I would not make that assumption in my own business case"[5].

United Aircraft Corporation would become the GTF's third customer. UAC was created by the Russian government in 2006, effectively merging many famous manufacturers including Tupolev, Ilyushin, Mikoyan, Sukhoi, and Irkut. The latter would take the lead for a new aircraft to replace the country's aging Tu-154, Tu-204, and Tu-214 models. The 160-seat MS-21 would be Russia's most advanced civil jetliner ever. It would feature extensive use of advanced materials, including a composite wing, and numerous components and systems from Western suppliers. Aimed squarely at the A320/737 segment, three variants of the MS-21 would seat between 132 and 178 passengers in a two-class configuration. Entry into service was targeted for 2016.

In 2009 came the news that the PW1400G had been selected as the engine supplier for the aircraft. Pratt & Whitney had its third customer for the geared turbofan. Although Western commentators held little hope that the aircraft would sell in significant numbers outside of Russia, the program provided Pratt & Whitney development funds to expand the GTF's thrust range to 28,000–34,000 lb, identical to the thrust range of the 737 and A320. This would yield important dividends to Pratt & Whitney in the future.

CFM's LEAP Ahead

As the undisputed leader of single-aisle aeroengines, CFM International did not remain idle while Pratt & Whitney was perfecting the GTF. In the late 1990s, it launched the TECH56 program to develop and mature technology that could be used on future CFM56 designs. This included new wide-chord fan blade designs and compressor and turbine module concepts that leveraged new three-dimensional aerodynamic computer modeling capabilities. This initiative was inspired in part by Pratt & Whitney's new PW6000 engine

[17]Scott interview, 21 Sept. 2015.

on the A318, which demonstrated that Pratt & Whitney was willing to step outside of its International Aero Engines joint venture to compete with CFM.

Opportunities to apply the new technology in the early 2000s were few because CFM was firmly established on the A320 and 737. It did propose a CFM56-7 for Bombardier's original C Series in 2004; that opportunity went away when Bombardier paused the program.

As fuel prices began their inexorable rise in the mid-2000s and the GTF became a larger threat, it was clear that CFM would need to up its game in defining its next-generation CFM56. In 2005, it launched the Leading Edge Aviation Propulsion (LEAP) development program. It assembled two teams. The first would focus on keeping the CFM56 program going with selective technology integration. The second team's focus was creating the next-generation CFM56. It invested more than $100 million per year into the program. LEAP's timing was fortuitous because several new technologies were maturing in the mid-2000s.

The first breakthrough technology was ceramic matrix composites (CMCs). Originally developed in GE's Global Research Center in New York, CMCs offered the promise of being just one-third the weight of conventional nickel alloys while being able to withstand temperatures that were 500°F higher. The material was composite of intertwined ceramic silicon carbide fibers embedded in and reinforcing a continuous silicon carbide–carbon ceramic matrix. GE first introduced CMCs in industrial gas turbines in 2000, and then focused on transferring the technology to aeroengines in 2006 [6].

Ongoing research by Snecma showed that composite fan blades, with over a decade of experience on the GE90, appeared ready for demanding single-aisle conditions. One of the biggest challenges, harkening back to Rolls-Royce's experience with Hyfil, was being able to survive bird strike testing. The breakthrough was to create a three-dimensional carbon fiber fan blade preform, and then fabricate the part using advanced resin transfer molding (RTM) technology.

Finally, GE gained valuable experience in additive manufacturing, or 3D printing. Although the technology had been around for several decades, it advanced to a point where it could be applied to metals at a commercial scale. GE collaborated with Morris Technologies, a Cincinnati-based additive manufacturing specialist, to determine if additive manufacturing could be applied to aeroengines. A breakthrough came in 2005, when Morris and GE demonstrated that highly finished components could be produced from cobalt–chrome superalloy using direct metal laser sintering [7]. Parts were built up layer by layer by melting powdered metal with a laser beam. This meant that engineers could conceive of wholly new part designs unencumbered by traditional "subtractive" manufacturing techniques. Cooling channels, for example, could be introduced into a part that would be impossible to make via traditional manufacturing techniques. Another potential advantage: several components could be consolidated into a single integrated piece. By the mid-2000s, it became clear that CMCs, composite fan blades, and additive manufacturing were maturing to a point where they could be introduced into a new aeroengine. All CFM International needed was an application.[18]

With Pratt & Whitney's GTF wins, CFM knew that the timing of a new engine would be sooner rather than later. It conducted extensive discussions with Boeing and Airbus and held customer symposia. It also initiated an online market research campaign, where it asked operators to rate the most important characteristics in a new engine. The answers clearly favored prioritizing reliability, fuel burn, quality, and time-on-wing; noise and nitrous oxide emissions received low ratings [8].

At the 2008 Farnborough Air Show, CFM launched an all-new turbofan, the LEAP-X, which would draw on technology from the LEAP56 initiative as well as GE's TAPS II advanced combustor design, and "blisks"—a combined compressor blade and disk. It also planned to integrate CMCs and RTM carbon-fiber composite fan blades. CFM reckoned that CMCs combined with reduced blade count had the potential to reduce the engine's weight by 175 lb. CFM would also integrate a new high-pressure core with two high-pressure turbine stages—a significant shift away from the compact, one-stage high-pressure turbine (HPT) design that was a hallmark of the CFM program. As noted in Chapter 5, GE didn't create new engine modules unless necessary, and CFM's core was originally derived from its 1960s-era F101 military engine. To address the GTF challenge, it concluded that it would need to drive a much bigger pressure ratio in its high-pressure compressor (HPC), which necessitated the move to two HPT stages. Thus, the LEAP transitioned from being a derivative engine to a genuine white sheet. Veteran journalist Guy Norris, co-author of *CFM: The Power of Flight*, summarized the transition: "The LEAP morphed from an advanced CFM56 to a mini-GEnx."[19]

GE would soon announce another technology risk: it would make the LEAP's fuel nozzles using additive manufacturing, replacing a vital component made up of 20 disparate parts with a single assembly. GE would ultimately form a joint venture with Parker Aerospace to make the new-age fuel nozzles (Fig. 6.6).

At the same air show where International Aero Engines members were embroiled in infighting, GE and Snecma extended the CFM International joint venture (JV), set to expire in 2012, to 2040. By renewing this incredibly successful partnership four years before it was up for renewal, they were sending a message that they were committed to the future. CFM International planned to invest in the LEAP-X demonstrator engine until 2012 and then "wait for the airplane guys" before taking the next step, per CFM56 leader

[18]Bill Klapper (former Vice President—CFM56), interview with author, 15 March 2016.
[19]Guy Norris, interview with author, 11 April 2017.

Credit: CFM International

Fig. 6.6 The LEAP fuel nozzle.

Chaker Chahrour. Should Boeing and Airbus not require a new engine, plan B would be to use the LEAP-X core for a new open rotor engine [9].

Enter the Dragon

As China became the largest single market for jetliners, its dreams of being a major aerospace supplier mushroomed. The approach of two massive Beijing-directed companies, AVIC 1 and AVIC 2, clearly wasn't working because China's penetration into the jetliner business was limited to aerostructures contracts, and its ARJ21 regional jet development program was years behind schedule. A new approach would be required.

In May 2008, Beijing announced the creation of the Commercial Aircraft Corporation of China (COMAC), a new aerospace company that would focus on civil jetliners. Its mandate would be to break China's dependence on Airbus and Boeing by creating its own large jetliners for domestic and international customers. Though state owned, COMAC was an attempt to reboot China's approach to civil aerospace by creating a "white sheet" company. To promote a more commercial and less political approach, it would be based in Shanghai rather than Beijing. COMAC would have a national monopoly on civil jetliners larger than 70 seats, while AVIC would focus on everything else—military aircraft, helicopters, turboprops, aeroengines, aerostructures, and systems. Its first task would be to take over the floundering ARJ-21 program from AVIC 1. This was hardly a recipe for displacing the duopoly, and concurrent with its founding it announced an ambitious new single-aisle aircraft spanning 130–200 seats. Later renamed the C919, it would enter service in 2016 (Fig. 6.7). The C919 would make limited use of Western suppliers and would feature a Chinese powerplant. China committed a whopping 30 billion yuan ($4.5 billion) to the project, and state media indicated that as much as 200 billion yuan would be available

Credit: Weimengat www.airliners.net

Fig. 6.7 The COMAC C919.

for the C919 project ($29.5 billion) [10]. This was not small beer—it was a bold gamble by China that it would establish itself as a major aerospace technology powerhouse on the world stage.

In 2009, COMAC had a change of heart regarding the use of a domestic powerplant—China simply wasn't ready—and initiated discussions with Western manufacturers to supply the C919's powerplant. Given Pratt & Whitney's three wins on the MRJ, C Series, and MS21, the C919 suddenly became a "must win" program for CFM. It made a major push for the program, including involvement by top executives from both GE and Snecma. Both companies were well-positioned. GE supplied the engine to the ARJ21 regional jet and had a broad range of investments in China in its aerospace and non-aerospace businesses. Snecma also had a solid history of collaboration with Chinese partners. "We went after this hard," recalled CFM Vice President Bill Klapper. "We even held annual operator conferences in Zhuhai to better understand the Chinese market."[20] After a one-year selection process, which included Pratt & Whitney's GTF and two engine proposals from Rolls-Royce (which ultimately withdrew from the competition), COMAC declared CFM the winner in December 2009. The LEAP-X1C would be the sole powerplant for the C919.

COMAC bought more than just the engine; it opted for a completely integrated propulsion system that included the nacelle and thrust reverser. GE also won significant avionics content on the new aircraft. To GE Aviation CEO David Joyce, the logic for pursuing the C919 was clear: "Emerging markets matter. We hunt worldwide…without overseas orders, our business would be 27% of its current size."[21]

[20]Klapper interview, 15 March 2016.

[21]David Joyce (CEO—GE Aviation), interview with author, 16 March 2016.

Western commentators publicly doubted COMAC's forecast of 2000 C919s over 20 years. No matter—CFM had its launch customer for the LEAP-X program. And Airbus and Boeing were paying attention. "Once we launched," Joyce observed, "it gave both Airbus and Boeing the confidence that we were committed to bring the technology to market, and, therefore, they could design their schedules and their airplanes around a guarantee that we would be in the market with a new engine in 2016" [11].

The Shot Heard 'Round the World: The A320neo

The new competition commanded Airbus's attention as fuel prices remained high. "There's no reason for the CSeries if you reengine the A320 and 737… there's no reason for the COMAC C919 and all other new programs," said Airbus Chief Commercial Officer John Leahy in mid-2010. Airbus's trade studies indicated that new aeroengine technology combined with advanced winglets, called "sharklets," could reduce fuel burn per seat by 13–15%. It appeared that economics favored reengining over a white-sheet approach, which could cost €12 billion ($15.2 billion) or more [12].

Despite Leahy's ruminations, the case to reengine was considered risky. Could Airbus really count on Pratt & Whitney's geared turbofan following the PW6000 debacle on the A318? Was the technology in CFM's LEAP-X ready for prime time? Lessors were generally opposed to the idea. And if Airbus reengined, would it leave itself open to a white sheet competitive response from Boeing that could quickly make obsolete its best-selling and most profitable aircraft?

Still, with four new aircraft using the geared turbofan and LEAP-X, the case for reengining strengthened with surprising speed. At the March 2010 International Society of Transport Aircraft Trading (ISTAT) meeting, BOC Aviation CEO Robert Martin captured the prevailing sentiment when he commented, "Even three months ago, this [reengining] wasn't on the table." [13] Now many expected an announcement as soon as the 2010 Farnborough Air Show. While Airbus evaluated its options, Bombardier pursued A320 customers with its groundbreaking C Series, including Air Baltic and Lufthansa. One customer group that wasn't enthused about the prospect of reengining was lessors, who had massive positions on A320s and 737s and were worried about the impact of new technology on the value of their assets [14]. Gerard Theron, Airbus's Head of Powerplant, would play a key role in the evaluation. Convinced that reengining was a possibility, he sent a letter to engine manufacturers in 2007 asking for A320 reengining ideas. With the C Series and MRJ launches, Pratt & Whitney's GTF was front and center. The GTF concept wasn't new to Theron. He had evaluated the Superfan proposal for the A340 in the late 1980s and had participated in demonstration programs in subsequent years. In 2007, he witnessed ground tests of a GTF, and in 2008 Airbus and Pratt & Whitney mounted a GTF to an A340 for flight tests. The results were impressive. The reduction gear performed flawlessly, and fuel burn met or exceeded expectations. "The tests gave us a sense that this would not be another PW6000," according to Theron. These test results, coupled with Pratt & Whitney's methodical 20-year development effort, won him and the Airbus engineering team over (Fig. 6.8). They would embrace the gear.[22]

In December 2010, Airbus rocked the aerospace industry with the announcement of the A320 new engine option, dubbed the A320neo (Fig. 6.10). Airbus would invest a modest €1 billion ($1.3 billion) in the program, which would offer 15% fuel burn savings, a 10- to 15-dB noise reduction, and a 500-nm range improvement with initial deliveries in 2015. It would shelve its plans for a white-sheet A320 replacement for later. Airbus launched the A320neo without a customary launch customer. Still, it was bullish and believed that the sales potential could reach up to 4000 neo-type aircraft over the next 15 years. Like the legacy A320, the neo would offer two engine options: Pratt & Whitney's PW1100G geared turbofan and CFM's LEAP-1A. For Pratt & Whitney, its new engine would feature an 81-in. fan and more importantly would be sweet vindication after several decades of work on the GTF. For CFM, this opportunity allowed it to maintain its position on the A320 and provided a crucial platform for the LEAP.

In a Corner: Boeing's Launch of the 737 MAX

With the stunning A320neo news, the aerospace world turned its attention to Boeing. How would it respond? Boeing had been studying a 737 replacement that could integrate 787 technologies since 2006. In a January 2011 interview, Boeing CEO Jim McNerney reiterated the company's bias to create a white-sheet aircraft, but the company left room to wait and see, to observe how the A320neo would do in the marketplace. Complicating the decision was the red-hot activity on the current model. In 2010, Boeing delivered 376 737s while booking orders for 508. Airbus's John Leahy predicted that Boeing would launch a white sheet, then abandon it and reengine when the "low level of maturity" for the "radically new powerplants" needed for a replacement became clear [15]. He would prove to be partially correct.

January 2011 began with the news of Virgin America's launch order for 30 A320neos. Indian low-cost carrier (LCC) Indigo also signed a nonbinding memorandum of understanding for 150 neos. Still, Boeing remained circumspect. In early February, CEO James McNerney told analysts again that its bias was to move to an all-new aircraft and not to reengine. Moreover, Boeing was concerned that it didn't have the engineering bandwidth to create a new single-aisle and potentially update the 777 at the same time [16]. Then there was the geometry of the 737 itself. Thanks to its lineage as an aircraft

---

[22]Gerard Theron (former Head of Powerplant—Airbus), interview with author, 10 June 2016.

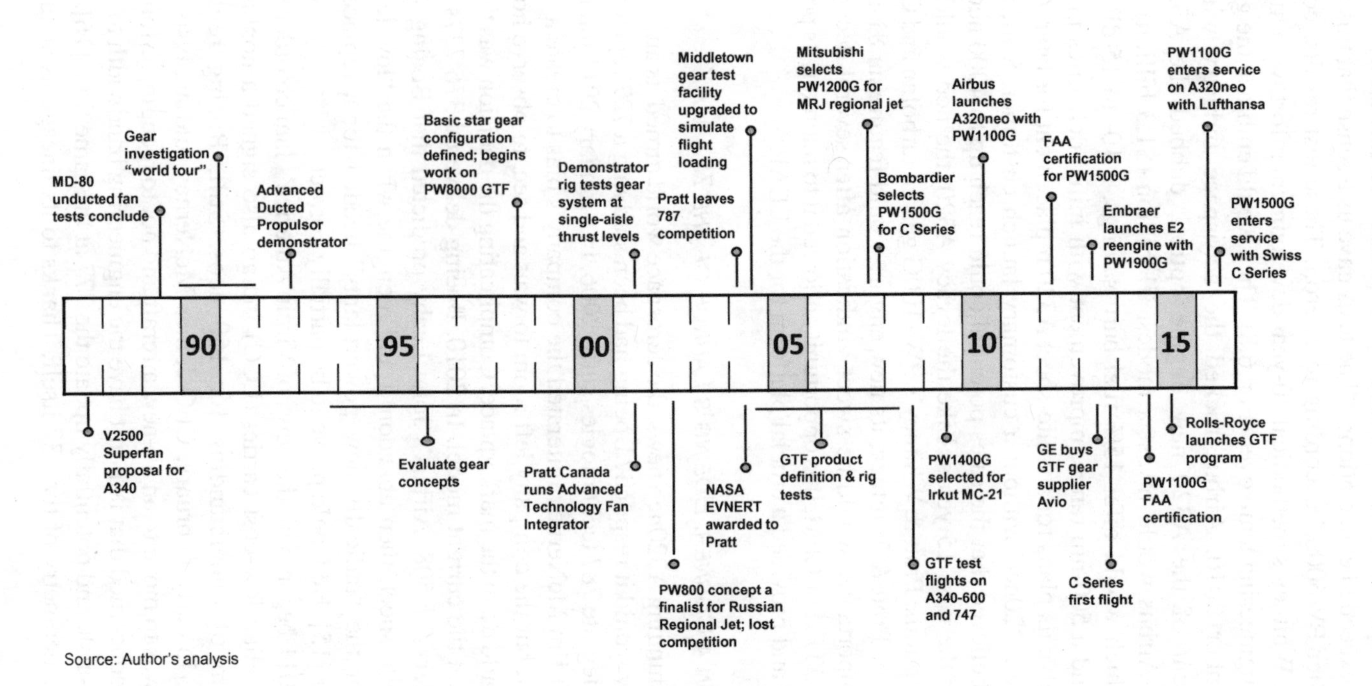

Fig. 6.8 GTF milestones.

originally designed in the 1960s, it had a low wing clearance relative to the A320. Placing a significantly larger engine under the wing would necessitate new landing gear, which in turn would drive the need for a new wing box and probably a new wing. This would add billions to the development cost. If Boeing was going to reengine the 737, it would need to accept an engine with a smaller diameter than the A320neo. It did perform a trade study of engine alternatives including the LEAP and GTF. Logic suggested an advantage to the LEAP given its incumbent status on the 737, GE's strength in compressor design, and the need for higher thermal efficiency (and less dependence on propulsive efficiency) given the 737's size constraints. There was also an exclusivity clause, which gave CFM preference on 737 derivatives.

In March, lessor International Lease Finance Corporation announced an order for 100 A320neos, with the GTF powering at least 60% of them. In April, Lufthansa followed through on its promise to Airbus after its C Series launch decision, and ordered 30 A320neos with the GTF.

The June 2011 Paris Air Show belonged to the A320neo. Air Asia, the largest low-cost airline in Asia-Pacific, ordered 200 neos powered by the LEAP-X worth $18 billion—the largest order in the 23-year history of the A320. Indigo firmed up its MOU for 150 neos and 30 A320s worth $15 billion. It was an order bonanza. A total of 667 commitments for the A320neo were announced during the event, representing a value of $60.9 billion. This brought its total orders to more than 1000. CFM International and Pratt & Whitney shared the riches. CFM brought in orders for a staggering 910 LEAPs worth $11 billion at list prices [17].

Unbeknownst to Boeing, American Airlines—a stalwart customer—became very interested in the A320neo and negotiated a potential deal for the aircraft. The prospect of American Airlines flipping to Airbus would be devastating. With this leverage in hand, it then went to Boeing and demanded a reengined 737.[23] "In a broader sense, Boeing realized that it was hard to justify a 'clean sheet' aircraft in a commodity market where prices hadn't increased in years," said industry analyst Richard Aboulafia. "The technologies weren't there to justify the investment."[24] Boeing had no choice but to acquiesce to reengining.

Boeing moved quickly, and invited GE to make a presentation to American Airlines in mid-July. The meeting went well. In *CFM: The Power of Flight*, GE Aviation executive Chaker Chahrour recalled:

> That was a great day for us, because it took us exactly where we wanted our strategic vision to take us. I called David Joyce from the airport in Dallas and rang Olivier Savin, my opposite number in France. Everyone was elated, and it was amazing how, literally in a few days, things had turned around [18].

---

[23]Anonymous airline executive, interview with author, 18 July 2016.
[24]Richard Aboulafia (VP-Teal Group), interview with author, 16 April 2017.

Credit : CFM International

Fig. 6.9 CFM's LEAP-1B.

The relationship between Joyce and his former boss, Jim McNerney, now the Boeing CEO, contributed to Boeing's change of heart.[25] On 21 July, American Airlines announced a massive order for 460 single-aisle aircraft including the A320, A320neo, 737NG—and 100 reengined 737 aircraft. In the space of a few weeks, Boeing pivoted from a new single-aisle aircraft in the late 2010s to a launch order for a 737 derivative. The aircraft would be called the 737 MAX (Fig. 6.10), and it would be powered by two LEAP-1B engines (the "B" was for Boeing) with 68-in.-diameter fans (Fig. 6.9). Although this was a 7-in. larger diameter than the CFM56-7 on the 737NG, it was 10 in. smaller than CFM's LEAP-1A on the A320neo and 13 in. smaller than Pratt & Whitney's PW1100G on the same aircraft. Boeing would later increase the LEAP-1B's fan size to 69.4 in. to close the gap slightly. It would enter service in 2017, which would give the A320neo (slated for a late 2015 entry) an 18-month head start.

By September, Boeing had expanded its order book to 496, but would only publicly disclose American as a customer. The A320neo had double the orders at the time. This led to some doubts regarding the MAX's competitiveness. Those doubts were erased on 13 Dec. 2011 when Southwest Airlines—arguably the world's most famous 737 operator—ordered 150 MAXs and 58 NGs valued at nearly $19 billion at list prices. This was Boeing's largest order ever in both dollar value and number of aircraft. "Southwest is a special Boeing customer and has been a true partner in the evolution of the 737," said Boeing Commercial Airplanes President and CEO Jim Albaugh. "Throughout our 40-year relationship, our two companies have collaborated to launch the 737-300, 737-500 and the Next-Generation 737-700—affirming the 737 as the

[25]Rick Kennedy (Manager—GE Aviation Media Relations), email to author, 26 Feb. 2018.

Credits: Don-vip, derivative work Lämpel, Clemens Vasters, Embraer

Fig. 6.10 A320neo, 737 MAX, and E2.

world's preferred single-aisle airplane. As launch customer for the 737 MAX, Southwest, Boeing and the 737 continue that legacy".[26]

REGIONAL REENGINING: THE E2

With the decisions set by three of the "big four" jetliner manufacturers, the industry's attention then shifted south to Brazil. What would Embraer do? "We saw Bombardier embarking on the C Series. We viewed it as one foot in the regional market and one foot in the air transport market," former Embraer executive Horacio Forjaz recalled.[27] Would Embraer follow Bombardier into the surging single-aisle market or stick to its knitting in the niche regional market? The Embraer market research team shifted into high gear evaluating customer needs and economics, the anticipated performance of the neo and MAX, and available technology. One incisive question from Embraer's Chairman—Board of Directors and former-CEO Mauricio Botelho guided its strategy: Why should a customer buy from us versus the two huge established companies? "We couldn't answer the question of why the customer would buy from us if we created a new single-aisle," he recalled. "Our technology was similar. We had to be pragmatic about our strengths and weaknesses. This was the basis of our decision."[28] The choice was clear—Embraer would reengine. The next question was which powerplant? Embraer had a long-standing history with GE because its CF34-8 and CF34-10 powered the E-Jet series. At the same time, Embraer's CEO Frederico Curado had publicly expressed skepticism about the GTF.

---

[26]Boeing Corporation, "Boeing 737 Max Logs First Firm Order From Launch Customer Southwest Airlines," Boeing Press Release, 13 December 2011.

[27]Horacio Forjaz (former Executive Vice President—Embraer), interview with author, 24 Sept. 2015.

[28]Mauricio Botelho (former CEO—Embraer), interview with author, 24 May 2016.

In what could be termed a mild upset, Pratt & Whitney was named winner of the engine sweepstakes in January 2013. Embraer would have three variants: an 80-seat E175-E2, a 90-seat E190-E2, and a 108-seat E195-E2 (Fig. 6.10). The PW1700, powering the E175-E2, would use the same architecture as the MRJ's PW1217G. The E190-E2 and E195-E2 would feature the PW1900, replicating the C Series's PW1524G. Pratt & Whitney not only scored a decisive win, but also executed its strategy of module reuse. Former Pratt & Whitney President Dave Hess recalled:

> The E2 was a tough competition. We had to fight the incumbent GE plus a very aggressive Rolls-Royce proposal. We had strong relationships with Embraer via Pratt & Whitney Canada for turboprops. The fact that we had hardware and test data also helped. It was a chain reaction. Each win begets the next win. The MRJ, C Series, A320neo, and MC21 all set up the E2.[29]

Pratt & Whitney's win may have been helped by General Electric's lack of bandwidth; it had three major development programs (GE9X, LEAP, and Passport) at the time, and limited capacity to execute and finance another new engine program.[30]

Embraer would go much further than simply reengining the aircraft with the GTF. It added improved wings, avionics, and systems. The E2 would be scheduled to enter service in the 2018–2020 timeframe, led by the E190-E2. The aircraft would offer a 16–23% operating cost advantage and a 15% maintenance cost advantage versus existing aircraft. Embraer estimated that it could capture 40–45% of a market sized at 6400 aircraft over 20 years [19]. Embraer received good news when it learned that the Mitsubishi Regional Jet EIS had been delayed until 2017.

Pratt & Whitney: Rolls-Royce Rapprochement?

Noticeably absent from this chapter is any discussion about Rolls-Royce. What was Rolls-Royce doing as its three competitors were creating revolutionary new single-aisle aeroengines? First, according to former Airbus executive John Leahy, they were focused on Boeing. "They [Boeing] had Rolls-Royce convinced they were going to be on the all-new single aisle." He added, "Rolls-Royce had stars in their eyes and they blocked us from doing the A320neo with IAE. I desperately wanted to do the neo with IAE and CFM, not with a brand-new guy" [20]. Rolls-Royce opposed a proposal to integrate Pratt & Whitney's gear into the architecture of the new IAE engine. So, it let IAE and the V2500 family generate cash to fund ongoing development of its Trent family.

[29]Dave Hess (former President—Pratt & Whitney), interview with author, 13 Nov. 2016.
[30]Kennedy email, 26 Feb. 2018.

The situation changed in 2011. First, Rolls-Royce and Pratt & Whitney dropped lawsuits against each other over disputes surrounding Rolls-Royce's Trent 900 and 1000 engines and the GP7200 engine. Pratt & Whitney had accused Rolls of copying turbine blade technology; Rolls claimed that Pratt & Whitney violated its intellectual property on the GP7200. Then, in October, Rolls shocked the industry when it agreed to sell its 32.5% stake in IAE to Pratt & Whitney parent company United Technologies Corporation (UTC). At the same time, it announced a new JV with Pratt & Whitney to focus on geared turbofans for future medium-thrust aircraft. Rolls-Royce conceded that gears would be required for future-generation high-bypass-ratio jets. It effectively chose to sit out the next round of single-aisle development and focus on larger aircraft. Pratt & Whitney now had control of the legacy V2500 and its new GTF engines. Airbus Executive Vice President Tom Williams opined on the advantages: "Putting Pratt & Whitney in charge of IAE is better…for customers that want the standard A320 (with V2500s) and the A320neo, CFM was able to offer package deals; now IAE can do the same. Up to now, Pratt & Whitney was at a disadvantage" [21].

Rolls and Pratt & Whitney would abandon their joint venture in 2013. For now, Rolls-Royce would focus on low-volume, high-value Trent twin-aisle engines.

A CHANGED LANDSCAPE

In the 2007–2013 timeframe, the jetliner industry turned on its head because of aeroengine innovation. The catalyst of it all was Pratt & Whitney and its geared turbofan. The GTF not only paved the way for reengining the three major aircraft programs (A320, 737, and E-Jet) in the single-aisle fleet, but also enabled four new aircraft programs: the MRJ, C Series, C919, and MC21. Three new competitors from Japan, China, and Russia threw their hats into the ring. And an astonishing 5473 net orders were booked for these aircraft over six years (Fig. 6.11).

The GTF not only reshaped aircraft product strategies, but also forced GE and Snecma to accelerate technology development. "The LEAP would have happened without the GTF," noted Guy Norris, "but it probably brought the timeframe forward by a decade."[31] It also motivated Rolls-Royce to sell its share of IAE and sit out the next round of single-aisle development.

One of the most surprising and unanticipated results of the GTF was that Pratt & Whitney took over the regional jet sector. It knocked GE off the E-Jet family in winning the E2 and locked up Mitsubishi, Embraer's new competitor. It also had Bombardier as the Canadian OEM attempting to cross the chasm from regional jets to single-aisles.

---

[31]Norris interview, 11 April 2017.

YEAR	MRJ	C Series	MC21	C919	A320neo	737MAX	E2	TOTAL
2007	15							**15**
2008	50							**50**
2009	50	47						**97**
2010	8	40	100	100				**248**
2011	100	43	75	92	1262	150		**1722**
2012		15		145	473	914		**1547**
2013		37		100	808	699	150	**1794**
TOTAL	**223**	**182**	**175**	**437**	**2543**	**1763**	**150**	**5473**

Source: Airline Monitor

Fig. 6.11 Net orders: GTF and LEAP aircraft: 2007–2013.

John F. Kennedy once said, "Victory has a thousand fathers, but defeat is an orphan," and there were several key leaders at United Technologies and Pratt & Whitney that deserve credit for bringing this disruptive innovation to market, beginning with UTC CEO Louis Chênevert. He was an unflagging champion of the program, allocated scarce development funds to keep it alive, and backed the controversial decision to walk away from the 787 and back an unproven design. Former President Steve Finger was also critical. When the financial crisis hit in 2008, many leaders would have cancelled the program. He stuck with it. Engineering SVP Paul Adams, who later became president, also played a pivotal role and embedded into Pratt & Whitney a philosophy of module reuse and scaling. And then there were scores of engineers including "the father of the gear," Michael McCune.

Then there was the LEAP, itself a breakthrough engine loaded with new technologies. It had two of the most successful companies in aerospace behind it in a JV that knew how to execute. Betting against GE or CFM was usually a fool's errand.

This set up a fascinating horse race for the mid-2010s. Could the aeroengine OEMs execute against the massive order book? Would new technologies prove their worth? Could the new aircraft OEMs succeed in entering the jetliner market? And would the twin-aisle segment follow the reengining trend? The jetliner industry girded for a pivotal period.

REFERENCES

[1] International Air Transport Association, *2009 Annual Report*, IATA, Montreal, Canada, 2009, p. 12.

[2] Parrett, B., "Carbon Fiber Competitor," *Aviation Week & Space Technology*, 18 June 2007, p. 76.

[3] Norris, G., "Pratt & Whitney's Geared Turbofan Moves Closer to Launch with Mitsubishi RJ Selection," *Aviation Week & Space Technology*, 15 Oct. 2007, p. 44.

[4] Anselmo, J., and Schofield, A., "CSeries Resurrected," *Aviation Week & Space Technology*, 21 July 2008, p. 38.

[5] Ibid.

[6] GE Aviation, "GE Aviation and the Ceramic Matrix Composite Revolution," YouTube [online video], 29 Oct. 2015, https://www.youtube.com/watch?v=is1BBilkyUM. [retrieved 4 March 2017].
[7] "Technology Laureate: GE Aviation's Greg Morris, Early Additive Adapter," *Aviation Week & Space Technology*, 22 May 2015, http://aviationweek.com/advanced-machines-aerospace-manufacturing/technology-laureate-ge-aviation-s-greg-morris-early-additi, [retrieved 4 March 2017].
[8] Norris, G., and Torres, F., *CFM: The Power of Flight*, Orange Frazer Press, Wilmington, OH, 2016, p. 564.
[9] Mecham, M., "Beyond LEAP," *Aviation Week & Space Technology*, 25 May 2009, p. 52.
[10] Perrett, B., "Comac's Big Push," *Aviation Week & Space Technology*, 22 June 2009, p. 45.
[11] Norris and Torres, p. 590.
[12] Norris, G., Walls, R., and Flottau, J., "Power Plays," *Aviation Week & Space Technology*, 17 May 2010, p. 22.
[13] Mecham, M., Walls, R., and Anselmo, J., "Reengining Reflux," *Aviation Week & Space Technology*, 22 March 2010, p. 40.
[14] Ibid.
[15] Mecham, M., "The Next Move," *Aviation Week & Space Technology*, 21 Feb. 2011, p. 22.
[16] Ibid.
[17] Norris and Torres, p. 597.
[18] Ibid., p. 598.
[19] Anselmo, J., "Embraer's Flight Plan," *Aviation Week & Space Technology*, 23 Sept. 2013, p. 18.
[20] Flottau, J., "Leahy Looks Back," *Aviation Week & Space Technology*, 29 Jan.–11 Feb. 2018, p. 68.
[21] Norris, G., "Gear Shift," *Aviation Week & Space Technology*, 17 Oct. 2011, p. 23.

Chapter 7

RETRENCHMENT

RESTRUCTURING AIRLINES

As revolutionary aeroengine technology turned the single-aisle segment upside down, the global airline industry restructured. One of the catalysts was higher fuel prices. In summer 2010, airlines were paying around $1.89/gal for jet fuel; however, by February 2011, the cost had increased to nearly $3.00/gal, due in large part to heightening tensions in the Middle East [1]. According to the International Air Transport Association (IATA), by 2012, fuel comprised 33% of expenses for the world's airlines, up from 28% two years earlier [2].

The distress from skyrocketing fuel prices ushered in a new wave of consolidation in North America (see Fig. 7.1) when United Airlines and Continental Airlines—two of the world's largest airlines—announced in May 2010 that they would be merging, citing complementary route networks and $1 billion in annual operational savings [3]. Although United's name and Chicago headquarters survived the transaction, most of Continental's management team, led by CEO Jeff Smisek, stayed on to run the new airline. Next, Southwest Airlines announced the acquisition of low-cost carrier AirTran Airways in September 2010, followed by American Airlines entering Chapter 11 bankruptcy in 2011. Fuel became such a large part of airline costs that Delta Air Lines took an unexpected step—it purchased its own oil refinery.

At this point, the only two US-based legacy carriers untouched by the latest round of consolidation were American Airlines and US Airways, although they would not remain alone for much longer. On 14 Feb. 2013, American, which was in Chapter 11 bankruptcy at the time, and US Airways announced that they would be merging. Although the American Airlines name remained due to its wider brand recognition worldwide, the US Airways management team remained at the combined entity with Doug Parker as CEO.

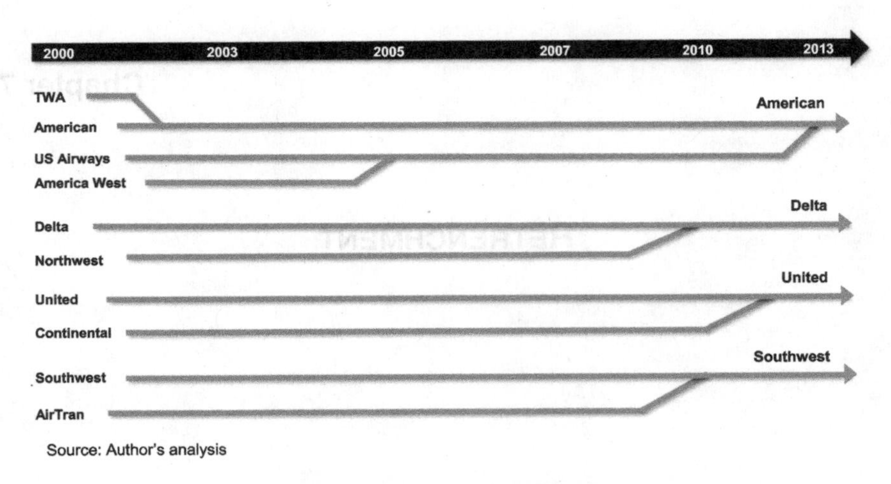

Fig. 7.1 US airline consolidation.

In the wake of this consolidation was four massive airlines—American, Delta, United, and low-cost carrier (LCC) Southwest—with $140 billion in revenue, more than 3200 jetliners, and 83% of US passenger traffic. Thus, more than 30 years after deregulation, the US airline industry came full circle back to a consolidated industry. And rather than mindlessly pursuing market share, which had been the experience in prior decades, airline executives focused on financial performance and capacity discipline. Profitability followed, with the four US majors posting profits of more than $20 billion. After decades of losses or paltry profits in the best years, airlines began to churn out double-digit profits. The remaining 17% of traffic was handled by regional airlines, low-cost airlines like Jet Blue and Spirit, and specialty operators including Alaska and Virgin America.

The story was much the same in the rest of North America. The Canadian market was dominated by Air Canada and low-cost carrier WestJet, and both were very profitable. In Mexico, flag carrier Aeroméxico was challenged by three low-cost carriers: Interjet, VivaAerobús, and Volaris.

In South America, Chile's LAN Airlines and Brazil's TAM Airlines joined forces in 2012 to create the largest airline in South America, aptly named LATAM Airlines Group. Transborder mergers were unusual in South America, but LATAM's scale—with more than 300 aircraft—made it by far the largest airline in the region. Other major carriers included Avianca, Aerolíneas Argentinas, and two Brazilian low-cost carriers—Gol and Azul—with the latter founded by JetBlue founder David Neeleman, as discussed in Chapter 3.

In the Middle East, high oil prices had the complete opposite effect and benefited the local economy and state-owned carriers as government oil revenues took off. The enormous growth in air travel demand in the Middle East was mainly fueled by three airlines: Emirates, Etihad, and Qatar, dubbed the "Middle-East Three." Air travel in the region grew an astounding 73% in revenue passenger kilometers (RPKs) between 2010 and 2015. Much of the growth was from passengers from outside the region as they transformed Dubai, Abu Dhabi and Doha into major hubs connecting traffic among Europe, Asia, Africa, and the Americas (see Fig. 7.2).

The success of these airlines, all government-owned and -supported, caused ripples within the airline industry. Emirates, for example, used the A380 to pick off lucrative routes from European carriers. US airlines also worried and filed complaints that United Arab Emirates and Qatar were unfairly providing subsidies to their airlines, putting US airlines at a competitive disadvantage.

Pressure ratcheted up on airlines in another mature air travel market—Europe—with the failure of four airlines in early 2012. Fuel costs for Europe's largest airlines increased by 18.9% and profits plunged 72% to just €69 million ($90 million) [4]. The economic pressure, coupled with full liberalization of the European market, meant that state-supported airlines like Alitalia teetered on the edge of insolvency. At the same time, low-cost carriers continued to expand, with their share of the European market growing from 24% in 2005 to 31% in 2012 (Fig. 7.3). Ryanair and easyJet thrived. A320 operator easyJet expanded its network throughout central Europe and tripled its revenue over the same timeframe. By 2017, it would reach 80 million passengers and 249 aircraft, with an astounding 92% load factor. Ryanair, operating an all-737 fleet, continued to emphasize underutilized airports and maintained its sharp focus on simplicity and low costs. It closed its airport check-in counters in 2009; everything would be done online except for baggage drop. It grew to be even larger than easyJet. By 2016, its fleet exceed 400 737s with revenue of €6.54 billion and after-tax profits of €1.24 billion ($1.38 billion), which made it one of the world's most profitable airlines.

Airline consolidation affected Europe just as it did the United States in the early 2010s. Following in the footsteps of Air France/KLM and Lufthansa Group, the aforementioned International Consolidated Airlines Group (IAG) was formed in January 2011 after British Airways and Iberia, the Spanish flag carrier, decided to merge. IAG then acquired Spanish low-cost airline Vueling in 2013 and Irish flag carrier Aer Lingus in 2015. IAG, Air France/KLM and Lufthansa Group would be ranked amongst the six largest passenger airlines in terms of revenue by 2016 (Fig. 7.4).

In Asia, fuel prices barely made a dent in the region's inexorable rise, with air travel demand in RPKs averaging 6.7% growth from 2011 to 2014. China led the way, and its jetliner fleet reached 2200 by 2014—an eightfold increase compared to 1990. The growth was underpinned by several factors. The size of the middle class in China and per capita disposable income grew precipitously. While this was occurring, the Chinese government embarked on an aggressive airport expansion project and provided generous support to its state-owned airlines. The largest were China Southern, China Eastern, and Air

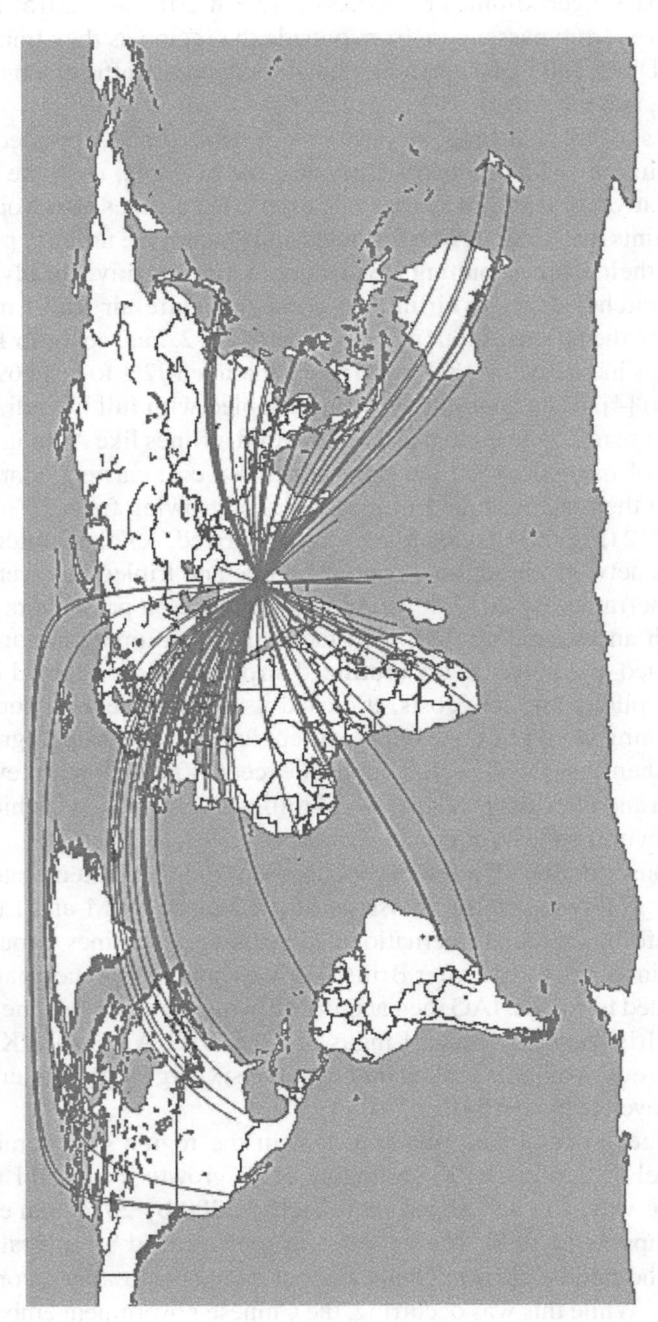

Source: Author's analysis

Fig. 7.2 The Emirates April 2015 route map.

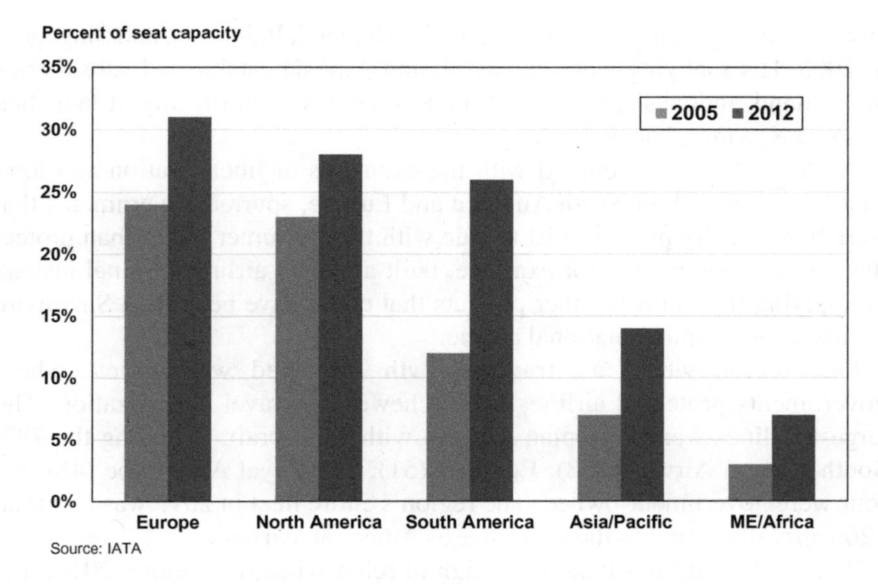

Fig. 7.3 LCC historical seat capacity share, by region.

China, which by 2016 were among the world's largest airlines. More than 80% of the fleet was single-aisles, nearly equally split between A320s and 737s [5].

Elsewhere in Asia, the liberalization of air travel led to an explosion in budget airlines, which in turn stimulated demand for additional air travel. Tony Fernandes, whose company purchased the failing Malaysian airline AirAsia in 2001, lobbied for liberalization of air travel in the region through

Rank	Airline	Country	Revenue ($B)	Profit ($B)	Fleet Size *
1	American Airlines Group	United States	41.0	7.6	946
2	Delta Air Lines	United States	40.5	4.7	852
3	United Continental Holdings	United States	37.5	7.1	750
4	Lufthansa Group	Germany, Switzerland, Austria, Belgium	35.5	1.9	656
5	Air France - KLM	France, Netherlands	28.9	0.1	342
6	International Airlines Group	United Kingdom, Spain	25.3	1.7	548
7	Emirates	United Arab Emirates	23.2	-0.7	241
8	Southwest Airlines	United States	20.2	2.2	720
9	China Southern Airlines	China	17.7	0.6	515
10	China Eastern Airlines	China	14.9	0.7	429

Sources: *Forbes*, company websites
* Fleet size excludes aircraft operated by regional partners
Excludes cargo operators; FedEx Express operated 688 aircraft

Fig. 7.4 World's largest airlines: 2016.

an open skies agreement among Malaysia, Thailand, Indonesia, and Singapore in 2003. His lobbying was successful, and AirAsia established subsidiaries in Thailand, Indonesia, India, and the Philippines. Additionally, it launched AirAsia X, a long-haul subsidiary.

AirAsia's success, coupled with the examples of liberalization and low-cost carrier growth in North America and Europe, spurred governments that were traditionally protectionist to side with the consumer rather than protect flag carriers. Singapore, for example, built a budget airline terminal instead of applying the funds to other projects that could have benefitted Singapore Airlines, the country's national airline.

One region where air travel growth stagnated was Africa, where governments protected airlines and eschewed air travel liberalization. The largest airlines were Ethiopian Airlines with 92 aircraft, including the 787; South African Airways (58); Egyptair (51); and Royal Air Maroc (46). All four were government-owned. The region's entire fleet in 2015 was less than 1200 aircraft—roughly the same size as American Airlines.

The world's airlines breathed a sigh of relief when, in summer 2014, fuel prices plummeted. This was driven primarily by increased production in the United States because of the "fracking revolution" as well as by the inability of the OPEC oil cartel to agree to cut production. Predictably, airlines around the world enjoyed a large tailwind in profitability.

Another major factor began to have a positive influence on airlines: the falling cost of capital (Fig. 7.5). In the wake of the Great Recession, the central

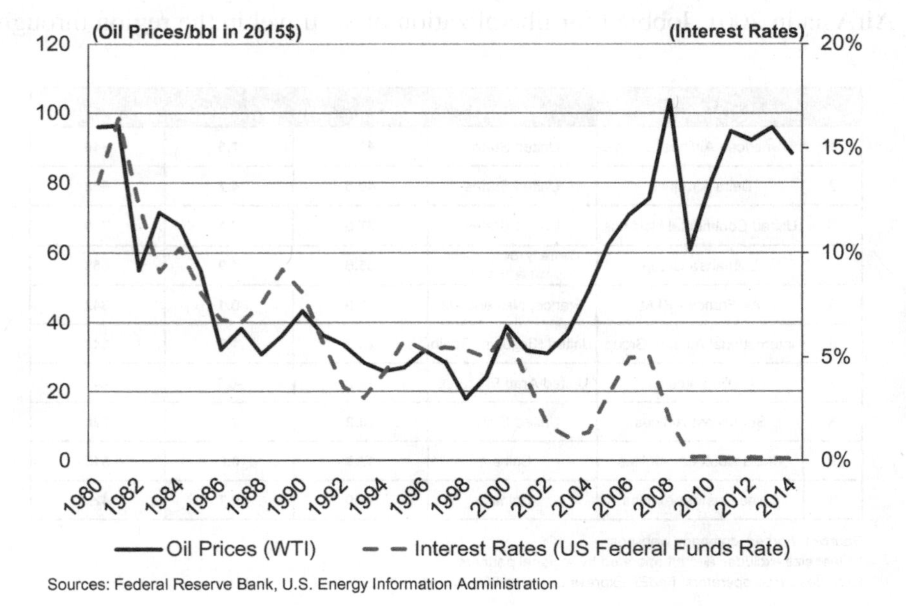

Fig. 7.5 Cost of capital and oil prices.

banks of leading economies lowered interest rates dramatically and flooded the global economy with capital. The influential US federal funds rate—the interest rate at which banks and credit unions lend reserve balances to each other—collapsed from 5.25% in June 2006 to effectively zero by December 2008 and would remain there for another seven years. This meant that the cost of financing aircraft plummeted. Southwest Airlines, for example, financed its 737-700s at a 6.11% interest rate in 2007, which equated to a $281,000 monthly payment per aircraft. By 2012, the same aircraft could be financed at a 4.17% interest rate, which reduced its monthly payments by $52,000 to $229,000 [6]. The Great Recession had reduced the cost of financing aircraft by 20% at a time when high fuel prices placed a premium on buying new jetliners. "For 30 years, the cost of capital and fuel traveled in tandem, but in the mid-2000s the fabric ripped," opined Teal Group Vice President Richard Aboulafia. "This set the stage for the largest orders boom in history."[1]

Jetliner orders, which stood at 734 in 2009, more than doubled in 2010, according to *The Airline Monitor*. In 2011, a record 2706 orders were booked. The jetliner buying bonanza continued. In 2012, another 2636 net orders were logged, followed by 3444 in 2013. And in 2014, customers bought 3748 jetliners. This brought the backlog for Boeing and Airbus to eight years of production. This was new territory.

While the backlog grew, airline leadership became more financially oriented. Historically, airlines rarely recovered their cost of capital. In the early 2010s, many airlines began to turn the tide by focusing on achieving high return on net assets (RONA) and return on invested capital (ROIC). Richard Anderson, the former CEO of Delta Air Lines, was particularly adamant about this, declaring that he aimed for 15% ROIC, just like other investment-grade transport businesses [7]. Rather than focusing on market share, he focused on growing profitable routes. Rather than buying the latest new-technology aircraft, he focused on asset productivity. Soon leading airlines—particularly network carriers—studied Delta Air Lines and adopted lessons learned to their own operations. The results were impressive. Pretax profitability for the world's airlines increased from 3.1% in 2011 to 9.2% in 2016, led by North America's impressive 14.4%. Airline return on net assets over the same timeframe more than doubled from 4.7% to 10.3% [6].

The ownership structure of aircraft also changed, as aircraft lessors expanded their influence. The growth of lessors was facilitated by signing of the Cape Town Convention in 2001, which created international standards for registration of contracts of sale, liens, leases, and legal remedies for default in financing agreements—including repossession of assets. This made it possible for lessors to recover assets from nonconforming customers anywhere in the world. The Cape Town Convention derisked the leasing business and set the

[1]Richard Aboulafia (Vice President—Teal Group), interview with author, 28 Nov. 2017.

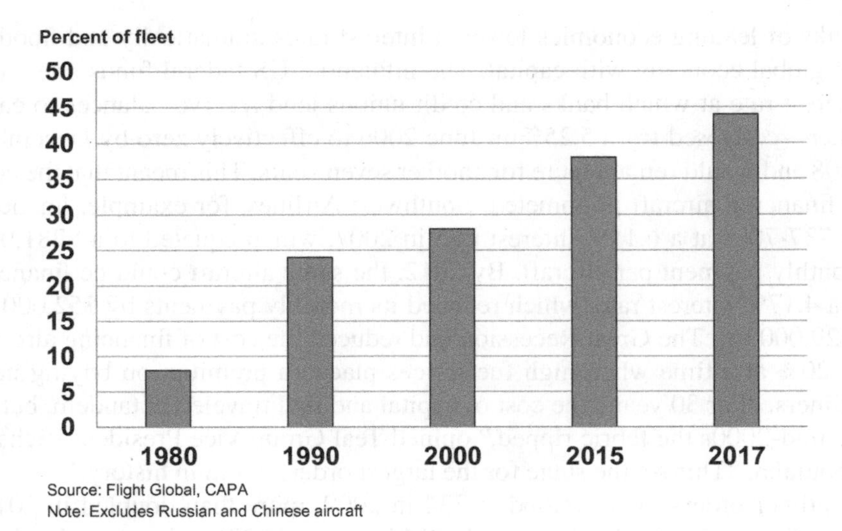

Fig. 7.6 Lessors' share of the fleet.

stage for significant growth of lessors (Fig. 7.6). "From early 1970s to the mid-1980s, leasing was a supplementary affair. Suddenly, the influence of lessors exploded in the late 1980s, and they accounted for 20% of orders," recalled AVITAS Senior Vice President Adam Pilarski. "They used to get worse deals than the airlines, now they get better."[2] By the mid-2010s, lessors owned nearly 40% of the jetliner fleet. The top 50 lessors alone owned more than 10,000 jetliners valued at $280 billion, with the two largest (AerCap and GECAS) owning a combined 2400 aircraft [8].

The heft and financial savvy of lessors drove the industry to standardization. "The best lessors are very good at not taking dogs," opined Pilarski. "They also prefer a single engine type on an aircraft, while airlines prefer multiple engines."[3] By making large orders of standardized aircraft, lessors could provide attractive lease rates to airlines. This was a godsend for start-up airlines and low-cost carriers, because lessors effectively lowered the entry barriers into the airline business. Rather than raising large amounts of capital to purchase jetliners, they could lease aircraft at attractive rates.

These structural changes in the global airline industry formed the context in which aircraft original equipment manufacturers (OEMs) sold aircraft in the 2010s. Leading economies liberalized aviation and privatized airlines, spurring air travel demand. Airlines consolidated and exercised more bargaining power. Low-cost carriers spread throughout the globe, expanding demand for reliable single-aisle aircraft. And a unique cocktail of low cost of capital with the lingering memory of a fuel price shock led to a tsunami of orders.

[2]Dr. Adam Pilarski (Sr. Vice President—AVITAS), interview with author, 7 Feb. 2018.
[3]Pilarski interview, 7 Feb. 2018.

TWIN-AISLE DELAYS

While the airline industry restructured, Boeing focused on bringing the 787 to market. The task proved to be a nightmare. Boeing's Tier 1 supply chain experiment blew up as some of its leading suppliers weren't up to the challenge (detailed in Chapters 8 and 10). The aircraft was overweight, which led to revised designs and material choices. Suppliers were drowning in red ink thanks to schedule changes. The challenge of introducing new aircraft technologies and a new supply chain model *simultaneously* was too much. After three delays to the schedule in 2008, Boeing announced another—its sixth—in June 2009. Entry into service slipped to late 2010. The 787 finally took to the air on 15 Dec. 2009—some two years after its splashy rollout ceremony.

The news for Boeing's other twin-aisle development program, the 747-8, wasn't much better. Boeing announced several delays to its schedule in 2008 and 2009, blaming redesigns, limited engineering resources, and the effects of an eight-week machinists strike [9]. It also took a $1 billion write-down for the programs due to difficult market conditions and higher internal and supplier costs. The 747-8 would not fly until 2011—nearly two years behind schedule [10]. This was surprising given the incremental nature of the updates (a modified wing with new engines) and the 40-year history of the program. Cargolux would ultimate take delivery of a freighter version of the aircraft, the 747-8F, in late 2011 rather than in 2009 per the original schedule. Lufthansa would take delivery of the passenger version the following year.

These delays contributed to another leadership change in late 2009, when Jim Albaugh replaced Scott Carson as the Commercial Airplane Group's CEO. Albaugh, the leader of Boeing's Integrated Defense Systems (IDS) business, wasn't steeped in the commercial jetliner business but was perceived as a steady hand known for executing complex programs. One of his IDS lieutenants, Pat Shanahan, was also brought aboard to run the 787. Boeing badly missed Alan Mulally, the leader of the highly successful 777 program in the 1990s, who was in the process of turning around Ford Motor Company.

Meanwhile, in Toulouse, Airbus ramped up deliveries of the A380 with its development issues behind it. It delivered 18 of its new superjumbos in 2010 and added Lufthansa as an operator. Initial routes from Germany included Tokyo, Beijing, and Johannesburg. In 2011, it delivered 26 A380s and added Korean Air and China Southern as customers. Initial feedback from customers was very positive; however, there was no hiding the fact that the number of orders was disappointing. From its 2001 launch to 2011, it garnered just over 250 net orders with 40% coming from one airline—Emirates. There were no North American customers, and new customers were slow in coming. Just over 60 aircraft were in operation in 2011—four years after its entry into service. Against tepid demand, the full magnitude of Airbus's investment wasn't clear but was assumed to be at least €10 billion ($11.5 billion in 2007).

Credit: Richard Aboulafia

Fig. 7.7 Richard Aboulafia.

Aviation industry guru Richard Aboulafia (Fig. 7.7) captured the sentiment of A380 critics when he labeled it as "the worst new product launch decision since the New Coke" [11].

REENGINING TWIN-AISLES

As jetliner orders skyrocketed, Airbus entered the final stretch of the A350 XWB development program. Unlike the A380, it was relatively drama-free and on schedule. Its first flight was in June 2013 with an anticipated entry into service in late 2014. Orders rolled in, including more than 230 in 2013, pushing the backlog to over 800. A noteworthy order that year came from Japan Airlines (JAL), a 777 operator and traditional Boeing customer, for 31 A350s. Speculation mounted that JAL's rival All Nippon Airways would follow its lead. One Bank of America Merrill Lynch analyst called the JAL order "a major upset" that "bodes poorly" for Boeing's market share and widebody strategy [12]. Customers had pressed Boeing since 2011 for an updated 777, and the pressure mounted for a response.

Boeing received approval to offer a reengined 777, the 777x, in mid-2013. The aircraft would feature a new 235-ft-long composite wing with an option for folding wingtips. The wings alone would yield a 7% fuel consumption improvement. New General Electric GE9X aeroengines would add another 10% fuel efficiency gain. The GE9X would feature a 134-in. (340-cm) diameter composite fan, a composite fan case, ceramic matrix composite parts in the hot section, and an overall pressure ratio of 61:1. Offsetting these gains was a 12-ton increase in the structural weight of the aircraft. The net fuel efficiency gain was projected to be 12–13%. The aircraft would stretch the legacy 777's fuselage, but thinner insulation would enable 10-abreast seating (instead of 9), which would push fuel burn per seat reduction closer to 20% [13]. Boeing appeared to have a hit on its hands given the popularity of the 777 coupled with the demand for fuel-efficient aircraft.

Boeing officially launched the aircraft at the November 2013 Dubai Airshow when it garnered an astounding 259 orders worth $95 billion at list prices. This was the largest product launch by value in jetliner history. Orders and commitments included Emirates (150), Qatar (50), Lufthansa (34), and Etihad (25). Two variants were offered: a 400-passenger 777-9X with a range of more than 8200 nm, and a 350-seat 777-8X with 9300-nm range. The 777-9X would have the lowest operating cost per seat of any jetliner, according to Boeing [14]. The irony of these orders was rich: The Big-Three Middle East carriers—all A380 customers—ordered the new twin-jet variant. Boeing not only was challenging the four-engine A380, but also was most likely shortening the life of its own 747. There was one final key takeaway: Boeing had achieved record orders without a white-sheet aircraft: no massive risks like the 787. Boeing CEO Jim McNerney summed up the prevailing sentiment when he said in 2014, "All of us have gotten religion…every 25 years a big moonshot and then produce a 707 or a 787—that's the wrong way to pursue this business. The more-for-less world will not let you pursue moonshots" [15]. Shortly after giving this speech, oil prices began to decline rapidly.

With the launch of the reengined 777 and the arrival of the 787, Airbus faced a dilemma: what to do with the A330? Airbus hoped that the smallest of the A350 variants, the 276-seat A350-800, would address the needs of A330 operators; however, it was more expensive and optimized for longer routes. Leading customers wanted more. At the 2014 International Society of Transport Aircraft Trading (ISTAT) conference, lessor CIT Aerospace—a leading A330 customer—called out Airbus. "We think that the A330 is at a crossroads," CIT President Jeff Knittel said, "Airbus needs to make some decisions." CIT felt that a reengined A330 would be ideal for the 250- to 300-passenger market for shorter routes—a niche where the 787 and A350 XWB were just too much aircraft [16]. Delta Air Lines CEO Richard Anderson and AirAsia X chief Tony Fernandes also called for reengining the A330.

Four months later, Airbus launched the A330neo (new engine option) at the Farnborough Air Show. Powered by Rolls-Royce Trent 7000 engines, it would deliver a 14% fuel burn per seat improvement. Two variants would be offered: a 252-passenger A330-800neo with a 7450-nm range, and a higher-capacity A330-900neo with 310 passengers and 1200 nm less range. In an ironic twist, the launch customer was Aircraft Lease Finance Company CEO Stephen Udvar Hazy—the same person who effectively killed the idea of reengining the A330 a decade earlier [17]. Airbus's head of product strategy, Bob Lange, explained the logic of the A330neo:

> We recognized that 80% of the operating cost improvements would come from the engine, and by virtue of higher-than-anticipated backlogs and production rates we had already embodied many airframe changes as continuous development. We were therefore well-placed to position an aircraft with very competitive capital costs as a complement to the A350 XWB.[4]

Moonshot Debacles

Boeing wrapped up flight testing of the 787 in mid-2011, and the aircraft entered service with launch customer All Nippon Airways later that year. Rival Japan Airlines followed. Both operators reported impressive fuel burn improvement. ANA's Trent-1000–powered 787s recorded fuel burn improvement of up to 21% compared to the 767-300ER. JAL's GEnx-1B–powered aircraft racked up similar results [18]. Passengers praised the travel experience, citing the air quality, large windows, and interior comfort. Other airlines lined up for the first wave of 787 deliveries. Ethiopian Airlines became the first airline outside of Japan to operate it; LOT Polish Airlines was the first European operator, and United Airlines led in North America. The aircraft's unique capabilities opened new routes as advertised. Air Canada, for example, offered a Toronto to New Delhi route profitably with the 787-9 after years of losses with the L-1011, 747, and A340 [19].

Still, there were teething problems with the first batches of 787s, including systems issues and software anomalies. It also suffered overheating incidents with its lithium-ion batteries, which led to a four-month grounding in 2013 and a retrofit program. Of the nine 787 operators in mid-2013, only one had dispatch reliability above 98%. Boeing worked feverishly to implement fixes, and by late 2015 its dispatch reliability was just under 99% [20]. Deliveries reached 114 in 2014 and exceeded 130 the following two years as a second production facility in North Charleston, South Carolina came on line. Boeing was seemingly executing its vision for the 787: sound economics, excellent reliability, and passenger satisfaction. And, as it predicted, it was beginning to fragment Pacific routes much as the 767 did to the Atlantic two decades earlier.

However, the financial results of the 787 were dire. Rather than development costs of $7–9 billion, the final total, thanks to numerous delays and redesigns, was in the $20–25 billion range [21]. And the "deferred production costs"—the amount Boeing was paid for each aircraft less its actual production costs

---

[4]Bob Lange (Senior Vice President – Airbus Product Strategy), interview with author, 16 June 2016.

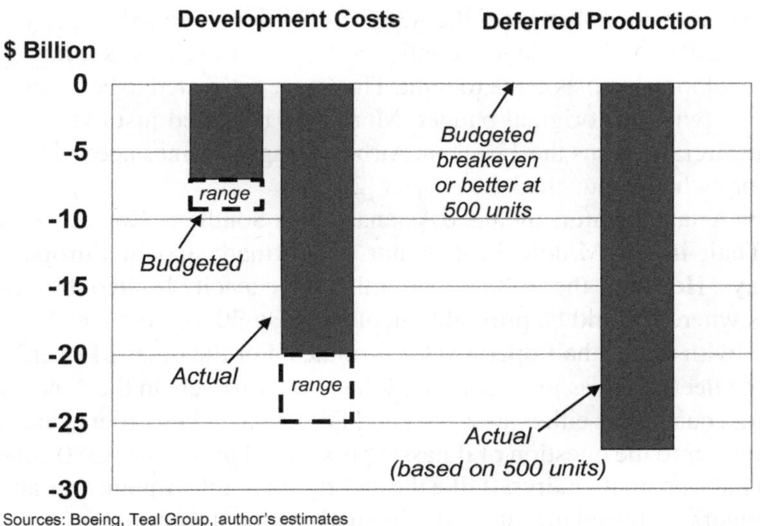

Fig. 7.8 787 losses as of 2016.

(including unamortized tooling)—were enormous. In 2012, it delivered 46 787s with a $5.6-billion loss—an average loss of $121 million each. Deferred production costs in subsequent years were $6.7 billion ($106 million per aircraft) and $5 billion ($43 million per aircraft). Boeing avoided adding these losses to its financial statements by using a program accounting method, where it calculated losses or profits by averaging 787-unit costs over a block of 1300 aircraft [22]. Factoring in $10+ billion in unamortized tooling, Boeing's deferred production costs would ultimately peak around $30 billion in 2017 before it started reversing the tide. Its original plan was to achieve breakeven after 500 units. Now it wasn't clear that Boeing could ever dig out of its $30 billion deferred production cost hole. The 787 was proving to be a commercial success but a financial debacle (see Fig. 7.8).

Boeing's costs for the 787 were more than financial; they also affected its product strategy. "The plan was for the 787 to come online 2008, and then Boeing would replace the 737 and 777," reasoned Scott Hamilton of *Leeham News*. "The late arrival of the 787 screwed up the whole product plan. Instead we got derivatives."[5] Boeing's scale and success in other businesses allowed it to weather the storm. It averaged nearly 98 deliveries of profitable 777s and 477 737s over the 2013–2016 timeframe, and its defense business was also successful. This experience spurred Boeing to rethink its commercial aircraft business—particularly its relationship with its suppliers. This is covered in Chapter 8.

[5]Scott Hamilton (Managing Director – Leeham Company), interview with author, 3 March 2017.

The news was no better for the A380. Deliveries averaged 28 per year from 2012 to 2016, and passengers continued to give it high marks; however, the true development costs came to light. They were estimated to be as high as $25 billion—twice the original budget. Moreover, it logged just over 300 orders for the aircraft versus the 1200 that Airbus thought airlines needed in that size category when it launched the project [23].

New Asian operators included Asiana, China Southern, Korean, Malaysian, and Thai; in the Middle East, Qatar and Etihad; and in Europe, British Airways. However, the A380's economics and capacity limited the number of routes where it could be profitably deployed. By 2016, over 200 A380s were active, with almost half operated by Emirates. For the other 11 operators, the average fleet size was just under 10. It had no customers in the Americas, and no cargo customers either after UPS and FedEx cancelled orders years earlier. The answer to the question of if passengers would prefer the A380 connecting mega hubs or smaller aircraft like the 787 flying point-to-point was answered. Passengers preferred the latter. Boeing had carried the argument.

On top of these challenges, the A380's engines were a generation behind the competition, and the 777x's seat mile costs would be comparable to the A380 when it came online in 2020. Flagship customer Emirates pushed for a reengined A380, but Airbus and aeroengine OEMs were loath to launch the variant based on niche demand and the needs of a single customer. In 2016, Airbus announced plans to cut back production to one A380 per month. The program appeared to be on life support.

Looking back, why did Airbus do it? "We saw a doubling of passenger growth, and we needed a complete product family. We were well aware of Douglas's history with just two big aircraft," said Airbus executive Christopher Buckley. "We also needed to attack Boeing's monopoly at the upper end. Throw in a bit of ego and we needed to have a big aircraft as well." He conceded, "Maybe we were a bit early and should have done the whole thing 10 years later."[6]

The A380 legacy also highlights an important difference from Boeing. "Boeing has done a better job at forecasting how air travel will evolve than Airbus," said Dr. Ron Epstein, Senior Equity Analyst with Bank of America Merrill Lynch. "The 787 was the right concept even though their industrialization plan was flawed and they didn't bet on the A380." Not every decision was good, though. "In the end, however, they did chicken out and launch the 747-8."[7]

THE A350 XWB: AN AIRBUS BRIGHT SPOT

While Airbus licked its wounds from the A380, the A350 XWB experienced early success. Qatar Airways took delivery of the first aircraft in December

[6]Christopher Buckley (Senior Vice President–Sales – Airbus), interview with author 10 June 2016.

[7]Dr. Ron Epstein (Senior Equity Analyst – Bank of America Merrill Lynch), interview with author, 20 Feb. 2018.

Credit: John Taggart

Fig. 7.9 Qatar Airways initiated A350 XWB service in 2015.

2014, and the first commercial flight took place the following month between Doha and Frankfurt. Three new operators—Finnair, Vietnam Airways, and TAM Brazil—came online in 2015 with another 14 deliveries that year. Initial XWB routes were an interesting mix of long-haul thin routes. Finnair flew it from its Helsinki base to New York, Beijing, Shanghai, and Bangkok. Qatar Airways (Fig. 7.9) operated it from its Doha hub to Adelaide, Singapore, Frankfurt, and New York. These were the very types of routes that the 787 targeted. In 2016, Airbus ramped up A350 XWB production rates in earnest with 49 delivered that year, and added Cathay Pacific, China Airlines, Singapore Airlines, Sri Lankan Airlines, Thai Airways, Air Caraibes, Lufthansa, and Ethiopian Airlines to its customer list.

The A350 XWB had minor teething problems, but dispatch reliability was generally good. It expected to reach 98.5% dispatch reliability by the end of 2016 with a goal of exceeding the reliability of the much more mature A330 just four years after entry into service, by early 2019 [24].

From a commercial standpoint, development costs for the A350 XWB were approximately $14 billion—about half of the A380's total [25]. It entered 2017 with a backlog of more than 700 aircraft, with strong sales on the horizon. This was no aircraft pursuing a niche market like the A380. In launching this aircraft, Airbus effectively hedged its bet that the future belonged to superjumbo aircraft; it also had a long-haul, medium-capacity jetliner like Boeing.

As the A350 XWB launched, its aeroengine OEM Rolls-Royce updated its product strategy. In February 2014, it announced a two-element product strategy in response to Pratt & Whitney's GTF, CFM International's LEAP, and General Electric's GE9X. The first element would be the *Advance*, a new three-spool aeroengine and a successor to its Trent family. It shifted work from the intermediate (middle) spool to the high-pressure spool by adding compressor and turbine stages. It also incorporated ceramic matrix composite parts into its hot section. The net result was a 20% improvement versus early-generation Trent engines.

The second element of Rolls-Royce's strategy was fascinating. It announced the *UltraFan*, an aeroengine that broke the mold in several respects. First, it incorporated a geared turbofan architecture that would operate at thrust levels up to 100,000 lbf—more than twice the thrust of Pratt & Whitney's PW1000 series. To make this happen, Rolls partnered with Liebherr to develop a 100,000-shp gearbox. Second, it would follow General Electric's lead and incorporate a composite fan, rather than its traditional hollow titanium approach. The fan would be variable pitch, meaning that it altered its angle of attack depending on engine operating requirements.

Third, the new engine incorporated titanium aluminide turbine blades. The plan was to introduce the UltraFan in the 2025 timeframe. The application, of course, was unknown. But Rolls-Royce threw down the gauntlet that it would respond to the architectural challenges from Pratt & Whitney and the material challenges from General Electric.

NEW REGIONAL JET COMPETITION

While Boeing and Airbus updated their twin-aisle range, the regional jet segment was somewhat stagnant. High fuel prices led operators to park regional jets or reduce their utilization. The smaller, sub-50-seat models were particularly affected, and in 2011 Embraer delivered its last EMB-145. Deliveries of new regional jets predictably suffered, with customers taking just 128 in 2012—a far cry from the prior decade, which averaged more than 270 deliveries per year (Fig. 7.10). Bombardier, focused on breaking into the 100- to 150-seat segment with the C Series, chose not to reengine the CRJ family. It would deliver just 265 CRJs over the 2011–2017 timeframe. This left the market to Embraer and its popular EMB-170/190 family, which straddled the regional jet market and the low end of the single-aisle mainline market. It would average 100 deliveries per year over the same timeframe. And to respond to the C Series, Embraer was updating its product line with

OEM	Model	2011	2012	2013	2014	2015	2016	2017	TOTAL
Bombardier	CRJ700/900/1000	46	14	26	59	44	46	30	**265**
	CS-100/300						7	17	**24**
Embraer	ERJ 135/140/145	2							**2**
	E170/190	103	106	90	92	101	108	101	**701**
Sukhoi	SSJ-75/95	5	8	15	27	14	26	34	**129**
Comac	ARJ21-700/900					1	1	2	**4**
	TOTAL	**156**	**128**	**131**	**178**	**160**	**188**	**184**	**1125**

Source: Teal Group

Fig. 7.10 Regional jet deliveries: 2011–2017.

the E-Jet E2 program. It started work on the geared turbofan–powered aircraft in 2013, and the EMB-190E2 took to the skies in May 2016 with entry into service planned for 2018. It had more than 230 firm orders including US regional airline SkyWest and leasing giant IFLC.

As the Bombardier–Embraer regional jet duopoly ran out of steam, three new competitors emerged. The first was Sukhoi, the famous Russian design bureau known mostly for high-performance fighters. After the collapse of the Soviet Union, Sukhoi diversified its revenue and created a civil aircraft division. After producing small turboprop utility and agricultural aircraft in the 1990s, it set its sights on the booming global jetliner business. In 2000, it announced a "Russian regional jet" initiative. After several years of trade studies, an 85- to 95-seat regional jet design emerged. It would be called the Sukhoi Superjet 100 (SSJ100). Unlike past Soviet aircraft, this would aim to be Western certified. Many of the aircraft's systems were made by European and North American OEMs, and a joint venture between Snecma and Russian OEM Saturn NPO would make the engine. In 2005, Sukhoi received its first orders—30 from Aeroflot and 10 from Russian lessor Finance Lease Corporation. Italian OEM Alenia Aeronautica took a 25% stake in the program in 2006 for $250 million, and in 2007, Alenia and Sukhoi formed Superjet International, a product support division headquartered in Venice. The joint venture would be responsible for aircraft completion and delivery to Western customers, including interior installation, customization, and painting. Alenia, the joint venture partner in the highly successful ATR regional jet program, brought credibility to the program.

The original plan was for a 2007 entry into service, but the program incurred several costly delays. It first rolled out in September 2007 and made its first flight in May 2008. It eventually received Russian certification and entered service with Armavia in April 2011. Orders for the Superjet were modest—just 170 through 2010—and mostly Russian. Many Western operators considered the aircraft to be too risky. Italian flag carrier Alitalia shied away from the SSJ100 in favor of EM190s despite Italian participation in the program [26]. A breakthrough came in early 2011 when Mexican operator Interjet ordered 15 SSJ100s. It would be another five years before the joint venture received another large Western order—15 SSJs from Irish airline Cityjet. Superjet International would deliver 95 SSJ100s through 2016—a far cry from its original aspirations—but Russia was in the global jetliner business on its own terms.

The second new entrant into the regional jet market was the Chinese. After spending much of the 1980s and 1990s engaging in trade studies and building 35 MD-80s under license, the Chinese government decided to pursue the regional jet segment with a 70-seat aircraft. This was a prominent part of its 10th Five-Year Plan. Aviation Industry Corporation of China (AVIC 1) would lead the project. The new aircraft, which resembled a shortened version of an MD-80—including rear-mounted engines and a *T* tail—would be called the ARJ21. Many attributed the MD-80 resemblance to China's licensing experience. Rumors circulated that drawings seen for the fuselage during development were simply McDonnell Douglas drawings with the identifying legend overlaid with a Chinese legend. This made the aircraft heavy. The 78-seat ARJ21 had an empty weight of 55,000 lb. In comparison, an EMB-170 with the same number of seats weighed 46,600 lb. The CRJ-700 was even lighter with an operating empty weight of 44,245 lb.

In 2002, General Electric won the competition to power the new jetliner with its CF34-10 and became a risk-sharing partner. First flight was planned for 2005 with entry into service in 2007. At least that was the plan. It incurred delays almost immediately from design changes, supplier issues, and inexperience. The first ARJ21 flew in 2008 with great acclaim in China. In 2009, responsibility for the program shifted to the Commercial Aircraft Corporation of China (COMAC) as part of broad restructuring of the Chinese aerospace industry. COMAC would be responsible for commercial jetliners, including the C919, while AVIC would be responsible for everything else—military aircraft, helicopters, commercial turboprop aircraft, engines, and systems. The program would eventually fall nine years behind schedule. In late 2015, COMAC delivered the first ARJ21-700 to Chengdu Airlines. The first commercial flight, from Chengdu to Shanghai, took place in June 2016. Chengdu took deliveries of two ARJ21s in 2016. The Civil Aviation Administration of China (CAAC) certified the aircraft for mass production in 2017, but its commercial prospects were weak. It was a heavy and inefficient aircraft that would likely only sell in China.

The third regional jet contender was the Mitsubishi Regional Jet—the original launch customer for the geared turbofan. After the euphoria of launch order from All Nippon Airways in 2008, the schedule started sliding to the right. In 2009, Mitsubishi announced that the wing box would be made of aluminum rather than carbon fiber composites, which delayed the entry into service to 2014. Three years later, it announced another delay attributed to technical programs with documentation, and in 2014 it shifted entry into service to 2017. Despite the delays, the MRJ landed some impressive customers, including US regional airline Tran States (50) and Sky West (100), in 2009 and 2012, respectively.

The MRJ first flew in in November 2015, and the program appeared to be gaining stability. Then in January 2017 it announced another delay—its fifth since 2008 – which it attributed to revisions of aircraft systems and electrical configurations. The new entry into service would be 2020—seven years behind schedule. Mitsubishi also faced an entry barrier in the US market, where most major airlines' pilot contracts have *scope clauses* that limit outsourcing of flying to regional airlines on aircraft exceeding 76 seats and a maximum takeoff weight of more than 86,000 lb. The MRJ90 launch platform exceeded both the seat and weight restrictions. The same was true of the Embraer E175-E2, which would be an important competitor. Despite the challenges, Mitsubishi maintained a backlog of 233 orders and 170 options [27].

The experiences of Sukhoi, COMAC, and Mitsubishi underscored the huge entry barriers to the jetliner business. OEMs from three of the largest economies and most technologically advanced countries attacked the regional jet sector with billions of dollars in investment and generous government support. Collectively, the three incurred 18 years of schedule delays and delivered just 133 jetliners from 2010 to 2017.

SINGLE-AISLE HIJINKS

While new regional jet entrants struggled, the single-aisle sector was awash in development activity, including three new designs and the A320neo and 737 MAX reengining programs.

Bombardier got to work on the C Series. In March 2009, Bombardier redesignated its two variants, the 108-seat C110 and 130-seat C130, as the CS100 and CS300, respectively (Fig. 7.11). Aerostructures and systems suppliers were selected, and the development program ramped up, with an entry into service date planned for four years later. Several delays were encountered, and the first flight of the CS100 took place in late September 2013. An uncontained engine failure of its new PW1000G geared turbofan engine slowed momentum, and the schedule slipped again. These delays increased development costs and began to stretch the financial resources of Bombardier, which was also suffering from a dearth of CRJ

Credits: Bombardier, Shimun Gu, Denis Fedorko

Fig. 7.11 The three single-aisle contenders: C Series, C919, and MC-21.

deliveries and a downturn in business jet demand. The company was running out of cash. By the time the CS100 was certified in December 2015, development costs exceeded $4 billion [28]. The year 2015 was pivotal for Bombardier for other reasons. In February it recruited Alain Bellemare to become its new CEO. A high-energy leader from United Technologies and a native Quebecer, Bellemare dove in to turn around a difficult situation. He recruited a strong leadership team that included Fred Cromer, the former president of leasing giant ILFC, and Nico Buchholz, the former Executive Vice President–Fleet Management for Lufthansa and the C Series launch customer. After reviewing the liquidity crunch facing Bombardier and the headwinds facing the C Series, the new leadership team approached Airbus to take a controlling interest in the program. Bombardier had been looking continuously for a strategic partner since the early days of the program. Airbus walked away from the negotiations; the timing and risks were not right for Airbus. The Quebec government then acted and paid $1 billion to take a 49% equity stake in a new joint company with Bombardier responsible only for the C Series. The C Series was effectively hived off from the rest of Bombardier with taxpayers owning nearly half of the new joint venture. Then Quebec's major pension fund, the Caisse de dépôt et placement du Québec, closed a deal to buy 30% of Bombardier's train business for $1.5 billion. This would give Bombardier liquidity to complete C Series development and fund the five-year turnaround plan launched by Alain Bellemare. Quebec Economy Minister Jacques Daoust explained the logic. "In the aerospace industry, we are at the top of the world," he said. "If we don't invest in businesses that we are good in, where should we invest?" Estimates were that Bombardier was responsible for 40,000 jobs and 2% of the province's gross domestic product (GDP) [29]. Proponents hailed the preservation of Quebec's most important industry. Critics believed that the government threw good money after bad, and that future bailouts would be required.

While Bombardier struggled to bring its new single-aisle to the market, China's COMAC faced challenges with its own single-aisle development, the C919. It received its first orders, for 55 aircraft, at the November 2010 Zhuhai Airshow from four Chinese airlines and two lessors, including GE Capital Aviation Services—not surprising given GE's significant content on the aircraft. Two delays came to light in 2013, which pushed back entry into service to 2017 or 2018. Despite its experience with the ARJ21, COMAC lacked the depth of experience and expertise required for the timely development of the ambitious program. Other issues included lack of delegation due to COMAC's government-influenced culture, poor interaction between engineering teams, late supplier selection, and working with an ultra-conservative regulatory agency—the Civil Aviation Administration of China (CAAC). The CAAC not only was inexperienced in certifying modern jetliners, but also was determined that the C919 would be a safe aircraft that would not humiliate the Chinese state [30].

More orders and delays followed, and in 2016 the aircraft's debut was delayed to 2019. The C919 first flew in 2017 (Fig. 7.11). By then, more than 700 orders were on the books—nearly all from Chinese airlines and many dubious in nature. Despite the likelihood that the C919 would not be competitive with the A320neo and 737 MAX, and was unlikely to sell many aircraft outside of China (it had no clear path for European Aviation Safety Administration [EASA] or Federal Aviation Administration [FAA] certification), it was a source of national pride. COMAC was intent on creating an "ABC" future: Airbus, Boeing, *and* COMAC.

Meanwhile, in Russia, Irkut—part of United Aircraft Corporation—pressed forward with its new entrant. The MC-21 (renamed from the prior MS-21 designation) would be loaded with new technology. Beyond the PW1400G geared turbofan, it would feature active sidesticks, a glass cockpit, and high composite content. Its composite wings would be fabricated using an out-of-autoclave resin transfer infusion process—a first for commercial aircraft. Irkut planned to eventually offer a second engine option, the Russian Aviadvigatel PD-14.

Two variants were planned: a 132-seat MC-21-200 and a 163-seat MC-21-300. Its situation was the polar opposite of the C919: Unlike COMAC, Irkut was loaded with engineering talent and systems integration experience, and it planned to certify the aircraft with EASA. Unlike COMAC, it lacked a large domestic market to support the new aircraft. The MC-21's list price of $91 million suggested a 15% lower acquisition cost than the A320, and it also expected a 12–15% operating cost advantage. Irkut expected to sell 1060 MC-21s over two decades [31].

Like the other new entrants, Irkut experienced delays in its development schedule. It first flew in May 2017 (Fig. 7.11). By then, it had 175 firm orders—nearly all from Russian lessors and airlines. Aeroflot would be the launch customer, with Azerbaijan Airlines the only foreign customer. This was hardly a harbinger for commercial success, but Irkut could at least count on lavish government support from Russian Federation President Vladimir Putin. Despite its notable use of technology, the MC-21 was—like the C919—a political aircraft. As Teal Group Vice President Richard Aboulafia opined, "The C919 and MC-21 are undermined by one crucial weakness: they are both being designed, built, sold, and supported by government-owned companies. Historically, government-owned aerospace companies do an extremely poor job of meeting market needs."[8]

THE MOTHER OF ALL JETLINER BATTLES

The jetliner sector has seen its share of battles over the years, including the DC-9 vs the 737, the 747 vs the DC-10 and L-1011, and the A330 vs the 767.

[8]Richard Aboulafia (Vice President—The Teal Group), interview with author, 7 Dec. 2017.

None compared to the competition between the A320 and the 737NG. In the 2000s and especially in the 2010s, each single-aisle transaction was a pitched battle between Airbus and Boeing that involved heavy discounting and other commercial concessions to win the deal. According to *The Teal Group*, Airbus and Boeing delivered more than 6,600 A320s and 737s over the 2011–2017 timeframe. The aggregate value of these transactions was more than $250 billion.

To the casual observer this would seem unusual given the duopoly status that Airbus and Boeing achieved and the massive entry barriers into the jetliner business. Surely, they could raise prices and maximize profit? Instead, the opposite happened. According to analysis by ICF International, Airbus received an average of $43–45 million for a new A320-200 in 2002 in then-year dollars versus a list price of $57 million—an average 23% discount. By 2015, the net amount received by Airbus was virtually unchanged, even though the list price increased to $97 million. Airbus increased its discounts each year and received the same net amount per aircraft. The story was the same for Boeing, where the comparable figures for a 737-800 were $44–46 million in 2002 (versus a list price of $61 million) and $47–49 million in 2015—a price inflation of just 0.6% per year (Fig. 7.12) [32].

Rather than maximizing margin, which most economists would predict, Airbus and Boeing were locked in a fierce market share battle. Why? "This is

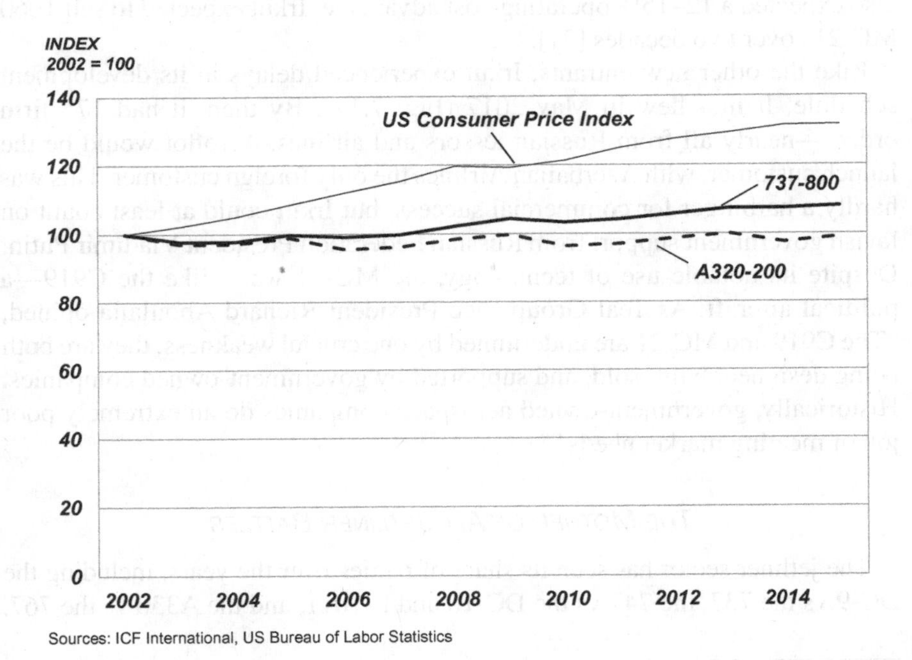

Fig. 7.12 Realized prices for the 737-800 and A320-200.

Credit: Airbus

Fig. 7.13 John Leahy: Airbus's trillion-dollar salesman.

not a cozy duopoly as some people charge," said Airbus VP-Product Strategy Bob Lange.

> The average order size increased due to airline consolidation and bulk buying, and with large orders we can reduce the product cost. The schedule of progress payments from these gives us predictable cash flow, so losing a deal can make a big dent in cash flow. We will compete to the wire if it is profitable.[9]

Airbus and Boeing have room to compete "to the wire" in competitions because the A320 and 737 are, by far, their most profitable aircraft due to the sheer volume of aircraft and the investments that have long since been amortized. Although neither OEM discloses aircraft profit margins, a reasonable assumption is that both aircraft yield margins of 20% or more. This means there is significant wiggle room on pricing to land important deals and still make a profit. "Airbus used to regularly undercut on pricing—they were very aggressive," recalled the CFO of a major US airline.[10]

Finally, there is the "John Leahy factor" (Fig. 7.13). The former Piper salesman rose through the Airbus organization to become Chief Operating Office–Customer, despite his US citizenship. Leahy was a force of nature and a tireless and ruthless competitor. "Under John's leadership we tried to win everything," recalled Airbus executive Bob Lange. Why was he so successful? "What set him apart was his ability to focus on *who* actually makes the decision. You don't do deals by focusing on the process."[11] More than 16,000 aircraft sales were completed under his leadership. "I told him that he was a trillion-dollar salesman," recalled Avitas's Senior Vice President

[9]Bob Lange (Senior Vice President—Airbus Product Strategy), interview with author, 16 June 2016.
[10]Anonymous US airline CFO, interview with author, 26 July 2016.
[11]Lange interview, 16 Jun. 2016.

Type	OEM	Model	2011	2012	2013	2014	2015	2016	2017	TOTAL
Single-Aisle	Airbus	A318/19/20/21	421	455	493	490	491	477	377	**3204**
		A319/20/21 neo						68	181	**249**
	Boeing	737-6/7/8/900	372	405	440	485	495	490	455	**3142**
		737-6/7/8/900 MAX							74	**74**
Twin-Aisle	Airbus	A330-2/300	87	101	108	108	103	66	67	**640**
		A340-5/600		2						**2**
		A350 XWB			.	1	14	49	78	**142**
		A380	26	30	25	30	27	28	15	**181**
	Boeing	747-8	9	31	24	19	18	9	14	**124**
		767-2/3/400	20	26	21	6	16	13	10	**112**
		777-2/300	73	92	94	99	98	99	74	**629**
		787-8/900	3	46	65	104	135	137	136	**626**
		TOTAL	**1011**	**1188**	**1270**	**1342**	**1397**	**1436**	**1481**	**9125**

Source: Teal Group
Excludes regional jets

Fig. 7.14 Jetliner deliveries: 2011–2017.

Adam Pilarski. "He corrected me that the right figure was $1.6 trillion!"[12] This may qualify Leahy as the most successful salesman in history in *any* industry.

The bare-knuckled Airbus–Boeing competition created a competitive hurricane that Bombardier, COMAC, and Irkut would confront if they hoped to become a major single-aisle player.

While Airbus and Boeing worked on their new reengined single-aisles, which were due to enter service in the 2016–2017 timeframe, they ramped up production to levels that were considered unfathomable in prior years. Airbus delivered 3,453 A320s from 2011 to 2017, an average of 493 per year. Boeing wasn't far behind with 3216 737 deliveries (Fig. 7.14). The memory of high fuel costs coupled with low cost of capital and the growth of low-cost carriers also bolstered the order book. Airbus would log more than 5800 A320 orders from 2011 to 2015, with the clear majority for A320neos. Boeing would bring in more than 4700 orders. Airbus enjoyed an advantage at the upper end of its product line because the A321neo parlayed several important advantages versus the MAX 9 including a newer design with fly-by-wire flight controls; a larger, more efficient aeroengine; and larger seating capacity. By early 2016, the A321 neo held a 5:1 order advantage versus the MAX 9 [33].

On 25 Jan. 2016, the first A320neo with Pratt & Whitney PW1100 aeroengines took to the skies in commercial service with Lufthansa. Predictably, the new engine encountered teething problems with Lufthansa and other early operators. Although it exceeded fuel burn expectations, it encountered issues with fan blade quality, carbon seal leakage, and combustor design in hot environments. Pratt worked feverishly to correct the issues. CFM International encountered is own issues with the LEAP introduction, including

[12]Pilarski interview, 7 Feb. 2018.

quality issues with low-pressure turbine disks and separation of coatings from its ceramic matrix composite turbine shrouds.

THE AIRBUS–BOMBARDIER SURPRISE

After the investment by the Quebec government into the C Series, CEO Alain Bellemare and the new leadership team pursued orders to solidify the future of the C Series. One came through in June 2016 when Air Canada ordered 45 CS300s with options for an additional 30. Critics wondered if this order was motivated by nationalism, because Air Canada was based in the same town as Bombardier. On the other hand, Air Canada previously selected the E-Jets over CRJs. There was more good news as launch customer Swiss took delivery of the first CS100. The following month, Boeing beat out Bombardier and Embraer in selling 25 737-700s to United Airlines. Boeing reportedly offered a $22 million-unit price—less than 30% of list price and well below what either Bombardier or Embraer could offer for their airplanes. An important motivation for the massive discount was to block the C Series from penetrating United [34].

Shortly after the sting of this disappointment, Bombardier broke through with a US major airline when Delta Air Lines bought 75 CS100s plus 50 options. Delta was the world's second largest airline, and the largest operator of the aging 110-seat 717. Bombardier appeared to be on its way to breaking the Airbus–Boeing duopoly. Then aeropolitics took hold. In April 2017, Boeing filed a petition with the US International Trade Commission (ITC) purporting that Bombardier "dumped" the C Series with United at $19.6 million each, below their $33.2 million production cost. A few months later, the ITC found that US industry could be threatened. The US Department of Commerce recommended duties of 300% on the C Series later that year. To many observers, this smacked of hypocrisy. Didn't Boeing benefit from US government support? If Bombardier was guilty of dumping, wasn't Boeing also guilty with its recent 70% discount to United? Didn't Boeing itself received government support via massive tax breaks from the State of Washington and many decades of lucrative US military programs? Add to this that Boeing had no competing aircraft in the capacity (100–125 seats) required by Delta. It didn't have a dog in the fight. From Boeing's perspective, Bombardier was using illegal government subsidies to threaten the low end of the 737 Max product family—its cash cow. It compared the situation to the early days of Airbus—government support of a much smaller competitor. "It will only take one or two lost sales involving US customers before commercial viability of the Max 7, and therefore the US industry's very future, becomes very doubtful," said Boeing Vice-Chairman Ray Connor to US trade officials" [35]. Then Bombardier's old nemesis Embraer entered the arena, when the World Trade Organization approved Brazil's request to investigate Canada's alleged use of "more than $3 billion" in government support.

Bombardier was backed into a corner. It was under attack by two major governments; major tariffs would be slapped on it signature Delta Air Lines order, which would likely scupper the deal; and it faced years of losses from the C Series to the detriment of the rest of the company—including its successful business jets franchise.

Bombardier put even more focus on its long-time effort to find a partner in the aircraft. Boeing, for reasons unknown, rejected another approach [36]. Bombardier was in talks with several potential Chinese investors, including COMAC. The Canadian government urged Bombardier to pursue a deal with Airbus [37], so in August 2017, Bombardier started discussions [38].

On 16 Oct. 2017 came the news that Airbus had acquired a 50.01% share of the C Series program (Fig. 7.15). The price for half ownership of a program that cost $6 billion to develop? Zero. Airbus paid nothing and took on no debt. Bombardier would retain 31% ownership, and the Quebec government the remaining 19%. The C Series would be integrated into the Airbus enterprise. And in a move clearly designed to address US trade concerns, Airbus said it would set up a second C Series final-assembly line in Mobile, Alabama (the site of its A320 factory) for delivery of C Series sold to US airlines.

The concerns of Canadian taxpayers aside, it was a strategic masterstroke. With the backing of Airbus, the future of the C Series was assured. It no longer risked being an "orphan" aircraft, and Airbus gained access to the most advanced 100- to 150-seat aircraft for free. Moreover, Airbus could, in the future, choose to pursue an even larger C Series variant—the 180-seat CS500—which would threaten Boeing's 737 MAX 8. The deal also was good

Credit: Airbus, Bombardier Inc. and its subsidiaries.

Fig. 7.15 Airbus takes control of C Series: Bellemare and Enders.

for Bombardier. "There was always the feeling that Bombardier had bitten off way more than it could chew, but this is a very fine jet; it just needed help," said Richard Aboulafia. "Bombardier no longer has a mountain of risk weighing on it, and it can manage itself like a normal aerospace company, not spending every moment wondering if it will be able to survive in one piece to see the break of dawn" [39]. Bombardier could now focus on regional and business jets as well as its train business.

Aviation journalist Scott Hamilton's assessment of the deal was colorful, "The stunning Airbus-Bombardier partnership for the CSeries program guarantees the future of the new airplane, kills off the A319 and thrusts a big stick up Boeing's tailpipe" [40].

Boeing predictably vowed to fight on and pursue its trade dispute, but the damage was done. Canada was now firmly in the Airbus orbit, and the aircraft it attempted to kill was stronger than ever. Its flagship 737 MAX was now caught in a pincer—pressure from below courtesy of a reenergized C Series and pressure from above from the A321neo. Left with a changed competitive landscape, it began to explore acquisition of Embraer. In July 2018, Boeing struck a deal with Embraer for an 80% state in its commercial jet business for $4.75 billion. Both companies expect the deal to close by 2019, pending regulatory approval.

Thus, the industry that John Newhouse labeled *The Sporty Game* in his classic 1983 book because of its high-stakes, winner-take-all nature consumed another new entrant. After billions in investment—some of it taxpayer funded—Bombardier succumbed to the inevitable and bowed out of the large jetliner segment.

Once again, the Airbus–Boeing duopoly proved its omnipotence.

REFERENCES

[1] Mouawad, J., "Airfares Are Chasing Oil Prices Higher," *The New York Times*, 23 Feb. 2011, https://www.nytimes.com/2011/02/24/business/24fare.html. [accessed 12 December 2017].

[2] International Air Transport Association, "Fact Sheet—Industry Statistics," Montreal, Canada, December 2017.

[3] Mouawad, J. and de la Merced, M., "United and Continental Said to Agree to Merge," *The New York Times*," BBC News, 2 May 2011, https://www.nytimes.com/2010/05/03/business/03merger.html. [accessed 10 Dec. 2017].

[4] Centre for Aviation, "European Airlines' Financial Results in 2012; Net Profit of Biggest 13 Down 72% for the Year," 26 March 2013, https://centreforaviation.com/insights/analysis/european-airlines-financial-results-in-2012-net-profit-of-biggest-13-down-72-for-the-year-102456 [retrieved 26 Nov. 2017].

[5] Centre for Aviation, "13 Chinese Airlines Could Each Have a Fleet of over 100 Aircraft by 2020," 25 May 2014, https://centreforaviation.com/insights/analysis/13-chinese-airlines-could-each-have-a-fleet-of-over-100-aircraft-by-2020-169778 [accessed 10 Dec. 2017].

[6] Epstein, R., "Commercial Aircraft Market Trends" [speech], Toronto, Ontario, 28 Nov. 2017.

[7] International Air Transport Association, "Fact Sheet—Industry Statistics," December 2017, http://www.iata.org/pressroom/facts_figures/fact_sheets/Documents/fact-sheet-industry-facts.pdf [accessed 30 March 2018].
[8] Hammacott, R., Vathylakis, A., and Wileman, A., "Top 50 Leasing Data," *Airline Business*, Jan.-Feb. 2018, pp. 28–33.
[9] "Boeing Announces Delay in Delivery of 747-8," *Seattle Times*, 25 Nov. 2008, http://old.seattletimes.com/html/boeingaerospace/2008393300_boedelays15.html [retrieved 26 Nov. 2017].
[10] BBC News, "Boeing Admits Further 747-8 Delay," 6 Oct. 2009, http://news.bbc.co.uk/2/hi/business/8293527.stm [accessed 26 Nov. 2017].
[11] Aboulafia, R., "September 2009 Newsletter," http://richardaboulafia.com/shownote.asp?id=302 [retrieved 26 Nov. 2017].
[12] Flottau, J., "Wide Opening," *Aviation Week & Space Technology*, 14–21 Oct. 2013, pp. 32–33.
[13] Norris, G., and Anselmo, J., "777X Production Investments Bolster Boeing," *Aviation Week & Space Technology*, 31 Oct. 2016, http://aviationweek.com/commercial-aviation/777x-production-investments-bolster-boeing. [retrieved 24 Nov. 2017].
[14] Boeing Corporation, "Boeing Launches 777X with Record-Breaking Orders and Commitments" [press release], 27 Nov. 2013, http://boeing.mediaroom.com/2013-11-17-Boeing-Launches-777X-with-Record-Breaking-Orders-and-Commitments [retrieved 26 Nov. 2017].
[15] Gates, D., "McNerney: No More 'Moonshots' as Boeing Develops New Jets," *Seattle Times*, 22 May 2014, https://www.seattletimes.com/business/mcnerney-no-more-lsquomoonshotsrsquo-as-boeing-develops-new-jets [retrieved 26 Nov. 2017].
[16] Norris, G., "Momentum Movers," *Aviation Week & Space Technology*, 24 March 2014, p. 33.
[17] Flottau, J., and Norris, G., "Second Life," *Aviation Week & Space Technology*, 21 July 2014, pp. 22–23.
[18] Norris, G., "Operators Reporting Positive 787 Fuel-Burn Results," *Aviation Daily*, 26 June 2012, http://aviationweek.com/awin/operators-reporting-positive-787-fuel-burn-results [retrieved 16 Dec. 2017].
[19] Summers, B., "Air Canada's 787 Expansion Plans Still in Play," *Aviation Week & Space Technology*, 22 Jan. 2016, http://aviationweek.com/commercial-aviation/air-canada-s-787-expansion-plans-still-play [retrieved 27 Dec. 2017].
[20] Norris, G., "With Better Dispatch Reliability, Boeing 787 Deliveries Reach 350," *Aviation Week & Space Technology*, 24 Nov. 2015, p. 58.
[21] Aboulafia, R., *The 2017 World Civil & Military Aircraft Briefing*, The Teal Group, Arlington, VA, March 2017, p. 528.
[22] Trimble, S., "Boeing 787 Unit Losses Decline, but Deferred Costs Rise," FlightGlobal, 22 April 2015, https://www.flightglobal.com/news/articles/boeing-787-unit-loss-declines-but-deferred-costs-rise-411502 [retrieved 27 Dec. 2017].
[23] West, K., "Airbus's Flagship Plane May Be Too Big to Be Profitable," *The Guardian*, 18 Dec. 2014, http://www.businessinsider.com/airbuss-flagship-plane-may-be-too-big-to-be-profitable-2014-12 [retrieved 4 Jan. 2018].
[24] Flottau, J., "Airbus, Operators Improving A350 Dispatch Reliability Two Years Into Service," Inside MRO, 29 Jan. 2016, http://aviationweek.com/caring-engines-today-and-future/airbus-operators-improving-a350-dispatch-reliability-two-years-servi [retrieved 4 Jan. 2018].
[25] Aboulafia, R., *The 2017 World Civil & Military Aircraft Briefing*, May 2017, p. 419.

[26] Nativi, A., "Superjet Scores Interjet," *Aviation Week & Space Technology*, 24–31 Jan. 2011, p. 34.

[27] Toh, M., "Mitsubishi Delays MRJ Deliveries By Two Years," FlightGlobal, 23 Jan. 2017, https://www.flightglobal.com/news/articles/mitsubishi-delays-mrj-deliveries-by-two-years-433402 [retrieved 3 Dec. 2017].

[28] Lu, V., "Bombardier's CSeries Jet Certified for Commercial Service," *Toronto Star*, 18 Dec. 2015, https://www.thestar.com/business/2015/12/18/bombardiers-cseries-jet-certified-for-commercial-service.html [retrieved 5 Dec. 2017].

[29] Valiante, G., "Bombardier CSeries, Aerospace Industry Too Important to Give Up: Quebec Government," *Toronto Star*, 29 Oct. 2015, https://www.thestar.com/business/2015/10/29/quebec-government-says-aerospace-industry-bombadier-cseries-too-important-to-give-up.html [retrieved 6 Dec. 2017].

[30] Perrett, P., "Not Just Inexperience," *Aviation Week & Space Technology*, 19 Aug. 2013, pp. 39–40.

[31] Polek, P., and Karnozov, V., "Russia's Irkut MC-21-300 Performs First Flight," AIN Online, 30 May 2017, https://www.ainonline.com/aviation-news/air-transport/2017-05-30/russias-irkut-mc-21-300-performs-first-flight [retrieved 7 Dec. 2017].

[32] Michaels, K., "The Flat Pricing Phenomenon," *Aviation Week & Space Technology*, 25 April–8 May 2016, p. 12.

[33] Flottau, J., "New Era," *Aviation Week & Space Technology*, 15–28 Feb. 2016, pp. 48–49.

[34] Hamilton, S., "Boeing Gives United A Smoking Deal on 737s to Block Bombardier from Gaining Traction," Forbes.com, 8 March 2017, https://www.forbes.com/sites/scotthamilton5/2016/03/08/united-boeing-and-the-competitors/#60dfc4130daf, [retrieved 8 Dec. 2017].

[35] John Hemmerdinger, "CSeries Prices Threaten 737 Max 7 'Viability': Boeing," FlightGlobal, 25 May 2017, https://www.flightglobal.com/news/articles/cseries-prices-threaten-737-max-7-viability-boein-437624 [retrieved 9 Dec. 2017].

[36] Morrow, A., and Van Praet, N., "Boeing Abandoned CSeries Talks Weeks Before US Duties Imposed," *The Globe and Mail*, 23 Oct. 2017, https://www.theglobeandmail.com/report-on-business/boeing-bombardier-c-series/article36699916 [retrieved 9 Dec. 2017].

[37] Lampert, A., and Hepher, T., "How Canada Pushed Bombardier Towards Airbus Instead of China Deal," *Financial Post*, 25 Oct. 2017, http://business.financialpost.com/news/how-canada-pushed-bombardier-toward-airbus-instead-of-china-deal [retrieved 9 Dec. 2017].

[38] Tomesco, F., Wingrove, J., and Clough, R., "Airbus Snaps Up Bombardier Jet in New Challenge to Boeing," Bloomberg, 16 Oct. 2017, https://www.bloomberg.com/news/articles/2017-10-16/airbus-to-buy-majority-stake-in-bombardier-c-series-jet-program [retrieved 9 Dec. 2017].

[39] Aboulafia, R., "Winners and Losers As Airbus Bails Out Bombardier's C-Series," *Forbes*, 17 Oct. 2017, https://www.forbes.com/sites/richardaboulafia/2017/10/17/bombardier-airbus-cseries-boeing/#f783c64491e9 [retrieved 9 Dec. 2017].

[40] Hamilton, S., "Airbus-Bombardier CSeries Deal Means No Tariffs on US-Assembled Aircraft, Says CEO," Leeham News, 16 Oct. 2017, https://leehamnews.com/2017/10/16/airbus-bombardier-cseries-deal-means-no-tariffs-us-assembled-aircraft-says-ceo [retrieved 9 Dec. 2017].

Chapter 8

SUPPLY CHAIN 2.0

Shortly after becoming CEO of Ford Motor Company, former Boeing Commercial Airplanes chief Alan Mullaly was asked if he was ready for the complexity of the automotive industry. "An automobile has about 10,000 moving parts, right? An airplane has two million, and it has to stay up in the air," he reminded his Detroit audience [1].

Indeed, the jetliner supply chain is complex. The large number of parts referenced by Mullaly must meet exacting standards and regulatory protocols. Globally there are thousands of suppliers, but for individual parts there may be just one or two firms with the capability or required certifications. On top of this, sourcing must comply with national security laws such as International Traffic in Arms Regulations (ITAR).

With about 1500 jetliners delivered annually, it is a high mix/low volume supply chain compared to other industries, which creates its own set of challenges. Then there is the aftermarket. Jetliner original equipment manufacturers (OEMs) are required to support their aircraft programs for life. To put this in context, Boeing must still concern itself with supporting 707 operations, 60 years after its introduction. Finally, there are massive financial implications of supply chain operations, because jetliner manufacturers typically source 60–70% of the value of an aircraft.

In the 1990s and 2000s, there was a fundamental shift in the way that jetliners were developed, sourced, and produced. Supply chain strategies changed significantly as aircraft and aeroengine manufacturers embraced Tier 1 supply chain models, which reduced the number of suppliers—sometimes dramatically—and delegated significant engineering and supply chain responsibilities to a select group of aircraft systems, interiors, and aerostructures suppliers. There was also a significant change in where activities took place as globalization and low-cost sourcing took hold. By the 2010s, the jetliner supply chain looked different. It became a global ecosystem. To understand how and why it changed, it is useful to recall how aerospace OEMs were organized as well as their supply chain practices in the 1970s and 1980s—the "premodern" era.

THE WAY THEY WERE: EARLY JETLINER SUPPLY CHAINS

The economist Ronald Coase won a Nobel Prize in 1991 for his Theory of the Firm. Coase posited that the boundaries of a firm were determined by transaction costs; when the cost of production, planning, transportation, and coordination were less within the boundaries of a firm than sourced from an outside supplier, a firm tended to internalize these activities (see Fig. 8.1).

For much of the jet age, most aircraft OEMs were vertically integrated. In-house production of aerostructures, aircraft systems, and interiors fed the final assembly line. Except for aeroengines and avionics, supplier activity was typically confined to individual parts and subassemblies. Engineering was also mostly captive and usually co-located with production. A typical manufacturer might have its engineers working on Mylar drawings and computers in a large "pen." Boeing was a good example. Tens of thousands of engineers were in the Puget Sound region within driving distance of its two

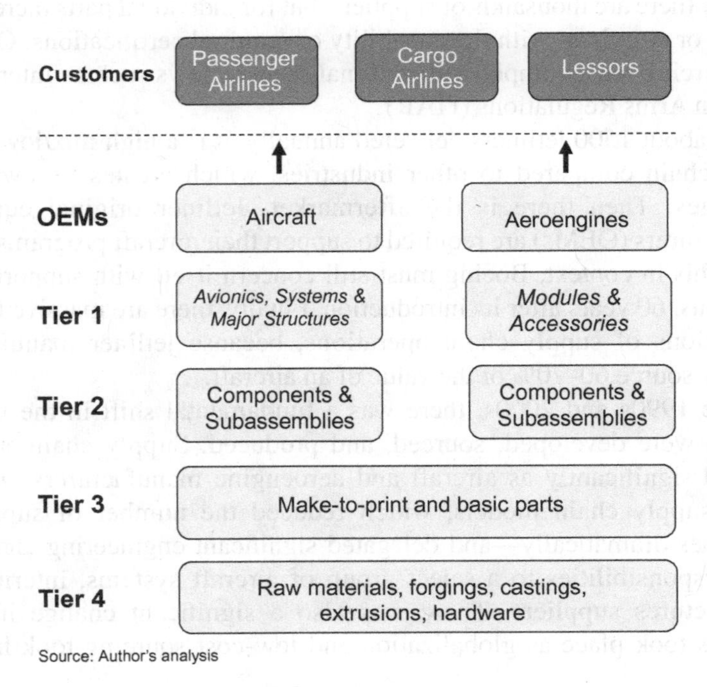

Fig. 8.1 Jetliner ecosystem.

major factories in Renton and Everett. The same was true of Douglas Aircraft in Long Beach and Fokker in the Netherlands.

Airbus Industrie began to chip away at this paradigm in the 1970s as major sections of its aircraft were manufactured in four countries. Like incumbent OEMs, each Airbus location performed most activities in-house, but this new supply chain paradigm required enhanced coordination to manage geographic dispersion. This included flights on grotesquely large transport aircraft called *Super Guppies* to move parts from decentralized production facilities to the final assembly plant in Toulouse (Fig. 8.2).

In the early 1990s, several major environmental changes reduced Coase's transaction costs and enlarged the possibilities of aerospace industrial organization. The first was the end of the Cold War and Chinese economic liberalization. This created a single market for aircraft sales opportunities, which were previously bifurcated by Western and Communist zones of influence. Now Western jetliner OEMs could aggressively pursue sales in China, the Commonwealth of Independent States (CIS; the former USSR), India, Eastern Europe, Central Asia, and Africa. For the supply chain, it meant that jetliner OEMs could tap into human capital and manufacturing capability in these regions.

Trade also liberalized during this time, resulting in lower tariffs (import taxes) and nontariff barriers for trade in aerospace goods and services. This happened at both the global and regional levels. The World Trade Organization's Plurilateral Agreement on Trade in Civil Aircraft, originated in 1980, eliminated tariffs for civil aircraft, aeroengines, components, and parts for 32 countries. The steady expansion of the European Union and regional

Credit: SrAStormy D. Archer

Fig. 8.2 The Super Guppy linked Airbus's supply chain.

trade agreements like the North American Free Trade Agreement (1994) and European Union's Association Agreement (2000) with non-EU members also played an important role in enabling aerospace trade.

The early 1990s also brought major changes to the way aircraft were designed and produced, shaped by globalization. As discussed in Chapters 1 and 2, the Boeing 777 was the world's first all-digital jetliner design. This meant that new aircraft designs were conceived in shared three-dimensional computer models that could be accessed by teams anywhere. At the same time, plummeting communications costs and the emergence of the Internet dramatically reduced collaboration costs. This meant that OEMs could send large files to remote engineering teams or suppliers, or even work around the clock as designs shuttled from one location to the next—analogous to a relay race in athletics. They could also stay connected to their suppliers via an electronic data interchange (EDI). Transportation costs also fell as major third-party logistics suppliers such as UPS, FedEx, DHL, TNT, and Kuehne & Nagel completed global networks that could run far-flung and complex supply chains. These changes created the potential to unbundle the aerospace value chain. Rather than doing everything in-house, why not assign work packages to talented and productive suppliers?

Toyota Motor Company revolutionized automotive supply-chain management in the 1970s and 1980s by appointing certain suppliers as the sole source of components, leading to intimate collaboration with long-term partners. A new breed of supplier emerged with the responsibility of delivering entire systems and assemblies to automotive OEMs. These Tier 1 suppliers possessed systems engineering, production, and supply chain responsibilities, alleviating the need for Toyota to be good at everything. Their customer relationships were based on the spirit of partnership rather than arms-length transactions. This novel supply chain approach enabled a "just in time" approach of delivering components to the assembly plant [2].

Historically, Western carmakers tended either to source in-house or award short contracts to the lowest bidders. They worked with as many as 2500 suppliers whereas the Japanese "lean" approach typically involved fewer than 300 suppliers [3]. After ceding competitiveness, Western automotive OEMs adopted the Toyota supply chain approach, and the Tier 1 supply chain model quickly spread through the global automotive industry in the 1980s.

The jetliner business seemed to be well-suited to the Tier 1 model as aircraft OEMs struggled to keep up with the technology development and investment requirements of new aircraft. Which company would go first?

BOMBARDIER AND EMBRAER: TIER 1 PIONEERS

In the early 1990s, Bombardier faced a dilemma. As it was ramping up deliveries of its new Canadair Regional Jet, it needed to determine how to fund an ambitious new business jet, the Global Express. "I told CEO Laurent Beaudoin that it would cost $1 billion in nonrecurring engineering, and he made it clear that this was unacceptable and we had to find a way to significantly reduce the investment," recalled former engineering chief John Holden.

> We came up with a new approach to risk sharing where we gave them requirements rather than detailed specifications. We brought in 10–12 risk-sharing partners and gave them large work packages with the goal of minimizing interfaces between us and smaller suppliers. And we would only reimburse them for nonrecurring engineering (NRE) if the program was a success. It worked—we reduced our NRE from $1 billion to $600 million.[1]

Tier 1 partners would fund roughly half of the Global Express's development. The largest partner was wing and center fuselage supplier Mitsubishi, which funded 20% of its development [4]. Other major system suppliers included Messier-Dowty (landing gear), Rockwell Collins (avionics), Parker Aerospace (hydraulics, flight controls, and fuel), Liebherr (air management), and Lucas Aerospace (electrical generation). Bombardier would encounter significant delays in developing the Global Express, but a new jetliner supply chain model was born: It would be called *Tier 1* (see Fig. 8.3).

Bombardier's regional jet competitor Embraer also decided to change its supply chain model, but for different reasons. Like Bombardier, it needed risk-sharing partners to share development costs as it contemplated a larger 70- to 100-seat regional jet family in the mid-1990s (the E-Jet). However, its financial situation was a mess. In 1994, the company had revenue of $250 million, a loss of $330 million, and a backlog of $150 million. Its debts were $400 million.[2] At the same time, the E-Jet would involve significantly more

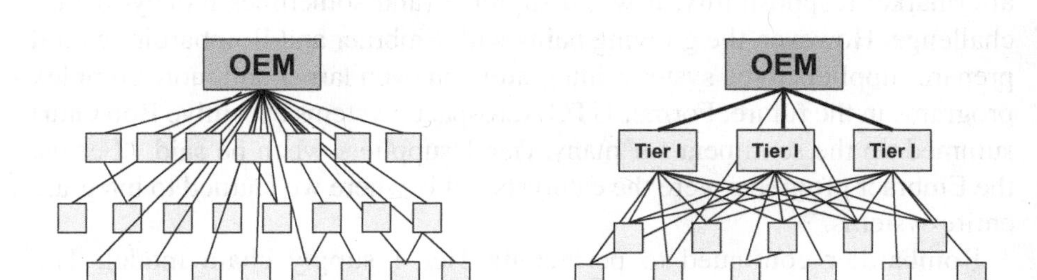

Fig. 8.3 Supply chain models.

[1]John Holding (former Vice President–Engineering—Bombardier), interview with author, 29 July 2016.
[2]Mauricio Botelho (former CEO—Embraer), interview with author, 27 May 2016.

complex systems than its 35- to 50-seat regional jet, the ERJ. Honeywell would supply an integrated modular Primus Epic avionics suite while GE would provide new aeroengines. A new fuselage and wings would also be required, and this would not be a simple stretch like the CRJ-700. Adding to the challenge were significant systems engineering requirements. Embraer's capability at the time was modest, and its procurement and supply chain teams were already stretched by the ERJ's 350 suppliers.

In the late 1990s, Embraer decided to make a bold break with the past and reduce its supplier count by an order of magnitude on the E-Jet; it would work with less than 40 primary suppliers [5]. "There were hundreds of suppliers on the ERJ, necessitating large procurement and supply chain teams, and we realized we needed a more efficient approach," recalled former executive Satoshi Yokota.[3] Gamesa, Kawasaki Heavy Industries, and Latecoere would be major aerostructures suppliers. Parker Aerospace would assume system responsibility for the flight control and fuel systems. Liebherr-Aerospace would supply the landing gear through a joint venture with Embraer. Many of these suppliers would be risk sharing, which allowed Embraer to close its funding gap and move forward with the E-Jet. One could assume that Embraer was visionary in moving to its tiered supply chain model—perhaps after benchmarking the automotive industry or its competitor Bombardier—but this wasn't the case. "We didn't adopt this supply chain model out of choice," former executive Horacio Forjaz recalled. "We were forced into it!" He added, "Not all suppliers were ready for the challenge."[4] Embraer would learn many of the same lessons that Boeing would learn 15 years later with the 787. Indeed, many E-Jet suppliers would experience growing pains as they adopted the role of systems integrator. Their supply chain and procurement organizations had to improve as they sourced components from many new suppliers—including direct competitors. Engineering needed to develop new systems integration skills. Program management had to cope with new levels of risk, and the customer support organizations had to take on much greater aftermarket responsibility. It was a supreme (and sometimes money-losing) challenge. However, the growing pains with Embraer and Bombardier would prepare suppliers to be systems integrators on even larger and more complex programs in the future. Former UTC Aerospace Systems executive Bob Guirl summed up the sentiment for many Tier 1 suppliers when he said, "For us, the Embraer programs were the catalysts—it is where we learned to integrate entire systems."[5]

Bombardier continued to perfect its Tier 1 supply chain model. The CRJ-700 would have just 40 major suppliers. Bombardier upped its supply

[3]Satoshi Yokota (former Vice President—Embraer), interview with author, 11 May 2016.

[4]Horacio Forjaz (former Vice President—Embraer), interview with author, 24 Sept. 2015.

[5]Bob Guirl (former Vice President—UTC Aerospace Systems), interview with author, 16 Oct. 2015.

chain prowess by bringing in supply chain executives from the automotive industry. Procurement chief Janice Davis came from Ford Motor Company, and Commercial Aircraft President Guy Hachey came from Delphi, a Tier 1 automotive supplier. This was the beginning of a migration of automotive supply chain executives into the aerospace industry. The reason for this migration was clear, according to Bombardier's Davis: "Margins were razor-thin, and the automotive industry had the most advanced supply chain practices."[6] Bombardier would later take its supply base down to 20 Tier 1 suppliers on the C Series.[7]

Bombardier and Embraer created a new architecture for aerospace supply chains. Jetliners are composed of two major, but separate, supply chains for aircraft and aeroengines. Only at final assembly do the two supply chains merge. Conceptually, each has four tiers of suppliers. For aircraft OEMs, Tier 1 suppliers include aircraft systems such as avionics, landing gear, and auxiliary power units; this supplier group generally owns their designs and retains access to their aftermarket. For aeroengine OEMs, Tier 1 suppliers make engine modules (e.g., low-pressure turbine modules) and major components (e.g., engine shafts and cases); most design rights remain with the aeroengine OEMs. There are several dozen Tier 1 suppliers for each supply chain. The next tier, Tier 2, is composed of hundreds of part and assembly suppliers. These firms generally retain some type of intellectual property—design and/or process—and sell to Tier 1 suppliers or directly to OEMs. Next are the Tier 3 suppliers, which provide "make to print" parts and components. As the name implies, these suppliers do not design their products. They are frequently referred to as "job shops," and there are thousands in the global ecosystem. The fourth supplier group, Tier 4s, include raw materials, forgings, castings, extrusions, and hardware.

This framework does not connote the *importance* of the suppliers. Indeed, suppliers of Tier 4 castings, forgings, and raw materials supply critical and high value jetliner inputs, which can reach millions of dollars per aircraft. They are among the most important suppliers; nonperformance can shut down a jetliner production line. Suppliers sometimes occupy more than one tier. A single company may be a Tier 1 supplier to one company and a Tier 2 supplier to another company. Tier 4 suppliers of raw materials, forgings, castings, and the like often sell to all other tiers as well as OEMs.

Harnessing Globalization: Low-Cost Countries

While regional jet OEMs gained experience with tiered supply chain models, another supply chain initiative was unfolding: sourcing from

---

[6]Janice Davis (former VP–Supply Chain—Bombardier), interview with author, 21 Sept. 2015.
[7]Rob Dewar (C Series Program Manager—Bombardier), interview with author, 28 June 2016.

low-cost countries. Major aerospace manufacturers scoured the globe to find new suppliers that could reduce costs or improve access to fast-growing jetliner markets.

In Europe, the obvious place to start was Eastern Europe and CIS countries to leverage their significant aerospace engineering and manufacturing capabilities. Several notable investments followed the 1993 Boeing engineering investment in Moscow highlighted in Chapter 1. United Technologies companies Pratt & Whitney and Hamilton Sundstrand made manufacturing investments in Poland and Russia in the mid-1990s. Snecma and General Electric also made engine-related investments in Russia in the late 1990s. Other popular low-cost locations included the Czech Republic, Romania, Turkey, Tunisia, and Morocco.

In the Americas, low-cost sourcing was focused on Mexico. Honeywell, Rockwell Collins, and Gulfstream Aerospace (business jets) made early investments in the 1980s, but Mexico was relatively inactive until the passage of the North American Free Trade Agreement (NAFTA) in 1994. The pace of investment then increased considerably when high tariffs—as much as 20%—were phased out. Aerospace investments for labor-intensive components such as wire harnesses and machined parts followed.[8] A trickle of investments in the mid-1990s became a flood in the late 1990s and 2000s as Mexico became an aerospace manufacturing hot spot. (Mexico's story as a rising aerospace cluster is covered in Chapter 14.) Brazil was also considered a low-cost location in the 1990s, and Embraer's final assembly facility in São José dos Campos acted as a magnet; Liebherr-Aerospace, Sonaca, and Kawasaki were early investors.

Globalization followed a different pattern in Asia. Numerous maintenance, repair, and overhaul (MRO) facilities were established there in the 1990s as the in-region fleet began to expand. Singapore, China, and Hong Kong were popular locations. In manufacturing, component manufacturer Moog was a pioneer in establishing a low-cost manufacturing center in Baguio, Philippines in 1984. In 1992, Rolls-Royce started a joint venture in Xian Aircraft Company to make turbine rings and compressor seals; competitors GE, Pratt & Whitney, and Snecma would all follow with their own manufacturing facilities later in the decade. In the 2000s, manufacturing investments in China exploded with dozens of major investments by a broad range of OEMs. This included two jetliner final assembly facilities. In 2002, Embraer established a joint venture with Chinese supplier AVIC 2 (China Aviation Industry Corporation 2) to assemble ERJ145s in Harbin. AVIC 2 was a Chinese consortium of aircraft manufacturers established in 1999. And in 2008, Airbus opened an A320 final assembly line in Tianjin—its first outside of Europe. Its motivation was access to the burgeoning Chinese market. The investment wave hit a crescendo in the late 2000s when, as discussed in Chapter 7, Comac demanded numerous

[8]Benito Gritizewski (President—FEMIA), interview with author, 12 July 2016.

in-country joint ventures (JVs) as the price of admission for participation on its C919 program.

China wasn't the only venue for Asian low-cost manufacturing. Boeing, Airbus, Spirit AeroSystems, and Honeywell all invested in manufacturing facilities in Malaysia. Singapore transitioned from a hub of low-cost MRO activity to a center for advanced manufacturing and research and development (R&D). Its headline investment was a Rolls-Royce Trent final assembly facility that opened in 2007—the first major aeroengine final assembly facility in Asia. India and Thailand were also popular aerospace investment destinations.

By the mid-2000s, two new key performance indicators (KPIs) entered the lexicon of aerospace executives: (1) the percentage of employees in low-cost countries (LCCs), and (2) the percentage of supplier spending in LCCs. In just over a decade, aerospace industry executives had changed their perspective to embrace a global marketplace for employees and suppliers. To paraphrase Shakespeare, the world was their oyster.

TIER 1 EXPANDS

While Embraer and Bombardier gained experience with the Tier 1 supply chain model in the 1990s, other major OEMs embarked on a similar journey in the 2000s. Rolls-Royce was one of the early aeroengine OEM adopters. "In the 1980s, we were very much a British company with many local suppliers," said former supply chain executive Alun Hughes. "In the 1990s we brought in more risk-sharing partners, but the 1998–2000 timeframe was a turning point in supply chain strategy. We injected a whole host of talent into the supply chain function and brought in executives from the automotive industry including Toyota, GM, BMW, and Delphi." The results were significant as Rolls-Royce downshifted from more than 500 suppliers on RB211 programs in the 1980s to about 100 suppliers on later Trent models in the early 2000s, to 50–75 suppliers on the Trent XWB[9] (Fig. 8.4). Twelve of the Trent XWB suppliers were risk-sharing partners, with 70% of the engine's value coming from outside suppliers, compared to 60% for prior engines [6]. Pratt & Whitney, General Electric, and Snecma also engaged in limited supplier reduction, but not to the same extent as Rolls-Royce.

Rolls-Royce went a step further in supplier cooperation in pioneering the risk and revenue sharing partner (RRSP) contracts on the Trent 500. In these arrangements, Rolls-Royce asked each of its RRSPs to make its own investment in aeroengine development programs. In exchange, partners received a proportional amount of the aeroengine's aftermarket revenue. This way, Rolls-Royce could reduce its investment requirements. In time,

[9]Alun Hughes (Sr. VP–Supply Chain—Rolls-Royce), interview with author, 20 Sept. 2016.

OEM	Model	Entry into service	Major Suppliers
Airbus	A380	2007	200
	A350XWB	2015	70
Embraer	EMB 145	1997	350
	EMB 170/190	2004	38
Rolls-Royce	Trent 700	1994	500
	Trent XWB	2015	50-75

Sources: Rolls-Royce interviews, AeroStrategy, Oliver Wyman, *Businessweek*, Author's analysis

Fig. 8.4 Supplier reduction.

RRSPs took on 30% or more of program risk for Rolls-Royce. Pratt & Whitney, GE, and Safran also used RRSP contracts to manage risk.

Meanwhile, the A380 project provided Airbus a unique opportunity to update its supply chain organization. Historically, each of Airbus's member companies had its own procurement and supply chain organizations. The formation of the European Aeronautic Defence and Space Company (EADS) in 2000 changed this, and it created a single supply chain organization for systems, engines, and interiors. Each company, however, retained its own airframe procurement organization. The workshare split on the A380 aerostructures was like other programs—major sections were built by Airbus, with targeted procurement from outside suppliers. Airbus worked with hundreds of suppliers, in contrast to Bombardier and Embraer, but gave significant responsibility to system integrators. Major supply chain difficulties emerged, but ironically, these were between Airbus companies. This included the A380 wire harness issue, which led to €6 billion ($8.3 billion) in lost revenues in the 2007–2010 timeframe alone [7].

Boeing's Supply Chain Nightmare

The 787's huge development costs following its launch in 2004 presented Boeing with a major dilemma: how to bring an aircraft loaded with new technologies to the market in four years for a development cost lower than its initial estimate that approached $10 billion. This challenge was compounded by a shift in corporate culture since the 1997 merger with McDonnell Douglas. As discussed in Chapter 7, new CEO Harry Stonecipher established return on net assets (RONA) as a key corporate measure that often favored short-term profits over long-term investment. This meant that if Boeing was going to launch the Dreamliner, it would need to take outsourcing to a whole new level. Many Boeing executives perceived this as dangerous, including senior technical fellow L. J. Hart-Smith, who in 2001 produced a prescient analysis projecting that excessive outsourcing would raise its costs and steer profits to its subcontractors [8].

Nonetheless, Boeing would embrace an aggressive Tier 1 model. Instead of outsourcing 5% of the aerostructures as it did for the 747 and 35–50% of the aerostructures' value for the 737, Boeing would outsource 70% for the 787. Rather than relying on hundreds of suppliers shipping parts and components for assembly, Boeing would work with approximately 50 Tier 1 strategic partners on the Dreamliner. Major risk-sharing partners would serve as integrators of major systems and aerostructures, and would not be paid until the 787's first delivery. Boeing's vision was that major systems and structures would arrive from around the world for final assembly in just three days. It estimated that this radical supply chain reboot would reduce the 787's development time from six to four years, and reduce its development cost from as much as $10 billion to $6 billion [9]. The other expectation was that Boeing would reduce the redundancy between its own engineering workforce and its suppliers. One example: although Boeing didn't manufacture landing gear, it had 140 design engineers looking over suppliers' shoulders as they executed make-to-print contracts [10].

The aerostructures supply chain approach was particularly radical (Fig. 8.5). In 2006, Boeing sold its Wichita operations, which provided most of its fuselages. Instead, the 787's composite fuselage would be shared by aerostructures suppliers Vought, Alenia, and Kawasaki. The aircraft's nose section would be made by Spirit AeroSystems, the new name for Boeing's former Wichita division. Alenia would make the horizontal stabilizer and Boeing the tail fin. Boeing would even outsource the composite wing, with Mitsubishi fabricating the wing and Fuji Heavy Industries making the center wing box. This was a surprise, because the wing is considered by executives to be the "crown jewel" of a jetliner's design. In aggregate, the three Japanese "Heavies"—Mitsubishi Heavy Industries, Fuji Heavy Industries, and Kawasaki Heavy Industries—would make 35% of the 787's structure and would receive $1.6 billion in Japanese government support to underwrite nonrecurring engineering, including capital equipment such as giant autoclaves [11]. Alenia and Vought would later combine to form a new company, Global Aeronautica, which produced more than 20% of the fuselage. Boeing would transport the parts to final assembly via the Dreamlifter, a modified 747. Thus, Boeing would adopt the same supply chain architecture as Airbus—airlifting major components between geographically separated production sites. Ironically, 35 years earlier, when discussing the Super Guppy connecting the A300's supply chain, the Boeing sales team used to superimpose a map of Europe over America to show distances travelled by the Airbus manufacturers and then pose the question: Is this really an efficient and modern airplane production system [12]?

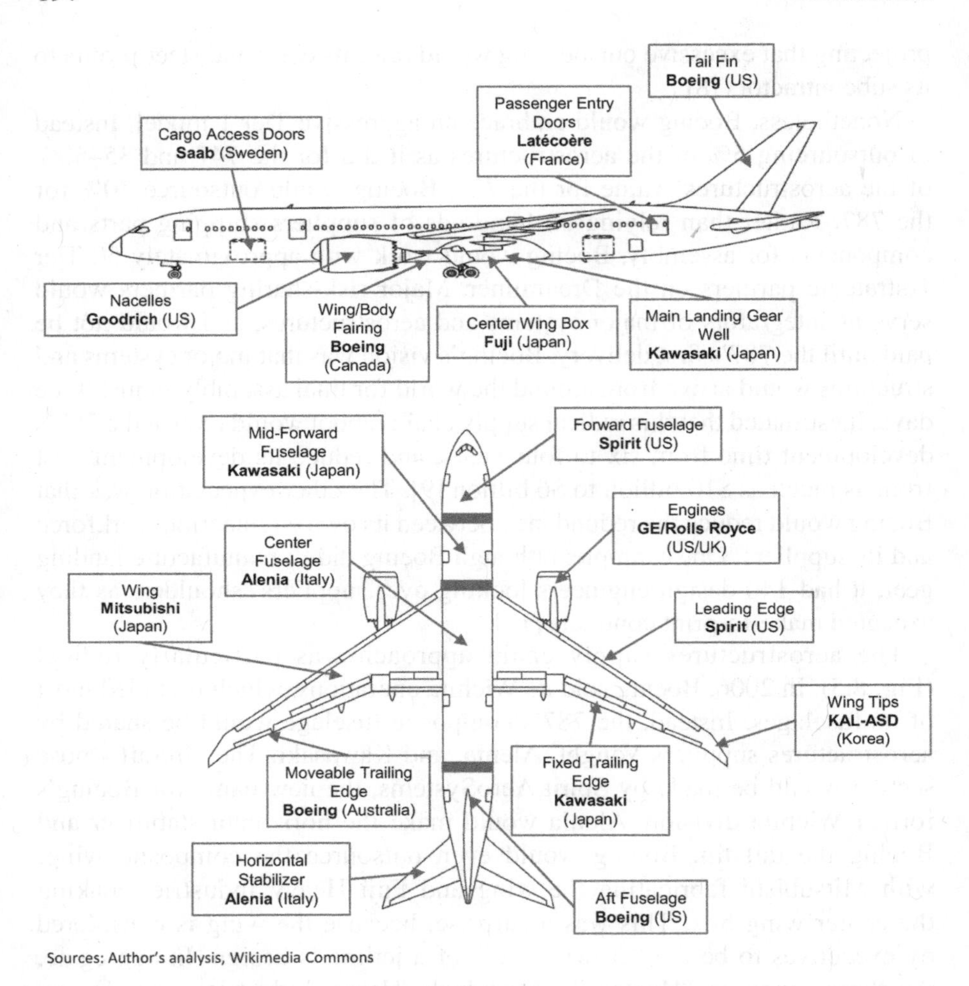

Fig. 8.5 787 major aerostructures suppliers.

Following the Tier 1 philosophy, the 787's aircraft system contracts were large and concentrated with relatively few suppliers. Hamilton Sundstrand captured major contracts to supply the Dreamliner's power generation and distribution, air management, and auxiliary power unit—an integrated approach for the Dreamliner's electric architecture, which comprised more than 600 line-replaceable units (LRUs). Messier-Dowty would provide the landing gear, and Rockwell Collins, Honeywell, and GE Aviation would provide major avionics systems. This shift in contracting approach contributed to significant consolidation of aircraft component suppliers, which is discussed in Chapter 12.

Boeing ripped up a supply chain playbook that had worked well for 50 years for an aircraft loaded with new technology. It was a supreme and audacious gamble. Asked in 2006 about the risk of this approach, Boeing

CEO Jim McNerney stated, "It's a good model in terms of development, and it's a good financial model, too. I think what we learn from the 787 program will impact the way we do every plane from now on" [13]. In Toulouse, Airbus supply chain executive Albert Varenne had a different reaction: "When I saw what Boeing was planning with new technology and the supply chain, I asked myself is this a bit too much?"[10]

Signs of trouble first emerged in October 2007 when Boeing announced a six-month delay to the program. By 2008, Boeing was more than two years behind schedule and needed to address the problems of underperforming suppliers. In March 2008, it purchased Vought Aircraft's 50% share of Global Aeronautica, a joint venture with Italian supplier Alenia, which produced the center fuselage. In July 2009, it acquired Vought's aft fuselage facility in North Charleston, South Carolina for just over $1 billion. And in late 2009, it bought Alenia's share of Global Aeronautica. In between this activity, Boeing announced Charleston as the location of its second 787 production facility. In the space of just under two years, Boeing took control of much of the 787's fuselage and gained two production facilities in the process. Ultimately, Boeing's supply chain debacle would contribute significantly to Boeing's NRE overrun well over $10 billion versus its original estimate, as well as its massive deferred production cost tab, which would peak at more than $30 billion by the mid-2010s.

Reflecting on lessons learned, Boeing CEO Jim McNerney opined, "I do believe in the global model that leverages engineering and manufacturing capability. But we drew the line too aggressively on the 787.... We bit off a little more than we could chew" [14] Many concluded that the 787 debacle sprang from placing too much faith in financially oriented leaders at the expense of competent program management and engineering executives. Boeing's program management competence, which was so evident on the 777 development program in the early 1990s, seemed to have disappeared.

In 2006, Airbus took on its own new challenge—the A350 XWB. It was determined to avoid the supply chain difficulties that plagued the 787. One change was a 24-month gap between design freeze and production—six months longer than the 787. This would give Airbus more time to identify potential production kinks. Unlike Boeing, Airbus would keep most key aerostructures in-house at its production facilities in France, Germany, Spain, and the United Kingdom; this would help to offset some of the risk associated with a high-composite aircraft. Under new procurement chief Klaus Richter, recruited from automaker BMW, it simplified its supply chain by reducing the number of major suppliers from 250 on prior models to 70 [15]. Work packages would be cut from 100 on the A330 to 20, with 50% risk-sharing

[10]Albert Varenne (Sr. Vice President–Supply Chain Strategy—Airbus), interview with author, 9 June 2016.

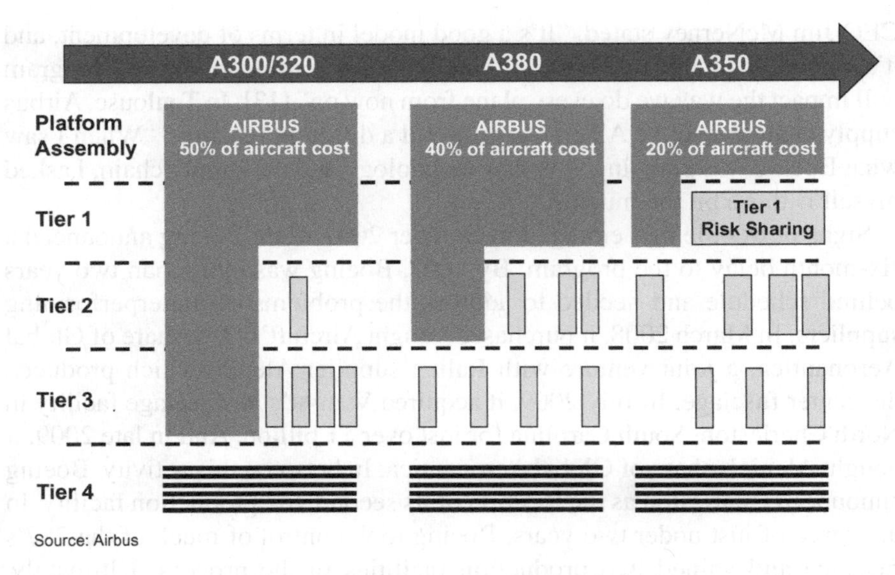

Fig. 8.6 Airbus supply chain evolution.

partnerships [16] (Fig. 8.6). Major system partners included Safran, Thales, Honeywell, Eaton, Moog, and Goodrich. Airbus would soon eat up the slack that it built into the schedule, and ultimately deliver its first aircraft to Qatar Airways nearly 18 months later than originally planned. Compared to the 787 and A380, however, this was a major step forward.

The A350 XWB's supply chain was a clear break from the past. In contrast to the A300 and A320 supply chains, where Airbus was responsible for 50% of the aircraft's value, Airbus had accounted for just 20% of the value of the A350 XWB, with Tier 1 partners—often sharing risks—playing a major role.[11]

The More for Less Era

The retrenchment following the A380 and 787 moonshots led OEMs to pursue cost reductions in their supply chains. Airbus faced significant pressure from a strong Euro and A380 losses. Its financial situation was tenuous. "Our long-term future is at stake," said Airbus President and CEO Louis Gallois. "We cannot continue to produce at Europe's Euro prices and sell at Boeing's dollar prices"[17]. Airbus announced a major cost-cutting initiative in February 2007: Power8. The goals of Power8 were ambitious: significant cost-reduction and cash-generating efforts leading to earnings before interest and taxes (EBIT) contributions of €2.1 billion from 2010 onwards and an

[11]Joe Marcheschi, "Be An Airbus Supplier—The Pacific Northwest Region," speech to Pacific Northwest Aerospace Alliance Conference, Seattle WA, 16 Feb. 2017.

additional €5 billion of cumulative cash flow from 2007 to 2010. Savings would come from reduced overhead (32%), lean manufacturing (16%), smart buying (31%), restructuring/focusing on core (12%), and faster product development (6%).

The plan also included an Airbus rarity—a workforce reduction of 10,000 employees, with 50% from full-time staff.[12] Labor strife followed, and Airbus was unable to reach its headcount reduction goals. But there were several important outcomes. Airbus spun out part of its aerostructures manufacturing capacity to create two new companies (Aerolia and Premium Aerotec), which it planned to sell—a move analogous to Boeing's Spirit AeroSystems spinout. It also peddled its Filton, United Kingdom wing structures manufacturing facility to GKN. To achieve "smart" buying, it created a single indirect purchasing function in 2007 and centralized all aerostructures procurement into a single organization in 2008.[13]

Boeing launched a major supply chain initiative in 2013. Unlike Power8, this initiative wasn't to address its crisis; rather, it was aimed at fundamentally restructuring Boeing's relationship with its suppliers and, ultimately, the jetliner OEM business model itself. It was frustrated by the fact that its Tier 1 suppliers earned double-digit margins—sometimes more than 20%—while Boeing Commercial Airplane Group earned mid-single-digit margins (Fig. 8.7). Boeing challenged the situation where it took most of the risk yet

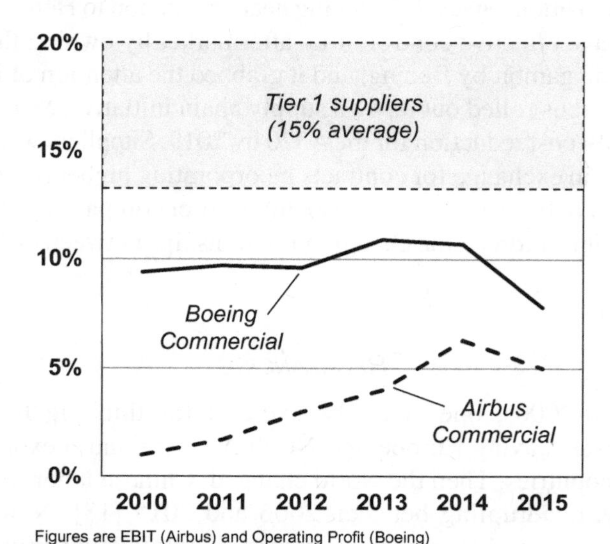

Fig. 8.7 OEM and supplier profit margins.

[12]Airbus press release, "Power8 Prepares Way for New Airbus," 28 Feb. 2008.
[13]Varenne interview, 9 June 2016.

suppliers reaped most of the rewards. The initiative was called *Partnering for Success*. Its goal was to achieve 15–20% cost savings via unilateral supplier price reductions, part redesigns, and lean initiatives. Controversially, it rolled out a "new business withhold" list, also known as the *No Fly List*, for suppliers unwilling to cooperate. Leading suppliers were invited to attend a meeting at Boeing's Chicago headquarters where McNerney personally introduced the program. Boeing's dramatic goals were met with shock and cynicism by many suppliers. Some had already leaned and globalized their operations in the prior decade; they viewed the initiative as a simple profit grab. Others in less profitable segments, like aerostructures, questioned the reasonableness of a 15% price reduction in a segment that routinely achieved mid-single-digit profit levels. *Was Boeing serious?* Would it really walk away from its critical suppliers?

Suppliers received the answer to this question in December 2013, when Boeing awarded the landing gear for the 777-X to Héroux Devtek, a small Canadian supplier that had never undertaken the development of such a large and important landing system. In making this choice, Boeing shunned UTC Aerospace Systems (formerly Goodrich) and Safran, the two largest landing system suppliers. It was a stunning move. Boeing awarded a major system on its newest aircraft to a supplier with $233 million in revenue compared to UTC Aerospace Systems' more than $13.3 billion. To add insult to injury, Boeing would also transition legacy 777 landing gear production to Héroux Devtek and would also take effective control of its aftermarket by owning the designs. It was a stunning gambit by Boeing, and it grabbed the attention of its suppliers.

In 2014, Airbus rolled out its own supply chain initiative, *SCOPE +*, which targeted a 10% cost reduction for the A320 by 2019. Suppliers were required to reduce prices in exchange for contracts incorporating higher A320 production rates. It was a narrower focus than Boeing's all-encompassing Partnering for Success, but it reinforced the changing relationships between OEMs and their suppliers.

REBALANCING

In the mid-2000s, one could be excused for thinking that aerospace production was leaving Europe and North America and inexorably shifting to low-cost countries. Then the world changed. Chinese labor rates increased significantly, quadrupling between 2006 and 2015 [18]. New, less labor-intensive manufacturing processes emerged. Intellectual property control became more important. And in the wake of the B787 development debacle, OEMs questioned the wisdom of depending on loosely knit, far-flung global supply chains. Thus, supply chain strategies morphed yet again in the 2010s.

The first big change was a redefinition of core activities and the appropriate balance of insourcing and outsourcing. This was particularly true at Boeing as

Aircraft System	777	787	777-X
Wings	Insourced most manufacturing	Outsourced to Japanese Aircraft Development Co.	Insourced at new composite wing factory in Everett
Fuselage	Outsourced to Japan Aircraft Development Co.	Outsourced to four risk sharing partners; later insourced*	Outsourced to Japan Aircraft Development Co.
Nacelle	Insourced systems integration and most manufacturing	Outsourced systems integration and manufacturing to Goodrich	Insourced systems integration; outsourced manufacturing
Flight controls	Insourced systems integration; outsourced manufacturing	Outsourced to Moog and other Tier 1 integrators	Insourced systems integration; outsourced manufacturing
Landing gear	Insourced systems integration; outsourced manufacturing	Outsourced systems integration and manufacturing to Safran	Insourced systems integration; outsourced manufacturing to Héroux Devtek

Source: Author's analysis

* Forward fuselage initially in-house until spin-off of Spirit Aerosystems in 2005; center and aft fuselage later insourced when Boeing purchased Global Aeronautica

Fig. 8.8 Boeing's supply chain evolution.

it digested the lessons of the 787 (see Fig. 8.8). "As the OEM, you don't have to do everything, but you need to understand everything," said former Boeing CEO Jim Albaugh.

> On the 787, we outsourced too much. We had five different sections of the fuselage all with their own processes. In hindsight, we should have built one section, qualified a process, and then outsourced. We won't outsource the wing again—you are giving someone else insight on the most important part of the aircraft. We were driven too much by return on net assets—we didn't balance it with risk.[14]

Boeing's 777-X supply chain decisions reflected these lessons. Wings that were outsourced on the 787 were brought in-house via a major new composite wing factory in Everett, Washington. System integration for the nacelle, flight controls, and landing gear were also brought in-house. These moves meant that Boeing's aftermarket opportunities increased considerably as it gained access to lucrative thrust reverser, flight controls, and landing gear aftermarket revenue streams. It also announced plans to expand in-house avionics capability and aircraft seat manufacturing.

Boeing wasn't the only OEM rebalancing its outsourcing. Embraer took full control of its JV with Kawasaki Heavy Industries to produce E-Jet wings, while in aeroengines, GE gained control over engine accessories and fuel systems through a series of acquisitions and joint ventures. One example was a joint venture with Parker Aerospace to produce fuel nozzle tips for the LEAP

[14]Jim Albaugh (former CEO – Boeing), interview with author, 22 Sept. 2016.

aeroengine using GE's additive manufacturing technology. Formerly, these crucial components were sourced from Parker Aerospace on an arms-length basis.

A second change in supply chain practices was a shift to greater investment in developed economies—particularly the Southeastern United States—rather than the headlong rush to LCCs in prior years. Several factors lay behind this shift. New jetliner final assembly facilities in Charleston, South Carolina (Boeing) and Mobile, Alabama (Airbus) acted as magnets for suppliers to build facilities near these crucial customers. Other factors benefiting the region included low energy prices, flexible labor, and pro-business political leaders. Embraer even chose to locate its final assembly facility for its Phenom business jets in Florida rather than Brazil. By the early 2010s, the region became the new hotspot for global aerospace manufacturing investment. The emergence of the Southeastern United States aerospace cluster is addressed in detail in Chapter 14.

Although there was rebalancing in where manufacturing activities took place, several trends ensured that low-cost countries maintained an important and growing role in the jetliner supply chain. The first was the intense focus on cost reduction ushered in by Partnering for Success and Power8. The second was the growing role of some LCCs—particularly China and India—as major markets for jetliners. They demanded in-country manufacturing as a condition for buying aircraft. "The Chinese told me if you don't do work here, we'll buy from Airbus," said one former Boeing executive.[15]

Supply chain decisions became more nuanced. Labor-intensive manufacturing activities such as hand lay-up composites, simple sheet metal fabrications, and basic five-axis machining continued to shift to LCCs. At the same time, advanced manufacturing and final assembly remained, for the most part, in Europe, North America, and Japan.

LEANING OPERATIONS

Much more was happening in manufacturing than outsourcing and LCCs; major changes were happening on the factory floors *within* aerospace companies. One catalyst for change was the Toyota Production System (TPS). The origins of TPS were in post-World War II Japan, when Toyota executives Taiichi Ohno and Shigeo Shingo began to build on legacy Ford Motor Company production approaches to create an entirely new approach to production. They developed a *just-in-time* production system, which, in turn, necessitated the creation of a new *Kanban* inventory control system, which "pulled" inventory only when needed rather than stocking large

---

[15]Carolyn Corvi (former VP and General Manager Airplane Programs and Supplier Management—Boeing), interview with author, 11 April 2017.

batches. Toyota soon discovered that factory workers had far more to contribute than just muscle power and embraced team development and cellular manufacturing. Another key discovery involved product variety. The prevailing Ford mass-production system did not cope well with product variety and small quantity production. Shingo, at Ohno's suggestion, went to work on the setup and changeover problem. Reducing setups to minutes and seconds allowed small batches and an almost continuous flow like the original Ford concept. It introduced a flexibility that Henry Ford thought he did not need. All of this took place between about 1949 and 1975.[16] Western manufacturers were either in denial or unaware of their productivity gap with Toyota's amazing productivity breakthrough.

This began to change in the early 1980s, when Toyota established a joint venture with US car maker General Motors (GM) in California called NUMMI. GM saw the joint venture as an opportunity to learn about TPS, while Toyota gained its first manufacturing base in North America and a chance to implement its production system in a US labor environment, thus avoiding possible import restrictions. The JV was a success and an important platform for spreading awareness of TPS.[17] NUMMI engineer John Krafcik would later coin the phrase *lean manufacturing* while contributing to the International Motor Vehicle Program at the Massachusetts Institute of Technology (MIT). This research spawned the 1990 book, *The Machine That Changed the World*, which compared Japanese, American, and European automotive assembly plants and proved conclusively the superiority of TPS.[18] This seminal work, written by MIT academics James Womack, Daniel Jones, and Daniel Roos, began to put "lean" on the minds of aerospace executives.

A second catalyst for leaning aerospace operations was the Six Sigma movement. In the early and mid-1980s, engineers at US electronics manufacturer Motorola decided that the traditional quality levels—measuring defects in thousands of opportunities—didn't provide enough granularity. Instead, they wanted to measure the defects per million opportunities. Six Sigma aimed for 3.4 defects per million opportunities in each process. Motorola developed this new standard and implemented methodologies and cultural changes that created billions of dollars in savings. AlliedSignal CEO Lawrence Bossidy embraced Six Sigma in the early 1990s with impressive results. Quality improved significantly, supplier defects fell by 70% in a single year, and productivity increased 6.1% in 1992 and 7.3% in 1993. American Airlines CEO Robert Crandall lauded its turnaround after being very critical

---

[16]Strategos, "A History of Lean Manufacturing," http://www.strategosinc.com/just_in_time.htm [retrieved 18 March 2017].

[17]Toyota Global Corporation, "NUMMI Established," http://www.toyota-global.com/company/history_of_toyota/75years/text/leaping_forward_as_a_global_corporation/chapter1/section3/item2.html [retrieved 18 March 2017].

[18]Strategos.

Credit: Carolyn Covi

Fig. 8.9 Boeing's lean pioneer Carolyn Corvi.

of its performance years earlier [19]. AlliedSignal's turnaround grabbed the attention of General Electric CEO Jack Welch, who created a goal for GE to achieve Six Sigma goals by 2000. Welch required all exempt employees to undertake a 13-day, 100-hour training program and complete a Six Sigma project by the end of 1998. A new battlefront formed in aerospace—the race to quality leadership. Rumor has it that Bossidy and Welch were playing golf one day and Jack bet Larry that he could implement Six Sigma faster and with greater results at GE than Larry did at AlliedSignal.[19] United Technologies Corporation (UTC) was also in the race, and then-CEO George David recruited Matsushita's head of quality assurance to move to the United States from Japan and set up an internal "quality university." By 1998 UTC reached 4.5 sigma (1350 defects per million)—a level on par with AlliedSignal [20].

While AlliedSignal and General Electric implemented Six Sigma, adoption of lean principles by aerospace companies slowly gained traction. There were several conduits. One was Carolyn Corvi (Fig. 8.9), a young Boeing executive who received a fellowship to study management at MIT in 1987. Her time at MIT overlapped with that of James Womack, author of *The Machine That Changed the World*. This was her first exposure to lean.[20] After returning to Boeing, she toured the NUMMI joint venture in 1992. "That day changed my life," Corvi told the *Seattle Times*. "I'd never seen anything like it. My poor husband, I talked to him about it until 2 in the morning." Corvi returned as an evangelist for what she'd witnessed: a synchronized manufacturing dance that produced a new car every 59 seconds [21]. She would later see lean applied to aerospace manufacturing at the Snecma factory in Villaroche, France. At the time, Boeing faced increased competitive pressure from Airbus, and Boeing's CEO, Frank Shrontz, received many quality complaints from customers. He

[19]iSixSigma, "The History of Six Sigma," https://www.isixsigma.com/new-to-six-sigma/history/history-six-sigma/ [retrieved 18 March 2017].

[20]Carolyn Corvi, interview with author, 13 Feb. 2017.

supported Corvi and other change agents, and lean gained traction at Boeing in the 1992–1995 timeframe.[21]

James Womack played an important role in transforming another aerospace company, nacelle supplier Rohr in Chula Vista, California. In the mid-1990s, Rohr was bankrupt, poorly run, and hadn't won a new nacelle program in years. Around this time, its manufacturing leader, Greg Peters, attended a lean course, and soon the company began to adopt the philosophy with Womack's assistance. Rohr focused on standard work, linkage, and flow at its 14 manufacturing cells as well as its engine integration facility in Toulouse. When Goodrich bought Rohr in 1997, lean was being deployed across the enterprise. The business became profitable by 2000, with 100% on-time delivery and significantly improved quality. Womack would later call it "the best aerospace organization that he'd seen."[22]

Meanwhile, back in Seattle, Boeing asked its suppliers to adopt lean. "We were called to Seattle in 1998 for a briefing on lean," recalled former Rockwell Collins CEO Clay Jones.

> I was taken immediately. On our flight back to Iowa, we decided to extend lean to our world—we'd call it *Lean Electronics*. We selected our most seasoned program manager to lead the initiative. All senior leadership had to go through rapid process improvement seminars, and ultimately thousands of lean events took place.[23]

Lean and Six Sigma had the same goal—both sought to eliminate waste and enhance productivity—but they took different approaches in achieving these ends. Six Sigma focused on reducing waste from variations in processes. It was a scientific method. In contrast, lean was an all-encompassing process to eliminate unnecessary steps in the development and production processes that did not add value to the finished product. Lean had more of a cultural component than Six Sigma, and often inspired passion from its adherents.

Although there were islands of success, the overall impact of lean and Six Sigma on aerospace after a decade was mixed. In 2001, James Womack and Deloitte partner David Fitzpatrick gave the industry a *C* grade in becoming lean and believed that the industry needed to cut its assets, many underutilized, to improve. Still, they cited growing awareness and momentum throughout the industry [22]. Six Sigma report cards were also mixed. Standouts like Honeywell (which merged with AlliedSignal in 2001) and GE saved more than $500 million per year, but there were many laggards and companies with disappointing results. A 2002 survey by *Aviation Week & Space Technology*

---

[21] Ibid.

[22] Marc Duvall (President—UTC Aerospace Systems Aerostructures Division), interview with author, 29 Oct. 2015.

[23] Clay Jones (former CEO—Rockwell Collins), interview with author, 5 Nov. 2015.

found that less than 50% of the respondents were satisfied by the Six Sigma results [23].

Lean and Six Sigma gained momentum in the early 2000s. Boeing's Carolyn Corvi, who became the vice president and general manager of the 737 and 757 programs, continued to be a change agent. "The conventional wisdom at the time was that you could never build more than 21 aircraft per month in Renton," she recalled, "and I asked *why*?"[24] Corvi led an initiative to shrink the 737 production line to a single, moving assembly line—much like she had seen at the NUMMI plant a decade earlier. Known as the *Move to the Lake* initiative, it also relocated engineers and managers to the factory buildings. The results of leaning out one of Boeing's most critical production lines were breathtaking. By 2004, manufacturing floor space was cut by 40%. Inventory turns were increased from single digits to 27 for Boeing's parts, and the time to assemble a 737 was reduced from 22 days to 13 days [24]. Critical to the success of this transformation was the involvement of organized labor, which overcame fears of job elimination to embrace lean and ensure its success.

Airbus responded to the challenge and introduced a moving assembly line for its A320s in Hamburg and launched the Power8 initiative in 2006, which included lean production and facility rationalization. These important initiatives would lay the foundation for dramatic production rate increases in later years.

By the 2010s, the jetliner ecosystem fully embraced lean and Six Sigma. The results were sometimes breathtaking. Boeing's Renton production facility—the same facility that Corvi challenged to produce more than 21 737s per month—exceeded 50 per month by 2017, with a target of 60 or more.

Globalizing Engineering

Supply chain and operations weren't the only functions rebooted in the 1990s and 2000s; engineering was also transformed. Until the late 1980s, engineering activities in most firms were organized in functional silos (e.g., aerodynamics, structures, design, materials, manufacturing, electrical, and others). Engineers worked with colleagues from other functions, but their loyalty was to their functional supervisors and bosses, who were measured on the utilization of their employees and defended their function's sovereignty. Moreover, there was a strong divide between engineering and operations. Although this approach often forged strong teams and deep functional expertise, it led to sub optimization on development programs. Accountability also suffered. Too often, engineering created designs that were expensive to produce and expensive to maintain. And in the pre-Internet and predigital era, most engineering was co-located. Outsourcing was limited to "job shops"

[24]Corvi interview, 11 April 2017.

staffed with retirees to cope with surges in engineering demand. Employees from these firms generally worked onsite with their former coworkers.

In the mid-1980s and early 1990s, many aerospace companies pursued a better idea: Why not break down the functional silos and organize around *projects*? In this approach, a cross-functional engineering team reported to a lead project engineer, who optimized the design, manufacturing cost, and maintainability around customer needs. All of the functions were part of the same team, so accountability was increased and feedback loops became instantaneous. The approach became known as integrated product team (IPT). Engineers in IPTs were often part of a matrix organization composed of functional leaders and project leaders. Thus, an engineer might report to both functional and project executives.

Implementation of this new approach wasn't easy. Cultures, processes, and metrics had to change. Functional experts, often with large egos, needed to become team members. Rolls-Royce's experience is illustrative. "In the 1980s, we went from crisis to crisis in engineering. Ralph Robbins wanted this addressed, and I realized that our issues were more related to process than technology," said former Engineering Director Phil Ruffles.[25] In the early 1990s, he implemented an IPT approach called Project Derwent that brought together engineering functions and manufacturing. The initiative created gates between project milestones that required all outstanding engineering issues to be addressed before the program advanced to the next step. This was a dramatic challenge to the status quo that had prevailed in Derby for decades, and its engineers threated a strike. "Nineteen ninety-one was the worst year of my life with the Trent 600 and Trent 700 programs and Project Derwent," Ruffles recalled, "but as a result the Trent 800 development program went through like a dream."[26]

As the barriers fell between functional disciplines, a sea change emerged in the early and mid-1990s that would transform engineering: distributing work to low-cost countries via captive engineering centers or contracting with engineering service providers (ESPs). Boeing's move to open an engineering center in Moscow in 1993—just 18 months after the end of the Cold War—was an early move. Why did they do it? "The Cold War was ending, and literally thousands of highly qualified technical people with not a lot to do were suddenly available," recalled former Boeing CEO Jim Albaugh. "This was an opportunity to get the best and brightest at a reasonable rate structure—75% had PhDs and they are very bright and excellent at aerostructures."[27] Boeing's Moscow design center would make important contributions to the 747-800, 787, and the Dreamlifter—Boeing's answer to the Airbus Super Guppy.

---

[25]Phil Ruffles (former Engineering Director—Rolls-Royce), interview with author, 1 June 2016.
[26]Ibid.
[27]Albaugh interview, 22 Sept. 2015.

Honeywell was also a low-cost engineering pioneer. In the mid-1990s, it created Honeywell Technology Solutions, an engineering resource available to all Honeywell businesses (aerospace and nonaerospace) with engineering centers in India, China, and the Czech Republic. In a similar vein, General Electric created the Jack Welch Center in Bangalore, India. This was GE's first R&D center outside the United States. Both Honeywell and GE would ultimately employ thousands of engineers at their Indian engineering facilities. Other aerospace companies followed their examples.

Giving a boost to low-cost engineering outsourcing was the *Y2K Bug*, which arose in the late 1990s because programmers created software with years represented with only the final two digits, which made the year 2000 indistinguishable from 1900. This opened the door for engineering service providers (ESPs) from India and Eastern Europe to work with aerospace OEMs and suppliers to ensure their systems didn't crash. Once these relationships were established, some ESPs broadened their offerings to include engineering and design services.

The results of the first wave of engineering outsourcing were mixed. Although labor costs in low-cost centers were substantially lower, so was productivity. Moreover, engineering teams had to learn new ways of collaboration with company-owned low-cost centers and ESPs. Work needed to be packaged properly with the right amount of engineering oversight. Employee attrition was also an issue. Nearly all aerospace firms faced a steep learning curve in unbundling engineering. According to an industry survey conducted in 2014, it took an average of seven to eight years to achieve breakeven on low-cost engineering outsourcing. In the same survey, cost savings of 10% or more were typical [25].

A new wave of low-cost engineering took hold in the mid-2000s that was driven by access to critical skills rather than simple labor arbitrage (Fig. 8.10). This was a result of maturation of the distributed engineering model as well as larger, more sophisticated ESPs. Centers of excellence were developed as Russia and Eastern Europe became a hotbed for structural analysis while India took on more software development. Although most of the investment was aimed at LCCs, it wasn't exclusively this way. EADS, for example, set up an engineering center in Wichita, Kansas, in 2004 to access the business aviation talent pool there. It followed with new engineering centers in Singapore (2006) and India (2009). Aerospace OEMs also found better ways to manage the work packages, including the adoption of "standard work."

This set the stage for the next phase of contract engineering: outcome-driven partnerships. Here, LCC engineering centers took responsibility for complete part design and sometimes manufacturing. Pratt & Whitney's experience with Indian ESP Infotech is illustrative. In the 1990s, Infotech provided basic analytical work to Pratt & Whitney. This expanded to complex analysis—structures, aerodynamics—and then evolved into complete part

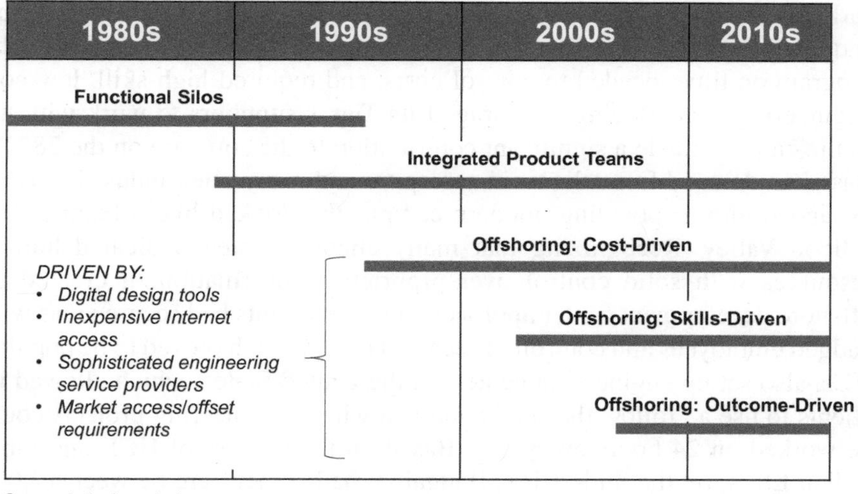

Source: Author's analysis

Fig. 8.10 Evolution of engineering outsourcing.

design. To provide support for military aeroengine programs, which generally require US citizenship, the two firms set up Infotech Aerospace Services, a joint venture in Puerto Rico with more than 600 employees. The move to outcome-driven engineering also changed the nature of contracts from time and material to fixed price. Now the ESP took on risk for the program. Some firms completed the journey by also taking on manufacturing responsibility. European OEMs led the way in embracing contracts that transferred responsibility and risk to ESPs.

Beyond Infotech, there were many other Indian ESPs. HCL Technologies Ltd. (Fig. 8.11) was the largest, and began work for aerospace clients in the

Credit: HCL Technologies

Fig. 8.11 HCL Technologies' Bangalore campus.

early 2000s. Its big break was with the 787. HCL's President, Engineering and R&D Services G. H. Rao recalled the situation: "Boeing wanted to do the program on time, needed to control costs, and required high-skill, low-cost engineering. And Boeing encouraged its Tier 1 suppliers to work with us. In the end, we made a significant contribution to the software on the 787."[28] Based on demand from clients in aerospace and many other industries, HCL expanded into a sprawling, modern campus that looked like it belonged in Silicon Valley. Recognizing that many clients wanted dedicated human resources with solid control over proprietary information, it created an offshore development center approach in which clients had secure areas with badged employees and controlled access. This approach proved to be popular. HCL also set up engineering centers in the United States, which allowed its clients to use a "follow the sun" approach where engineering projects could be worked on 24 hours every day. Based on the success of HCL and other Indian ESPs, by the mid-2010s, Bangalore had somewhere between 12,000 and 20,000 engineers working on civil aircraft programs, which qualified it as one of the largest concentrations of aerospace engineers on the planet.

The development of India as a major engineering hub underscored the extent that jetliner supply chains were transformed in the two decades following the Cold War. Manufacturing activities, once controlled by vertically integrated aircraft and engine OEMs, were dispersed to a global network of suppliers. The industry embraced the Tier 1 supply chain model with mostly positive results (apart from the 787 debacle). After a slow start, lean and Six Sigma movements took hold and underpinned improved quality and productivity. And engineering, once confined to functional silos, became integrated with manufacturing and then globalized. The jetliner supply chain, to borrow a term from the computer industry, effectively rebooted.

REFERENCES

[1] Kiley, D., "The New Heat on Ford," Bloomberg.com, 1 July 2007, https://www.bloomberg.com/news/articles/2007-06-03/the-new-heat-on-ford. [retrieved 16 Oct. 2016].

[2] "The Machine That Ran Too Hot," *The Economist*, 25 Feb. 2010, https://www.economist.com/node/15581072. [retrieved 3 March 2017].

[3] Womack, J., Roos, D., and Jones, D., *The Machine That Changed The World*, Macmillan, New York, 1990, p. 146.

[4] Aboulafia, R., *2016 World Civil & Military Aircraft Briefing*, The Teal Group, Fairfax, VA, 2016, p. 652.

[5] "Winds of Change: Get Ready for Seven Trends Transforming Commercial Aerospace," Commentary—AeroStrategy Management Consulting, Ann Arbor, Michigan, March 2007.

[28]G. H. Rao (President–Engineering and R&D Services—HCL Technologies Ltd.), email to author, 23 March 2018.

[6] Norris, G., "Powerful Partners," *Aviation Week & Space Technology*, 13 April 2009, pp. 47–48.
[7] Floteau, J., "Déjà Vu," *Aviation Week & Space Technology*, 15 Feb. 2010, p. 23.
[8] Hiltzik, M., "787 Dreamliner Teaches Boeing Costly Lesson on Outsourcing," *Los Angeles Times*, 15 Feb. 2011, http://articles.latimes.com/2011/feb/15/business/la-fi-hiltzik-20110215. [retrieved 17 Oct. 2016].
[9] Tang, C., and Zimmerman, J., "Managing New Product Development and Supply Chain Risks: The Boeing 787 Case," *Supply Chain Forum*, Vol. 10, 2009, pp. 76–78.
[10] Kemp, K., *Flight of the Titans*, Virgin Books, London, 2006, p. 145.
[11] Bowermaster, D., "Heavies Help Carry 787," *Seattle Times*, 13 May 2005, https://www.seattletimes.com/business/heavies-help-carry-787. [retrieved 19 July 2016].
[12] Kemp, p. 89.
[13] Velocci, A., Anselmo, J., and Asker, J., "Only the Paranoid Survive," *Aviation Week & Space Technology*, 26 June 2006, pp. 48–51.
[14] Velocci, A., and Anselmo, J., "Lessons Learned," *Aviation Week & Space Technology*, 14 July 2008, p. 74.
[15] Matlack, C., "What Airbus Learned from the Dreamliner," *Bloomberg Businessweek*, 17 April 2008, https://www.bloomberg.com/news/articles/2008-04-16/what-airbus-learned-from-the-dreamliner. [retrieved 26 July 2016].
[16] Wall, R., "Transition Time," *Aviation Week & Space Technology*, 7/14 May 2008, p. 79.
[17] "Power8 prepares way for "New Airbus," Airbus Press Release, 28 February 2007.
[18] Morris, D., "Will Tech Manufacturing Stay in China?" *Fortune*, 27 Aug. 2015, http://fortune.com/2015/08/27/tech-manufacturing-relocation. [retrieved 24 July 2016].
[19] Velocci Jr., A., "Turnaround Earns Allied New Credibility," *Aviation Week & Space Technology*, 15 Aug. 1994, pp. 38–39.
[20] Velocci Jr., A., "Pursuit of Six Sigma Emerges As Industry Trend," *Aviation Week & Space Technology*, 16 Nov. 1998, pp. 52–57.
[21] Gates, D., "Meet the People Who Helped Shape Boeing's 100-Year History—Part 2," *Seattle Times*, 15 July 2016, https://www.seattletimes.com/business/boeing-aerospace/meet-the-people-who-shaped-boeings-100-year-history-lopez-corvi-and-albaugh. [retrieved 2 August 2016].
[22] Velocci Jr., A., "Effective Application of Lean Remains Disappointing," *Aviation Week & Space Technology*, 22 Jan. 2001, p. 56.
[23] Velocci Jr., A., "Full Potential of Six Sigma Eludes Most Companies," *Aviation Week & Space Technology*, 30 Sept. 2002, p. 60.
[24] Mecham, M., "The Lean, Green Line," *Aviation Week & Space Technology*, 14 July 2004, p. 144.
[25] Ferguson, I., "The World Is Not Flat," ICF International/*Aviation Week & Space Technology* white paper, 2014.

Chapter 9

BRAINIACS

Avionics are the brains of jetliners. They consist of hundreds of components for critical aircraft functions including flight control, autopilot, communications, navigation, surveillance, and even aircraft health monitoring. The term *avionics* was coined by journalist Philip Klass as a linguistic blend of *aviation* and *electronics* [1].

Avionics are among the most sophisticated aircraft systems, and their significant capability improvements yielded important benefits for consumers including more direct routes, lower takeoff and landing minimum ceilings, and dramatic improvements in safety. Air travel deaths decreased 60-fold over a 50-year period, an achievement unparalleled in the history of transportation.

At one time, there were dozens of avionics suppliers that produced specific avionics "black boxes." By the 2010s, most jetliner avionics were produced by Honeywell, Rockwell Collins, and Thales, three huge, sophisticated suppliers engaged in intense competition. How did this vital equipment sector get started, what were the key developments, and how did it evolve into a consolidated equipment sector?

THE EARLY HISTORY OF AVIONICS

Electronics first found their way onto aircraft just years after the Wright Brothers first flew, and their use was accelerated by World War I. Much of the pioneering activity was in Europe. In 1910, Frenchman Paul Brenot of Société Francaise Radio-électrique (SFR) conducted transmission tests from an aircraft and produced 18,000 airborne radios by the end of World War I [2]. In the late 1920s, it was producing dual band radios covering both shortwave (3–30 MHz) and longwave (15–35 KHz) frequencies. Another early producer of avionics was CSF (Companie Générale de Télégraphie Sans Fil). In Great Britain, automotive supplier S. Smith & Sons expanded into aviation in 1915 after it won a War Office contract for aircraft accessories. By World War II, Smiths was a major manufacturer of aircraft altimeters and compasses for the Royal Air Force and Royal Navy. CSF and SFR came under the control of German firm Telefunken.

On the other side of the Atlantic, inventor Elmer Sperry started a firm in 1910 to focus on gyrostabilizers and gyrocompasses. His son, Lawrence, pioneered the idea that if the three flight axes of an aircraft—yaw, pitch, and roll—could be harnessed to the stability of a gyroscope, an automatic control system might be developed. By 1914, he had invented a crude autopilot and split with his father to form his own avionics company. After an untimely death on a 1923 flight over the English Channel, his firm was reunited with his father's, and the company would prosper during World War II making analog computer–controlled bomb sights, airborne radar systems, and automated takeoff and landing systems.

In 1929, Vincent Bendix, a successful automotive supplier, formed a firm focused on aviation. The Bendix radio division was established eight years later. The firm experienced dramatic growth during World War II and manufactured the majority of avionics for US aircraft, including crude radars. Another entrepreneur, Bill Lear, also made the transition from the automotive industry to aerospace. After inventing the first practical car radio, he turned his attention to aviation and founded Lear Developments in 1931. His first big breakthrough was the Learoscope, a radio compass that was a forerunner to today's automatic direction finder. He talked World War I ace Eddie Rickenbacker into putting it on an Eastern Airlines flight from Miami to New York. After the flight, Rickenbacker called Lear's direction finder "the most important air navigation aid developed to date" [3]. Recognizing the threat to his own business, Vincent Bendix made an offer to buy Lear Developments but was rebuffed; 20 years of intense competition would ensue.

Meanwhile, in Cedar Rapids, Iowa, a precocious teenager named Arthur Collins created innovative high-frequency radios. He formed the Collins Radio Company in 1933 and caught a big break when his equipment was used by Admiral Richard Byrd on an expedition to Antarctica. After surviving patent challenges by AT&T and RCA, he adapted his radios for airborne applications and invented *autotune*, which enabled pilots to switch frequencies by pushing a button. By 1937, Braniff was the first airline to equip its entire fleet with Collins Radio equipment. American Airlines followed suit [4]. Collins would produce some 26,000 "autotuned" airborne radios during World War II; wartime contracts totaled $110 million [5]. In the space of 12 years, Collins transitioned from homebuilt radios to creating a major avionics supplier.

Some of these pioneers of the field are shown in Fig. 9.1.

Landing aids became an area of focus in the early 1930s when airports installed the earliest forms of approach lighting. Tests of instrument landing systems (ILSs) that incorporated radio beacons began in 1929, and in 1938

Credits: Courtesy of The History Museum, South Bend, Rockwell Collins, and Kansas State Historical Society

Fig. 9.1 Avionics pioneers Elmer Sperry, Vincent Bendix, Art Collins, and Bill Lear.

the first landing of a scheduled US passenger airliner took place. After further development by the US Army in World War II, the International Civil Aviation Organization adopted the Army standard for all member countries.

Not surprisingly, World War II instigated other critical avionics developments. One was aircraft radars. The first airborne surveillance mission was carried out by a 776-ft German Graf Zeppelin blimp, which flew three surveillance missions along the British coast with a crude and very large radar set. In 1940, a British delegation led by Henry Tizard offered many of its most technological advances to the United States to bargain for scientific assistance. One of these inventions was a cavity magnetron, which made it possible for radar to operate at higher frequency bands. A joint British–American research team at the Massachusetts Institute of Technology Radiation Laboratory (Rad Lab) leveraged this invention to make centimeter-band radar practical, allowing detection of smaller objects such as submarine periscopes. These new radar sets were dramatically smaller than earlier generations of radar, which allowed them to be packaged in a size suitable for aircraft. American industrial might got behind the new invention. The Rad Lab would develop 150 models of radar systems that were designed for use in aircraft, submarines, ships, and coastal defense. More than 1 million radar sets would be produced during World War II [6]. At times, radar operators noticed some extraneous echoes showing up on their display. After investigating, it was discovered that the echoes operators were seeing on their display weren't aircraft or anything related to missions, but rather interference from weather. Following the war, radar would be adapted to a new function—weather detection.

Following the war, avionics developments would accelerate with the growth of civil aviation and the onset of the jet age. Communications systems expanded to include more and more channels. Very-high-frequency (VHF) radios, which used electromagnetic waves between 30 and 300 MHz, provided "line-of-sight communications," which meant that there needed to be a straight path between a transmitting and a receiving antenna unobstructed by the horizon. To communicate over the horizon, 3- to 30-MHz high-frequency (HF) radios, commonly known as shortwave radios, were used. Not surprisingly, Collins Radio was a leader in airborne HF communications.

Another major challenge was improving aircraft navigation systems. Specifically, how could an airplane navigate without reference to external radio-frequency aids? Here, another US Midwesterner, Charles Draper, played a major role. In the 1940s, Draper led an initiative at the Massachusetts Institute of Technology to make dramatic improvements in the accuracy of gyros so that they could become a reliable navigation system. After initially focusing on fire control systems for military aircraft, Draper turned his attention to aircraft navigation. In 1953, Draper flew in a B-29 from Boston to Los Angeles automatically guided by its inertial navigation system, which weighed 2800 lb. For the first time, the feasibility of an inertial navigation system was demonstrated outside the laboratory [7]. Significant accuracy improvements and weight reduction in inertial navigation systems would follow in subsequent decades. Draper became known as the "father of inertial navigation."

Avionics in the Jet Age

The jet age would usher in significant improvements in avionics as larger, faster aircraft flying longer routes emerged (Fig. 9.2). One challenge was that jetliners needed precise means of navigating transoceanic routes.

Segment	Innovation	Year
Flight Control	Fly-by-wire	1988
Navigation	Global positioning system	1983
	Inertial navigation system (mechanical)	1973
	Inertial navigation system (ring laser gyro)	1982
	Inertial navigation system (fiber optic guidance)	1995
	Flight management system	1982
Safety & Surveillance	Transponders	1960
	Ground proximity warning system	1969
	Enhanced ground proximity warning system	1995
	TCAS	1995
Displays	Cathode ray tube displays	1982
	Liquid crystal displays	1995
	Heads-up display	1975
Communications	Satellite communications	1979
	ACARS	1979
	Solid state radio	1966
Architecture	Integrated modular avionics	1995
	AFDX	2007
Recording	Flight data recorder	1950s
	Cockpit voice recorder	1960s

Source: Author's analysis
Note: Some innovations occurred earlier in business and military aviation

Fig. 9.2 Dates of important jetliner avionics innovations.

Inertial navigation systems, which estimated the aircraft's speed, direction, and orientation without external references, made their debut in jetliners in the 1950s. These systems, which were based on spinning mass (gyroscope) technology, were particularly important for aircraft navigation during transoceanic flights, when aircraft were out of range of conventional navigation aids. Long haul aircraft were typically equipped with three inertial navigation systems each.

Aircraft navigation capabilities continued to improve in the 1960s, when jetliners installed VHF omnidirectional range (VOR) systems. These enabled aircraft with a receiving unit to determine their position and stay on course by receiving radio signals transmitted by a network of fixed ground radio beacons. Soon, VOR networks became widespread, creating "Victor highways" to link VOR stations. ILSs continued to improve, and by the 1960s, the first instrument landing system equipment for reduced visibility landings were introduced. Avionics became smaller and much lighter thanks to solid state electronics; King Radio made the first solid state transceiver in 1966.

Inertial navigation systems continued to improve and build on Charles Draper's legacy. Shortly after the invention of the laser, it was posited that accurate navigation devices could be created by measuring two beams of light of the same polarization traveling in opposite directions. The first ring laser was built under Sperry scientist Chao Chen Wang in the early 1960s. This technology not only was more accurate than conventionally spinning-mass gyros, but also was lighter and more reliable. Soon, ring-laser gyro (RLG) technology would replace less-accurate mechanical gyroscopes and enter service on the 757 and 767. The accuracy of RLG inertial systems, measured in "drift rate," was 0.01 deg/hr. This would be improved upon by fiber optic gyroscope (FOG) technology in which an external laser injected counterpropagating beams into a coiled and long optical fiber ring. These systems were expensive, and their installation was generally limited to long haul aircraft. Sperry and Litton were the first major suppliers.

The stakes for aircraft safety were higher in the jet age, which meant that avionics needed to assist in determining the cause of incidents and create a feedback loop to improve safety. The Comet accidents of the 1950s inspired an Australian inventor, David Warren, to modify analog recording technology to a crash-survivable flight data recorder (FDR) in the 1950s. The United Kingdom was the first country to mandate FDRs on aircraft in the 1960s. Early FDRs recorded 24 parameters. Cockpit voice recorders (CVRs) would soon be added to the standard avionics equipage [8].

FDRs and CVRs made a significant contribution to accident investigations and jetliner safety, but controlled flight into terrain (CFIT) remained a major issue. "Scandinavian Airlines had three major crashes in the late 1950s and 1960s," recalled avionics pioneer Don Bateman, "[and I asked] can we do something about this? We got serious about developing a solution in 1965 and 1966."[1] Don Bateman's obsession with aviation safety could be traced back to his youth in Saskatchewan, Canada, when, in 1940, he broke his elementary school rules to visit the scene of a mid-air collision between two military trainers. As part of his punishment, his elementary school teacher ordered him to write a detailed account of what he had witnessed. When he handed in his assignment, she told him, "You sure can't spell—you're going to be an engineer" [9]. His teacher was prescient, and in 1958 Bateman joined Boeing to work on 707 avionics. He later joined a local start-up called United Electronics, which was later acquired by Teledyne. By 1969, Bateman's team collaborated with Boeing and SAS to demonstrate a ground proximity warning system (GPWS) that alerted pilots if their aircraft was in danger of flying into the ground or an obstacle. CFIT rates declined, and the Federal Aviation Administration (FAA) mandated that airlines use GPWSs in 1974 after a TWA 727 crash in Virginia.

Meanwhile, another surveillance capability—weather radar—became widespread on jetliners in the early 1950s. Bendix, Garrett, and RCA were early leaders, and some carriers used military equipment. For example, Braniff Airways and Panagra (a Pan Am joint venture) became the first airlines to use weather radar in South America in 1954 using a US Navy radar system. An added benefit was that it provided a navigation aid in areas where ground radar facilities were lacking [10].

Goodbye Third Pilot

The fuel crisis of the 1970s provided the drive for improved navigation capabilities and led to development of the first flight management system (FMS). Sperry Flight Systems' Tern-100 automatic navigation system was the first truly automatic FMS using external sensors. It was initially certified on the Boeing 727 in 1974, followed by the 707 in 1975 and the 747 in 1979. Sperry's breakthrough led to a 1978 contract to provide FMS as standard equipment on the 757 and 767. The FMS integrated input from the aircraft's sensors and databases to provide in-flight management of the flight plan. It also overcame the limitation of autopilots in not being able to handle multiple maneuvers at once. This would accelerate the elimination of flight engineers from the cockpit crew, an innovation that Airbus pioneered on the A310 when it introduced the forward-facing crew cockpit. Airbus chief engineer Bernard Ziegler (Fig. 9.3) made the decision to leverage FMS and pioneer a two-pilot cockpit in the A300B and A310. He was so despised by the French pilots' union that he received death threats and at one point was forced to live under police protection [11]. Honeywell executive Bob Witwer summarized the revolution when he said, "The FMS made the third

[1]Don Bateman (former VP—Honeywell), interview with author, 23 Aug. 2016.

Credit Airbus:

Fig. 9.3 Airbus's Bernard Ziegler: flight control revolutionary.

pilot go away" [12]. This reduced operating expenses for airlines and helped to lower the cost of air travel. The 737-300, 757, 767, and MD-80 would follow with two-person cockpits.

Another major avionics innovation promulgated by Airbus was the flight control computer, the brains of fly-by-wire flight controls pioneered on the A320 and discussed in Chapter 2. The A320 was also the first jetliner to incorporate full flight-envelope protection into its flight-control software, another breakthrough instigated by Bernard Ziegler. The goal was to build a jetliner that could never be stalled. Airbus included "hard" protection features in which the flight control computer prevents pilots from overbanking, applying either a too-high or too-low pitch angle, or overstressing the aircraft. Another breakthrough was that pilots controlled the aircraft by sidesticks, eliminating the traditional yokes in the cockpit. This not only allowed the pilot to sit closer to the aircraft displays, but also allowed aircraft designers to shorten the length of the aircraft and reduce its weight. Aerospatiale (predecessor to Thales) worked closely with Airbus engineers to create the flight control computers that translated movements of the sidestick into electrical signals that controlled actuators and flight control surfaces. Here, it leveraged its experience gained from the French Rafale fighter program to Airbus's advantage. "The A320 was a revolutionary aircraft," according to Thales executive Richard Perrot. "The fly-by-wire system was one of the most important avionics innovations in the last three decades. This was the first step in the development of [what would become] Thales as a major European avionics supplier."[2]

[2]Richard Perrot (VP-Marketing—Thales), interview with author, 20 Sept. 2016.

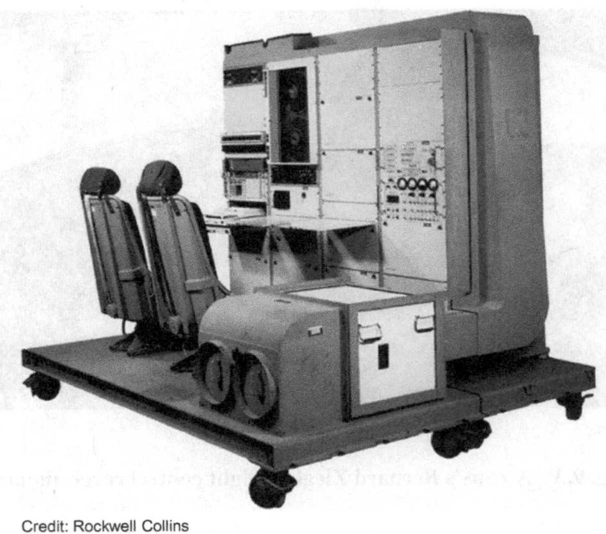

Credit: Rockwell Collins

Fig. 9.4 The first GPS receiver, used by Rockwell Collins, was 6 ft tall.

Arguably the single most important innovation in navigation systems was the development of the global positioning system (GPS) by the United States in the 1970s. GPS is a satellite-based navigation system that allows a user to determine its precise location by receiving signals from multiple satellites orbiting the Earth at an altitude of 12,000 miles. The program was initiated by the US Department of Defense in 1973 for the benefit of the US military, which at the time was fully engaged in the Cold War. On 19 July 1977, a Collins engineer named David Van Dusseldorp sat on the rooftop of a company building in Cedar Rapids, Iowa, and received the world's first GPS signal with a 6-ft-tall receiver that required two operators (Fig. 9.4).[3] Collins would shrink GPS receivers, and in 1983 demonstrated GPS for airborne navigation on a trans-Atlantic flight to Paris.

GPS was initially restricted to military use until 1983, when President Ronald Reagan issued a directive making GPS freely available for civilian use. This was in response to the downing of Korean Air Lines Flight 007, a Boeing 747 carrying 269 people, by Soviet fighters after it strayed into prohibited airspace. GPS would dramatically increase the accuracy of navigation systems as falling costs and smaller, lighter systems facilitated its introduction on most jetliners. And importantly, GPS was not dependent on ground-based navigation aids, which significantly improved navigation accuracy over the

[3]"Rockwell Collins Celebrates 40 Years Since Receiving the World's First GPS Satellite Signal," Rockwell Collins press release, 19 July 2017, https://www.rockwellcollins.com/Data/News/2017-Cal-Yr/GS/20170719-GPS-40-Anniversary.aspx [retrieved 8 April 2018].

ocean. By 1995, 10,000 air transport and general aviation aircraft would be equipped with GPS [13].

The era also saw advances in avionics surveillance capabilities. One vexing problem had bedeviled aviation since a 1956 mid-air collision between a DC-7 and an L-1049 Super Constellation over the Grand Canyon: *how to avoid mid-air collisions?* Avionics technology evolved to the point where an aircraft could interrogate all other aircraft in a determined range about their position. In 1981, the US FAA announced plans to implement an aircraft collision avoidance concept called the Traffic Alert and Collision Avoidance System (TCAS). Piedmont Airlines, United Airlines, and Northwest Airlines flew early versions of TCASs in the late 1980s. After observing the success of these trials, the FAA mandated that all turbine-powered aircraft with more than 30 seats be equipped with TCASs by January 1993. Europe, Australia, and Hong Kong followed with similar mandates for January 2000. Another surveillance system, weather radar, continued to improve as digital technology, color displays, automatic scanning capability, turbulence protection, and wind shear protection became integrated jetliner weather radar systems.

Displays evolved from mechanical devices and "steam gauges" to electromechanical displays, and by the early 1980s, cathode ray tubes (CRTs). CRTs offered several advantages including greater flexibility, better visibility, and significantly lower maintenance costs. Electronic flight instrumentation systems (EFISs) featuring CRTs made their debut on the 757, 767, and A310 in the early 1980s. Another advancement was the head-up display (HUD), a transparent display that presented data without requiring users to look away from their usual viewpoints. The Dassault Mercure was the first jetliner to demonstrate this technology in 1975; Sundstrand and Douglas would follow in the late 1970s. Alaska Airlines, which operated to remote airports in mountainous terrain and challenging weather in the Pacific Northwest, was a pioneer in adopting HUDs for its operations.

Aircraft relied on VHF radios for line-of-site communications (over or near ground) and HF for long-range and transoceanic flights. HF communication could sometimes be unreliable, and this weakness was addressed when satellite communications made its debut in the late 1980s. United Airlines was an early adopter to enable enhanced communications capability over the Pacific Ocean. Advances weren't limited to voice communications; in 1978 a datalink system called ACARS was introduced that enabled two-way transmission of operations, navigation, and maintenance information. This made it possible for passengers to know their connecting gate before they landed.

SUPPLIER CONSOLIDATION

As technological innovations gathered pace, consolidation of the fragmented avionics supply base also unfolded that would winnow a crowded avionics field to five major jetliner suppliers by 1990 (Fig. 9.5). In the United States, Allied Corporation purchased King Radio Company and Bendix Corporation in 1983. Two years later, it merged with Signal Corporation to create AlliedSignal. Signal's assets included Garrett Aviation, a major gas turbine and component original equipment manufacturer (OEM). By combining significant avionics capability with aircraft systems, AlliedSignal vaulted itself into the upper echelon of aircraft suppliers.

Honeywell also grew through acquisition when it bought Sperry Aerospace in 1986. The logic was clear according to Honeywell CEO Edson Spencer. "Combining Honeywell's very successful guidance and navigation systems for commercial and military aircraft with Sperry's highly regarded line of flight management systems and instrumentation will make Honeywell a major supplier for the cockpit of the future," he said [14]. The combination created more than broad capability; it created significant scale because the two firms had combined revenue of $2.6 billion and 30,000 employees.

The other big US avionics supplier, Collins Radio, grew throughout the jet age without the aid of major acquisitions through its strength in displays, navigation, and its bread and butter—communications. It fell into financial difficulty in 1971 and was purchased by North American Rockwell, whose name later changed to Rockwell International. Collins now had security and the financial backing of a successful prime contractor. In the late 1970s, it went on to win significant content in 757 and 767 common cockpits, and significant content on the 737. It would finish the decade by gaining a strong position on the 747-400, including the industry's first central maintenance computer. Collins would finish the 1980s as the largest jetliner avionics supplier with ~50% market share.

In Europe, Smiths Industries improved its position in 1987 when it acquired the avionics units of Lear Siegler, which included the avionics capabilities of Lear Developments, which Bill Lear sold to Siegler in 1962. Lear would go on to revolutionize business aviation with his eponymous Learjet. A crown jewel of the acquisition was the 737 flight management system.

Dramatic avionics consolidation took place in France in response to the changing landscape. In 1989 a new joint venture company, Sextant Avionique, was created by joining Thomson-CSF and Aerospatiale. This brought together four significant French avionics suppliers: Crouzet, Sfena, EAS, and the avionics division of Thomson-CSF. The new firm boasted nearly complete avionics capabilities and combined revenues of 5 billion francs ($830 million), ranking it as the largest European avionics supplier and the fourth largest globally [15]. Sextant's growth would be linked with Airbus. Beyond the aforementioned collaboration on the A320's flight control computer, Sextant provided approximately 70% of the cockpit for the aircraft.[4] This would translate into significant content on the A330 and A340 where Sextant would

[4]Perrot interview, 20 Sept. 2006.

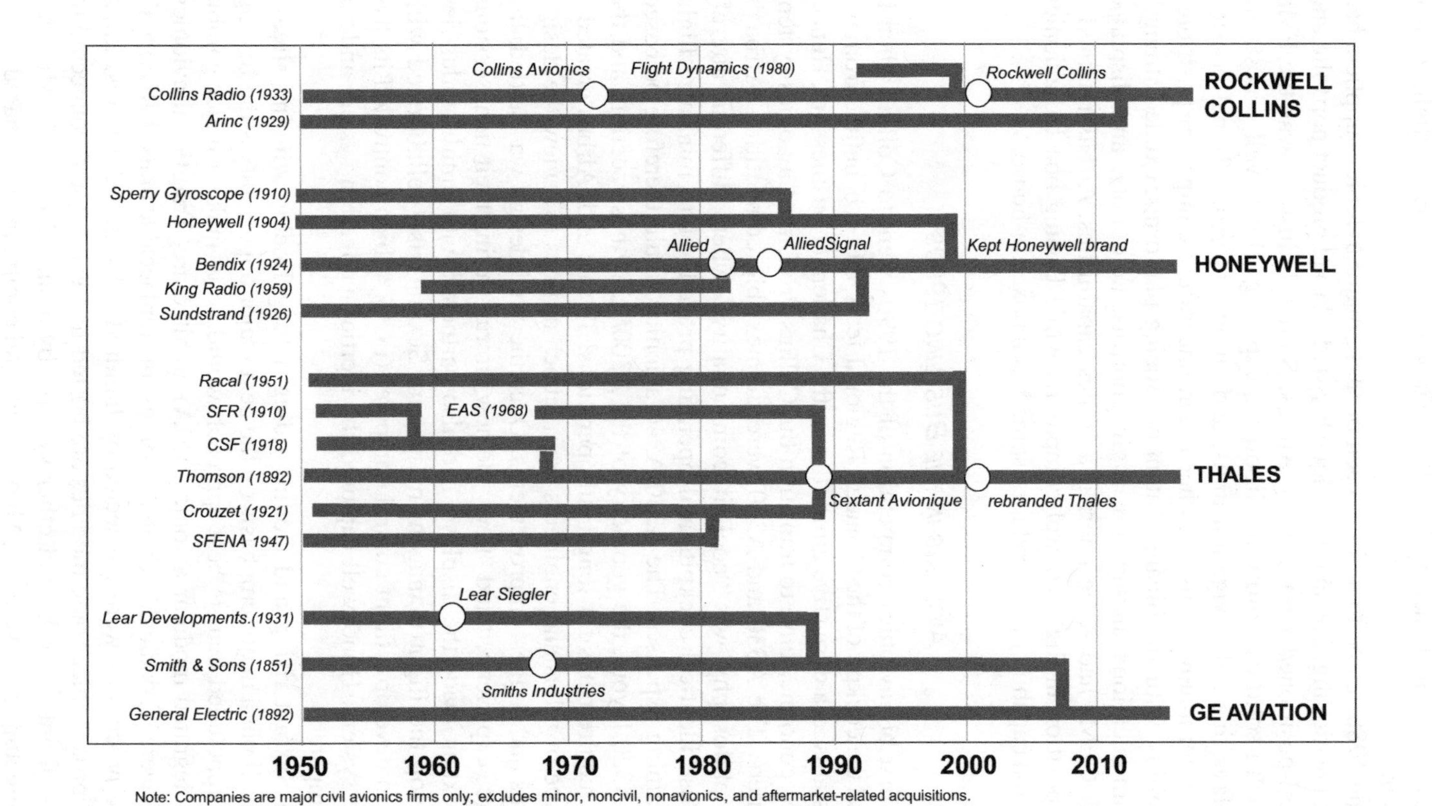

Fig. 9.5 Creation of major jetliner avionics suppliers.

win the displays and once again collaborate with Airbus on the flight control computer.

As the 1980s came to a close, each of these major avionics suppliers had distinct positioning. Sextant Avionique boasted a broad product portfolio and was well-positioned with a rising Airbus. Smiths Industries was the FMS leader with solid capability in airborne computing. Honeywell was strong in displays, inertial navigation units, and autopilots. AlliedSignal boasted excellent communications, weather radar, and safety equipment positions. And Collins, with significant content on Boeing platforms, exuded strength in communications, navigation, flight controls, autopilots, and displays. Aircraft OEMs had some internal avionics capabilities. Airbus retained a strong position in flight control computers, while Boeing boasted avionics integration capability and produced some black boxes in-house.

Avionics in the Bustling 1990s

The first big avionics competition of the 1990s went to Collins when it captured the cockpit of the Canadair Regional Jet (CRJ). It benefited from the CRJ's status as a derivative aircraft from the Challenger business jet. Hence, it made economic sense to transition the Collins Pro Line 4 avionics system to jetliners. The A330 and A340 were the next big competition. Airbus's design philosophy was "cockpit commonality," where different aircraft would use identical cockpits and operating procedures to minimize flight crew training expenses. The A330/A340 avionics would therefore be based on the A320 layout, the geometry of the A300/A310 nose section, and the unique requirements for long-haul operations. To this end, Airbus created a task force of pilots and engineers from launch airlines to review the design iterations and suggest improvements. Once the final design was settled, the task force experienced and "flew" the new aircraft during their final sessions [16]. Sextant made the displays, and also partnered with Smiths Industries to develop the flight management system. Sextant also collaborated with Airbus to develop a flight control computer (FCC), which controlled its fly-by-wire system. Honeywell captured the inertial navigation system and air data computer.

With Sextant rising and Collins riding a long Boeing winning streak, Honeywell put a major emphasis on the next major opportunity—the Boeing 777. It concluded that its best approach would be to make a major gambit on an integrated modular avionics (IMA) architecture. Traditional avionics systems were "federated," where numerous line replaceable units (LRUs) with separate processors were connected by hundreds of individual wires. In an IMA approach, electronic modules are hosted in a cabinet and connected to displays, flight controls, and other systems by a shared network. Modules were miniaturized, leveraging Moore's law, and computing was shared.

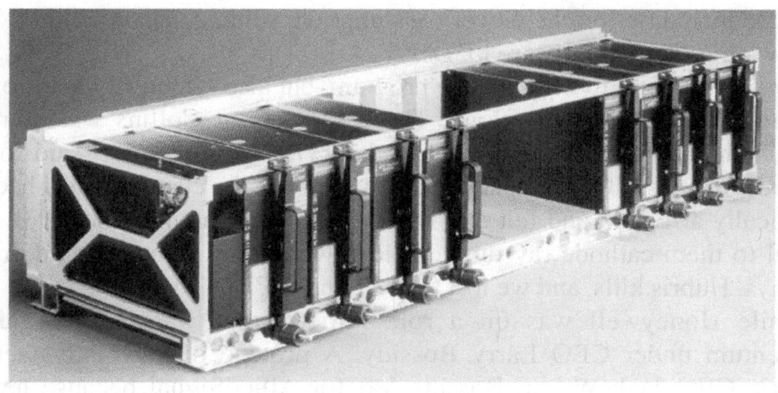

Credit: Honeywell

Fig. 9.6 Honeywell's AIMS.

The IMA concept wasn't completely new. It had been used on the US F-22 fighter, Boeing's C-130 Avionics Modernization Program, and some advanced business jets, but it had not been used for a major jetliner program—particularly a program as critical as the 777, which would enter service with a 180-min. Extended-range Twin-engine Operational Performance Standards (ETOPS) certification (a subject covered in Chapter 13).

Honeywell's proposal for the 777 was the airplane information management system (AIMS; see Fig. 9.6). Two cabinets would host the 777 modules, and the system would be knitted together with a shared ARINC 629 data bus with significantly greater capacity than the prevailing ARINC 429 twisted-wire standard. The ARINC 629 data bus was the airborne equivalent of a local area network—it provided much faster and higher capacity communications between aircraft systems, sensors, and the avionics cabinets. AIMS required a flawless introduction in an airplane loaded with new technology and a four-year development window. Joint Honeywell–Boeing teams worked around the clock in Phoenix and Seattle to define and develop AIMS hardware and software, which included 600,000 lines of code. Honeywell designed and built two full flight deck simulation facilities to allow integration and rigorous verification testing of the entire shipset of AIMS equipment.

The next big competition was the 737NG. Boeing chose to retain its "federated" architecture—avionics functions that are packaged as self-contained units—to avoid the huge nonrecurring engineering cost of a new avionics system. Honeywell won the displays and the air data computers, Smiths retained its position with the FMS, and Collins captured the autopilot and niche positions in radios and navigation sensors.

In 1994, Boeing announced another major Honeywell victory: It would provide active matrix liquid crystal displays (LCDs) for the 777 as well as the 737NG. LCDs offered several advantages over the CRT technology that they replaced including weight, power consumption, volume, and reliability. Soon all new aircraft would feature LCDs.

Honeywell not only won a huge program, but also effectively leapfrogged its competitors with technology. The leadership at Collins was stunned. "Collins had owned Boeing—we had the 737, 747, 767, 757," said former CEO Clay Jones. "The company had lost focus on the customer, had become technically arrogant, and felt that the business was an entitlement. We tried to sell to them cathode ray tube displays when they wanted liquid crystal displays. Hubris kills, and we lost the 777 and 737NG."[5]

While Honeywell was on a roll, AlliedSignal was quietly gaining momentum under CEO Larry Bossidy. A protégé of legendary General Electric CEO Jack Welch, Bossidy left for AlliedSignal because he was too close in age to be considered as Welch's successor. When he took over in 1991, he set about pulling AlliedSignal, then a motley collection of 52 businesses spanning aerospace, automotive, and specialty materials, from its corporate drift. One of Bossidy's key initiatives was implementation of total quality management, Six Sigma, and "black belt" manufacturing—well ahead of most other industrial firms and nearly everyone in the aerospace industry. Annual productivity improvements reached 6% per year in the mid-1990s, which led to a workforce reduction of 28%. By 1997, margins for the $14.5-billion company increased from 4.7% to 11.4%, and its stock price increased by 429% since he took over. Bossidy was named CEO of the year in 1998 by *Chief Executive Magazine*, beating luminaries such as Intel's Andy Grove and Dell Computer's Michael Dell [17].

AlliedSignal was more than lean; it was also driving avionics innovation through its acquisition of Sundstrand and its ground proximity warning system product line. In the early 1990s, Don Bateman (Fig. 9.7) concluded that he could improve the system that he had invented if he combined GPS and reliable terrain data for as much of the world's terrain as possible. AlliedSignal had good data for Europe and North America, but not the rest of the world. Bateman approached his manager with an interesting proposition: Why not fill the voids by buying detailed maps compiled by the Soviet-era military? "The Russians had compiled terrain information since the 1920s. They had the data and needed the cash after the end of the Cold War," recalled Bateman, "but we lacked a budget to purchase it, so we hid the expenditure in our test equipment budget."[6] This gave GPWS forward-looking capability and increased pilot warning time of controlled flight into terrain considerably. An *E* (for enhanced) was added to the GPWS, which became known as EGPWS. Several airlines, including United Airlines and British Airways, tested the new system, but the catalyst for

[5]Clay Jones (former CEO—Rockwell Collins), interview with author, 5 Nov. 2016.
[6]Bateman interview, 23 Aug. 2016.

Credit: Aviation Week & Space Technology Archive

Fig. 9.7 Don Bateman: inventor of GPWS and EGPWS.

widespread adoption sprang from a tragedy. In December 1995, 159 people died when an American Airlines 757 crashed into a mountain near Cali, Colombia. In response, American installed EGPWS on its entire fleet, and other airlines soon followed. The US FAA mandated that all production aircraft by 2000 include EGPWS and that in-service aircraft install it by 2003. Thus, Bateman's modest investment hidden in a test equipment budget would pay enormous benefits for his employer. AlliedSignal also enjoyed a strong position in TCAS, which eliminated mid-air collisions as it rolled out across the US fleet.

AlliedSignal had perfected a value-creation machine that combined productivity and innovation. What next? Its next move would forever change the avionics sector and the jetliner supply chain. After months of secret talks between Bossidy and Honeywell CEO Michael Bonsignore, the two firms announced their intent to merge in July 1999. This was a blockbuster deal, creating a firm with a combined $23.5 billion in revenue and $10.5 billion of aerospace business. The jetliner supply chain had never seen a Tier 1 merger of this magnitude, and the complementary nature of the two firms' avionics portfolios was clear: Honeywell, leading the charge with IMA, was the leading cockpit integrator; AlliedSignal was a powerhouse in EGPWS, TCAS, communications, and navigation.

Bossidy would become chairman of the new firm, while Bonsignore would become its new CEO. The Honeywell brand was retained for the entire firm, and the AlliedSignal brand would retire. The product overlap of the merger was minimal, but the US Justice Department asked Honeywell to divest its

TCAS 2000 product line. Defense electronics supplier L-3 Communications bought the product line and agreed to sell 30% to Thales. The firm would be called Aviation Communications & Surveillance Systems (ACSS). Although Thales owned just 30% of the equity, it was a major win because it filled a major gap in its product portfolio—TCAS.

Days after the megadeal was announced, Honeywell received more good news. Embraer had selected its Primus integrated modular avionics system for its ERJ-170/190 regional jet program. It would also be a risk-sharing partner. As the new avionics gorilla, Honeywell Aerospace ended the 1990s on roll. It had swept that decade's major competitions at Boeing and Embraer, its merger with AlliedSignal promised to unlock significant productivity gains, and it boasted unprecedented scale and breadth in civil avionics. What could go wrong? A lot, as it turned out.

Brussels Bombshell

If Larry Bossidy was CEO of the year in 1998, it would be fair to consider General Electric's leader Jack Welch as the CEO of the decade in the 1990s. Since taking the helm in 1981, Welch increased GE's market capitalization from $14 billion to more than $400 billion 20 years later, making GE the most valuable and largest company in the world.[7] In 1999, *Fortune* named him the Manager of the Century. He was more than the leader of one of the world's most admired companies; "Jack" was a capitalist rock star and global icon. In October 2000, while on the floor of the New York Stock Exchange, Welch was asked by a reporter his reaction to a proposed merger between United Technologies and Honeywell. He was floored—the news came as a complete surprise. GE had looked at Honeywell earlier in the year and concluded that although there was a solid fit, it was too expensive.

At the time Honeywell traded in the $50–60/share range. Now it was down to $36, and Larry Bossidy had announced his plans to retire. Welch and the GE leadership team quickly pulled an offer together and reached Honeywell Chairman and CEO Michael Bonsignore five minutes before the board was scheduled to vote on the UTC offer [18]. Honeywell dropped consideration of the UTC offer and agreed to a $44 billion offer from GE. This deal would require approval not only of the US Department of Justice, but also the European Commission (EC). The EC had approved the AlliedSignal–Honeywell merger in the prior year, but had scuttled major industrial mergers between telecommunications majors WorldCom and Sprint as well as the combination of media giants Time Warner and EMI. Both deals had significant product overlap. In contrast, there was little aerospace product

[7]"John F. Welch, Jr.," General Electric, https://www.ge.com/about-us/leadership/profiles/john-f-welch-jr [retrieved 27 Jan. 2018].

Credit: danacreilly

Credit: Hamlton83

Fig. 9.8 Mario Monti and Jack Welch.

overlap between GE and Honeywell. GE lacked a position in avionics. In aeroengines, GE was a leader in the large-thrust segment while Honeywell's strength was in business jet engines and APUs. Regulatory approval was an afterthought. "This is the cleanest deal you'll ever see," said Mr. Welch, who predicted closure of the deal by February 2001 [19].

In early 2001, the EC Competition Commissioner Mario Monti (Fig. 9.8) held several meetings with GE leadership, and it became apparent that there were concerns about the breadth of the new industrial combination. Could GE use its aircraft leasing business GECAS, which was at the time the world's largest lessor, to influence decisions on avionics selections—particularly given its role as a launch customer for some programs? Could the sheer size and breadth of GE–Honeywell hinder competition for avionics, aeroengines, and aircraft systems by creating new system bundles? Not surprisingly, Rolls-Royce, UTC, Thales, Collins, and other suppliers adamantly opposed the transaction and were pushing for outright rejection or significant divestitures—including, not surprisingly, attractive companies that they potentially could scoop up.

In May 2001, the US Justice Department weighed in. It approved the merger with the requirement that Honeywell sell its military helicopter engine business and open up its aftermarket for APUs and small engines. This was good news for GE—minor divestitures—and created hope that the EC would follow suit. Days later, the EC surprised GE when it issued a 155-page statement of objections. In June, GE offered $1.3 billion in divestitures including some avionics capabilities and later that month sweetened the offer to $1.9 billion. It soon received word that the EC demanded $5–6 billion in divestitures. GE made one final bid to complete the deal: It would also sell 19.9% of GECAS. On June 28, Commissioner Monti rejected GE's offer as insufficient. The world's largest industrial merger was finished [20]. This was a stunning development—not only for the avionics sector, but also for the global economy. *Time* magazine captured the zeitgeist following the deal's collapse:

> Welcome to globalization. The collapse of the GE-Honeywell merger shows that companies that benefit from a global market can now be governed in all they do by any of the countries or regions in which they do business. There's no settled code of rules in the global marketplace, just a haphazard collection of local practices and habits. Still, the GE case is extraordinary. Never before have officials outside the US nixed a merger between two giant American corporations already approved by the Department of Justice. Never before have US companies lobbied so ferociously against their US rivals in a foreign capital [21].

Momentum Shift

The reverberations from the Brussels surprise would continue at the Phoenix headquarters of Honeywell Aerospace. Anticipating approval of the deal, key personnel decisions were made and some GE executives moved into Honeywell offices. "We were basically operating as GE before the deal was approved," according to a former executive. "When the deal was scuttled, it had a huge impact on morale."[8] This fueled turnover of key executives. And on the heels of the AlliedSignal acquisition several years before, many of the leaders that had conceived and executed the integrated modular avionics strategy and had forged close relationships at Boeing were gone. There was more bad news that year when Honeywell paid Northrop Grumman (the new owner of Litton) $400 million to end an 11-year lawsuit on Ring Laser Gyro technology. Still, in 2002, it managed to win two work packages on the A380, including the flight management system and the aircraft environment surveillance system (AESS), which combined EGPWS, TCAS, Mode S, and weather radar into a single modular avionics suite.

Meanwhile, change was afoot in Cedar Rapids, Iowa, at the Rockwell Collins headquarters. Clay Jones (Fig. 9.9), an executive from the firm's DC office but technically an outsider, became the new leader of the air transport division in 1995. A former fighter pilot, Jones exuded confidence and was committed to regaining the leadership position in civil avionics that Collins had enjoyed in the 1980s. One of his first big initiatives was a commitment to enhanced productivity and Six Sigma principles; Collins called this *lean electronics*. Hundreds of lean events were launched around the company as Collins followed in the footsteps of its competitor AlliedSignal—but 5 to 10 years later. Jones also focused on breaking down barriers between

---

[8]Jeffrey Johnston (former executive—Honeywell), interview with author, 1 Feb. 2016.

Credit: Rockwell Collins

Fig. 9.9 Rockwell Collins CEO Clay Jones.

the air transport, business, and regional and government divisions in Cedar Rapids. C Avenue, the street separating the government systems team from the other divisions on the sprawling Cedar Rapids campus, was known as being "40 ft wide and 1 mile tall." The logic for tearing down the barriers was clear: with "clean sheet" avionics systems costing $400–500 million to develop, why not create an architecture that could be reused across multiple platforms and markets, saving time and money?[9] This meant technology sharing and common architectures. A division dedicated to services and the aftermarket was created, and R&D spending amped up. Collins would be spun off from Rockwell International in 2001 with Jones as the CEO. What the new company, called Rockwell Collins, needed now was a business development target. Enter the 787.

The 787 Dreamliner would be a program built on new technology, and the cockpit was no exception. But what exactly were the discriminators, and how could Collins break Honeywell's stranglehold on Boeing? The first step was to get close to the customer. A new cadre of leaders from the government sector, including Bob Chiusano, Kelly Ortberg, and Greg Irmen, reestablished the relationships at Boeing that had atrophied over the prior decade and held weekly 6 a.m. meetings to plot the 787 capture strategy.

[9]Kelly Ortberg (CEO—Collins), interview with author, 9 May 2016.

In parallel, Smiths Aerospace had developed an IMA system for the Boeing military C-130 AMP modernization program and adapted that system processor and remote data concentrators under IR&D to demonstrate the capability to host the first truly open system architecture. Smiths teamed with Rockwell Collins, who provided network architecture, to offer Boeing an IMA system like the 777's AIMS, but with a more flexible architecture that was lower cost and easier to update.

It wanted huge displays and distributed computing like the A380's common core system. And there were opportunities to create new products. "Honeywell's pressure on us forced us to innovate, which led to the funding and creation of new products. We knew we needed discriminators," recalls former Collins sales executive Randy Lincoln.[10]

Boeing partitioned the avionics system into seven major work packages. The headline package, the common core system (CCS), was awarded to Smiths Aerospace. The Smiths CCS open architecture approach allowed Boeing to procure more than 50 different software applications from more than 20 avionics and aircraft systems suppliers and integrate them in a common core resource cabinet of processors using a partitioned operating system. Another key package, displays, was awarded to Rockwell Collins. It included five 9-in. by 12-in. displays—double the display area of the 777—plus two head-up displays (Fig. 9.10). Boeing designed the 787 flight deck to be familiar to 777 pilots so that only five days of transition training would be required for crews to transition from one to the other. Collins also won the communications system and the integrated surveillance system (ISS). The ISS approach was similar to what was used in the A380 and combined terrain, traffic, and weather radar

Credit: Rockwell Collins

Fig. 9.10 787 cockpit with Collins displays.

[10]Randy Lincoln (former VP-Sales—Rockwell Collins), interview with author, 9 Oct. 2015.

functionality into a single cabinet. This saved 60% in size, 36% in weight, and 26% in power consumption. A key element of the system, the Collins multiscan weather radar, could automatically "hunt" for storms and turbulence up to 320 nm away, saving the crew from the tedious task of manually altering the tilt angle of the radar [22]. Additionally, Collins landed a bundled communications package for all radios, control panels, antennas, and audio systems. This was a significant win because these were traditionally buyer-furnished equipment where airlines could choose from several sources. Now they were in a single, integrated system. Collins also captured pilot controls and the AFDX communications network—a key element of the common core system. It was Collins's largest content ever on a jetliner.

Meanwhile, Honeywell won two packages—a far cry from its major win on the 777. The first, the inertial reference suite (IRS), combined its IMU strength with VOR, ILS, GPS, and other sensors. It also won the crew information system/maintenance system (CIS/MS), which gave the crew a constant update on the health of major aircraft systems. It was a great comeback for Rockwell Collins, a strong showing for Smiths Aerospace, and a humbling defeat for Honeywell. "We were full of ourselves," recalled a former Honeywell executive, "we thought we would win five of the work packages."[11]

What was notable about the 787 was the remarkable degree of avionics integration compared to past federated architectures (Fig. 9.11). Smiths's common core system consisting of more than 50 line-replaceable units (LRUs) provided the brains and central nervous system for the aircraft, a Collins integrated surveillance system included TCAS, TAWS, weather radar, and Mode S transponder. Honeywell's integrated navigation package included FMS, inertial reference, air data, automatic heading and reference system, and the entire host of navigation receivers including GPS. Like the A380, Boeing knitted the entire avionics suite together with a high-speed Ethernet based on the ARINC 664 standard as part of the common core system.

An important implication of the 787's integrated architecture was that its avionics were 100% supplier-furnished equipment (SFE). Boeing, rather than airlines, decided the avionics suppliers. An entire 787 shipset was worth approximately $2.5 million per aircraft. In contrast, the 737's federated architecture was worth approximately $1.5 million per aircraft.[12]

Collins would follow the 787 by winning the cockpits for the Mitsubishi Regional Jet (2007) and Bombardier C Series (2008) with its Pro Line Fusion avionics system. It also won the majority of the avionics content on the 747-800. It then displaced Honeywell displays again in winning significant content on the 737MAX and 777-X. "The 787 really laid the groundwork for future success;

[11]James Cudd (former VP-Strategy—Honeywell Aerospace), interview with author, 4 March 2016.
[12]Randy Lincoln (former VP-Sales—Collins), interview with author, 29 Sept. 2016.

737 (federated)	787 (integrated)

Buyer-Furnished Equipment (~35% value)
• VHF radio
• HF radio
• Satellite communications
• Multimode Receiver (VOR, ILS, ADF, DME, GPS)
• Inertial navigation system
• Flight data recorder
• Cockpit voice recorder
• Health monitoring recorder
• Weather radar
• Data loaders
• Air data computer
• Emergency locator transmitter
• Enhanced ground proximity warning system
• Traffic collision avoidance system
 	
Supplier-Furnished Equipment (~65% value)
• Cockpit displays
• Flight controls
• Flight management system
• Panels and switches
	Supplier-Furnished Equipment (100% value)
	• Cockpit displays
	• Head-up displays
	• Common computing resource (Common Core)
	• Common data network (Common Core)
	• Flight control computer & controls
	• Integrated surveillance system
	  • Traffic collision avoidance system
	  • Mode S transponder
	  • Terrain awareness warning system
	  • Weather radar
	• Crew information/maintenance system
	• Integrated navigation system
	  • Flight management system
	  • Inertial reference system
	  • Air data system
	  • Integrated navigation receivers (ILS, market beacons, VOR, GPS and GLS, ADF, DME, ELT)
	  • AHRS
	• Panels and switches
	• Data loaders
	• Flight data recorder
	• Cockpit voice recorder

Connected by individual wire pairs (ARINC 429)	*Connected by common data network (ARINC 664)*
Shipset value ~ $1.5 million	**Shipset value ~ $2.5 million**

Source: Author's analysis

Fig. 9.11 Avionics comparison: 787 vs 737.

it led to the MAX and the trifecta was the 777-X," said Kelly Ortberg.[13] Collins also won significant content on the A350 XWB, including communications, navigation equipment, and information management systems.

Beyond these programs, the vision of sharing avionics architectures came to fruition as the Pro Line Fusion integrated avionics system was selected for more than 15 platforms including turboprops, business jets, regional jets, helicopters, and tanker aircraft. Rockwell Collins attracted notice as one of the best managed aerospace firms. Much of the credit, according to former Boeing CEO Jim Albaugh, rested with Clay Jones. "I can make the case that over the last decade [prior to 2015] he did as good of a job as any other executive in the business in transforming his company and getting close to customers," he said.[14] Jones would retire as CEO in 2013 and hand over the reins to seasoned executive Kelly Ortberg.

One of Ortberg's first important acts was to acquire ARINC, a communications services supplier best known for its VHF air-ground data services. It was a bold "beyond the box" bet by the new CEO, and Collins's largest acquisition to date. Rather than depending solely on new equipment and aftermarket sales, Collins would tap into the fast-growing communications and information services markets. "There is no doubt the digital information

[13]Ortberg interview, 9 May 2016.
[14]Jim Albaugh (former CEO—Boeing), interview with author, 21 Sept. 2015.

exchange will continue to expand at a rapid rate," he said. Collins instantly became one of the airline industry's largest communications services suppliers. Ortberg turned to Collins executive Jeff Standerski to run the new business. "There was tremendous excitement around both companies," said Standerski. "The cultures matched, and we recognized the potential to create a new type of company with information management and communications solutions."[15]

Thus, Collins completed a 30-year round trip that saw it rise, fall, and regain its footing. How did a world leader in one of the most dynamic aerospace product segments flourish in Iowa—in the middle American heartland and off the beaten track? Here, the focus of having 10,000 employees in one somewhat isolated location made a positive difference. This facilitated architecture reuse as engineers and leaders moved back and forth between the civil and military divisions—and across C Avenue. Not to be overlooked were the soft factors: "The people and culture underpin creativity—not the location," said former executive Bob Chiusano.[16]

Thales Rising

Collins wasn't the only avionics OEM on the rise. Thales—the new name of Sextant Avionique after it rebranded in 2001—was also on a roll. It gained focus after jettisoning its Crouzet automation and control component company in 2000 to focus on avionics, cabin electronics, defense electronics, and air traffic management. Its beachhead in the jetliner business was Airbus. Riding the success of its strong positioning on the A320, A330, and A340, it would add important capability to its portfolio. The first was TCAS. In 2000, Thales became a joint venture partner with L3 Communications in Aviation Communication & Surveillance Systems—a company created when the US Department of Justice authorities forced Honeywell to divest its TCAS capability when it merged with AlliedSignal. Thales also developed flight management system capability by collaborating with Smiths Industries on multiple Airbus commercial transports in the mid-1990s, positioning it to compete with Honeywell. Thales developed its own FMS on the Airbus A400-M, a military transport aircraft, albeit with key challenges encountered during its development.

Its expanded avionics capability and close relationship with Airbus converged on the A380, when Thales collaborated with the aircraft OEM to create its first integrated modular avionics architecture. The system featured interactivity, which allowed pilots to use a cursor to input information or control flight. It also included Avionics Full-Duplex Switched Ethernet

[15]Jeff Standerski (former SVP, Information Management Services—Collins), interview with author, 11 Nov. 2017.

[16]Bob Chiusano (former COO—Collins), interview with author, 15 Oct. 2015.

(AFDX), supplied by Rockwell Collins, which preceded the aforementioned Common Core architecture used by Boeing on the 787. Thales also captured the displays and other important components on the A380.

Thales would go on to win the cockpit for the Sukhoi Superjet in 2003. Its role on the Russian program was substantial. It was the avionics systems integrator and produced the full avionics suite including displays, integrated modular avionics, and communication, navigation, and surveillance systems.

With two major integrated modular avionics under its belt, Thales won much of the 350 XWB cockpit in 2008 (Fig. 9.12). The new aircraft built upon the integrated modular avionics concept developed for the A380, and would also manage key systems on the aircraft including landing gear, fuel, brakes, oxygen, cabin pressurization, and pneumatics. The Thales IMA architecture replaced multiple separate processors and LRUs with around 50% fewer standard computer modules, known as line-replaceable modules. Like the A380, it ran on a 100-Mbit/s network based on the AFDX standard.[17]

Thales was the first company to harmonize and host an electronics flight bag—an electronic information management device that helped flight crews

Credit: Airbus

Fig. 9.12 The A350 XWB cockpit.

[17]David Learmont, "A350 Avionics to Expand on A380 Systems," FlightGlobal.com, 7 July 2007, https://www.flightglobal.com/news/articles/a350-avionics-to-expand-on-a380-systems-215493. http://www.flightglobal.com/articles/2007/07/24/215493/a350-avionics-to-expand-on-a380-systems.html [retrieved 27 Jan. 2018].

perform flight management tasks more easily and efficiently with less paper. Any of its six identical displays could display electronic flight bag information. Thales also supplied the head-up display, which was a customer option on the aircraft. And importantly, it captured the ADIRU—a subsystem that combined air data and inertial reference functionality. In all, it won 15 systems and had a significantly higher shipset value on the A350 XWB than it had on the A330. Still, it struggled with meeting its cost estimates for the program, and it suboptimized in sharing architectures between platforms—the same problem Collins faced before Clay Jones's arrival. "There was a policy to tailor the product to the needs of a specific aircraft," admitted Michel Mathieu, head of the Thales avionics division [23]. The value of the A350 XWB to Thales was $2.9 billion over 20 years [24].

As Thales grew and developed a complete product line, it expanded its global footprint to including major facilities in the United States and Asia. It moved into inflight entertainment, which is discussed in Chapter 11. In the space of two decades, Thales unified a fragmented French avionics sector and made jetliner competition a three-horse race with Rockwell Collins and Honeywell.

GE's Return to Avionics

It took several years for the dust to settle following the Honeywell debacle, but GE did not lose its appetite to be a player in avionics. It would need to enter the market through acquisition. Honeywell was clearly off the table, Thales wasn't politically feasible, and Rockwell Collins was too expensive. However, the #4 player, UK-based Smiths Aerospace, became available after the parent company shifted corporate strategy to deemphasize aerospace. GE acquired the Smiths aerospace business in 2007 for $4.8 billion.

The acquisition broadened GE's offerings by adding Smiths's flight management systems, electrical power management, mechanical actuation systems, and airborne platform computing systems to GE Aviation's aeroengine capabilities. The combined GE Aviation and Smiths Aerospace revenues in 2006 were $15.6 billion.

In avionics, Smiths had narrow capabilities—it was known for its FMS capability and its significant win of the common core system that was in development for the 787—but GE ownership enabled it to grow. In 2010, it won the avionics core processing system, display system, and on-board maintenance system for the newly launched C919 in partnership with AVIC Systems. The C919 was a GE corporate priority embraced by CEO Jeffrey Immelt to strengthen its position in China.

In business aviation, GE won another important competition on the Gulfstream G500/G600 with its Data Concentration Network, taking aspects of its 787 common core system and modifying them for Gulfstream. GE captured this work from what many at Honeywell had expected would be parts of its Primus Epic system. Although GE Aviation did not have the scale of the "big three" of Honeywell, Collins, and Thales, it served notice that it would invest in, and try to grow, its avionics position.

GE would later win the common core system for the 777-X. Due to the success Boeing had on the 787 in integrating the many software applications into GE's open-architecture CCS at a lost cost, it decided to forego the Honeywell AIMS architecture used on the legacy 777.

BRAINIACS

By 2017, the value of jetliner avionics production had reached \$3.7 billion, and supplier market shares shifted as a result of important program decisions for the 787 and A350 XWB. After losing share throughout the 1990s, Rockwell Collins reassumed market leadership with a 35% share, followed by Honeywell (28%) and Thales (22%). The remaining 15% was composed of GE Aviation, BAE Systems, ACSS, Esterline, L3 Communications, and Teledyne. (See Fig. 9.13.) Avionics wasn't just a large business—it was lucrative, with leading suppliers regularly posting pretax margins of 15–20%.

In 2017, Boeing surprised many when it announced that it was setting up an avionics center of excellence and pulled some of the lucrative avionics work back in-house. The extent of Boeing's ambitions wasn't fully understood, but its announcement may have contributed to a second transformational event. In September 2017, United Technologies agreed to acquire Rockwell Collins for \$30 billion.

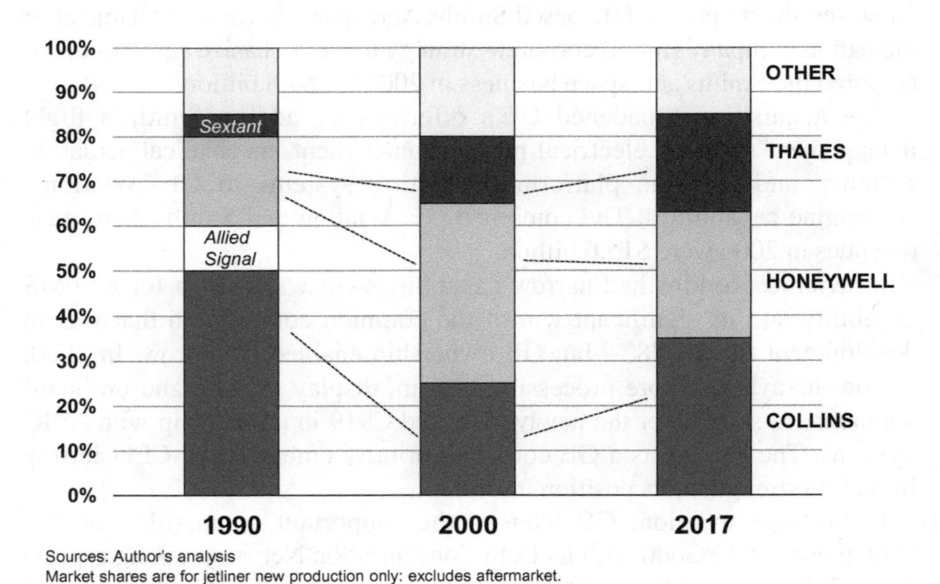

Fig. 9.13 The evolution of air transport avionics market shares.

From its humble beginnings in the early 20th century, the fragmented avionics sector consolidated with three highly capable, global suppliers and a fourth (GE Aviation) knocking on the door. And their impact on the safety of air travel was phenomenal. Consider this: By 2015, there were fewer than 2 passenger deaths for every 100 million passengers on commercial flights. By comparison, from 1962 to 1971 there were 133 passenger deaths out of every 100 million passengers. The odds of dying in a crash aboard an airplane in the United States or the European Union was just 1 in 29 million; the odds of being killed by lightning are three times greater [25]. This step-change in safety is one of the most important factors underpinning the massive growth in air travel during the jet age. A key enabler of this shift was avionics innovation. Indeed, avionics suppliers deserve to be known as the brainiacs of the jetliner ecosystem.

REFERENCES

[1] McGough, M., "In Memoriam: Philip J. Klass: A UFO (Ufologist Friend's Obituary)," *Skeptic*, 26 Aug. 2005, https://www.skeptic.com/eskeptic/05-08-26. [retrieved 28 July 2016].

[2] Pujes, J.-P., *A Century of Electronics*, Imprimerie Chauveau, Chartres, France, 2006, p. 21.

[3] Rashke, R., *Stormy Genius: The Life of Aviation's Maverick Bill Lear*, Houghton Mifflin, Boston, 1985, pp. 73–74.

[4] Bradband, K, and Smith, B., *Rockwell Collins: 75 Years of Innovation*, WDG Communications, Cedar Rapids, Iowa, 2010.

[5] Ibid., pp. 24, 31.

[6] Mansen, M., "Airborne Weather Radar," *Avionics News*, March 2011, pp. 100–104.

[7] Klass, P., "MIT Sparks Inertial Guidance Efforts," *Aviation Week & Space Technology*, 10 Aug. 1951, p. 71.

[8] Sweetman, B., "Aviation's Top Ten," *Aviation Week & Space Technology*, 18–31 Jan. 2016, p. 47.

[9] Gates, D., "Redmond Aviation Engineer's Lifelong Work Saved Thousands of Lives," *Seattle Times*, 4 Feb. 2012, https://www.seattletimes.com/business/redmond-aviation-engineers-lifelong-work-has-saved-thousands-of-lives. [retrieved 4 August 2016].

[10] "Race Ends in a Tie," *Aviation Week & Space Technology*, 31 May 1954, p. 54.

[11] Langewiesche, W., *Fly By Wire*, Picador, New York, 2009, pp. 102–103.

[12] Alcock, C., "Honeywell Centenary Highlights Technology-Rich Heritage," AIN Online, 21 May 2014, https://www.ainonline.com/aviation-news/aerospace/2014-05-21/honeywell-centenary-highlights-technology-rich-heritage [retrieved 27 Jan. 2018].

[13] Nordwall, B., "Broadening Base for Avionics," *Aviation Week & Space Technology*, 15 March 1993, p. 102.

[14] Hiltzik, J., "Honeywell Pays Unisys $1 Billion for Sperry Unit," *Los Angeles Times*, 15 Nov. 1986, http://articles.latimes.com/1986-11-15/business/fi-3495_1_honeywell-chairman. [retrieved 4 August 2016].

[15] Lenorovitz, J., "France Prepares to Merge Four State-Controlled Avionics Firms," *Aviation Week & Space Technology*, 30 Jan. 1989, p. 78.

[16] Potocki de Montalk, J. P., "New Avionics Systems—the A330/A340," *The Avionics Handbook*, edited by Cary Spitzer, CRC Press, Boca Raton, FL, 2001, pp. 30-2–30-3.

[17] "CEO of the Year 1998," *Chief Executive*, 10 July 1998, https://chiefexecutive.net/ceo-of-the-year-1998__trashed. [retrieved 31 July 2016].

[18] Welch, J., *Jack: Straight from the Gut*, Warner, New York, 2001, pp. 356–357, 360.
[19] Ibid., p. 363.
[20] Ibid., pp. 366–373.
[21] Elliot, M., "The Anatomy of the GE-Honeywell Disaster," *Time*, 8 July 2001, http://content.time.com/time/business/article/0,8599,166732,00.html. [retrieved 31 July 2016].
[22] Norris, G., Thomas, G., Wagner, M., and Smith, C. F., *Boeing 787 Dreamliner—Flying Redefined*, Aerospace Technical Publications International, Perth, Australia, 2005, p. 136.
[23] Taverna, M., "Prime Time," *Aviation Week & Space Technology*, 25 Oct. 2010, p. 54.
[24] Kaminski-Morrow, D., "Airbus Selects Thales for A350 XWB Cockpit Avionics," *Flight International*, 21 Jan. 2008, https://www.flightglobal.com/news/articles/airbus-selects-thales-for-a350-xwb-cockpit-avionics-220969. [retrieved 3 October 2016].
[25] Allianz Global Corporate & Specialty and Embry Riddle Aeronautical University, "Global Aviation Safety Report," December 2014, p. 4.

Chapter 10

SKIN AND BONES

Competition in the jetliner ecosystem takes place between firms, leaders, and programs. It also takes place between materials.

Aerospace has always been at the vanguard of advanced materials and structures. The demand for military aircraft in World War I was a catalyst for the development of the aluminum industry. The development of the SR-71 Blackbird by the Lockheed Martin Skunkworks helped to create the modern titanium industry. Aircraft were one of the first applications for composites. And then there is the incredible demands of aeroengines, which fostered the development of superalloys that operate at temperatures as high as 2800°F.

The selection of materials is particularly important to jetliners, where weight savings yield significant cost reduction to operators or can extend the aircraft's range. There are also other selection criteria including strength, impact resistance, and corrosion resistance. Former Boeing Commercial Airplanes CEO Alan Mulally explained:

> When it comes to weight, the materials that we select are really important. The requirements on the materials vary depending on where they are used. In some cases, the material properties are important for fatigue where the airplane structure moves, like in the cabin where we pressurize it to go up in flight and we depressurize it to land, so the structure is always moving. Whereas in the bottom of the airplane corrosion is important where water can get there of condensation. So, we're always looking for materials that satisfy many requirements but in the most weight-efficient manner [1].

Initially, aircraft original equipment manufacturers (OEMs) manufactured most of the aircraft's structure in-house with limited outsourcing to small "mom & pop" suppliers. In time, large, sophisticated aerostructures suppliers emerged with advanced manufacturing processes and scale.

As operating and manufacturing economics assumed growing importance in the jetliner business, decisions regarding the "skin and bones" of the aircraft assumed growing importance. It became clear that the interplay of materials, manufacturing processes, and suppliers could make or break an aircraft program.

How did the use of jetliner materials evolve, and who are the major suppliers of materials and structures?

Jetliner Materials: The Early Years

Although wood was the primary aircraft material until the 1930s, new lightweight alloys were developed just years after the Wright brothers' first flight. In 1909, Germany developed an early aircraft aluminum alloy, called *duralumin*, that contained aluminum, copper, magnesium, and manganese. The first mass-production aircraft to make extensive use of duralumin was the Junkers J.I observation aircraft in World War I. Around the same time, the Aluminum Company of America (Alcoa) introduced its first aerospace alloy dedicated to aircraft structures. By the end of the war, the Germans rolled out the Zeppelin D.I, the first production all-metal aircraft using a *monocoque* design, where the aircraft's skin carried loads in addition to its structures. In the 1920s, steel tubes gained popularity in aerostructures, although a leading publication indicated that the United Kingdom was the only nation that had seriously tackled the problem of using steel as a substitute for spruce [2]. Aluminum monoplanes eventually emerged, and by the 1930s sleek all-aluminum transports like the DC-3 had entered service.

World War II brought step-function increases in aircraft raw material demand—particularly for aluminum. From 1942 to 1945, Alcoa built and operated numerous plants for the US government including 8 smelters, 11 fabricating plants, and 4 refineries. It isn't surprising that in 1942 German saboteurs landed on Long Island, New York, and Jacksonville, Florida, on missions to destroy Alcoa plants. They were unsuccessful.[1] Alcoa's largest aluminum rolling mill, one of the largest buildings in the United States, was set up far away from the coasts in Davenport, Iowa (Fig. 10.1). Meanwhile in the Soviet Union, the State Defense Committee evacuated leading aluminum producer Verkhnaya Salda Metallurgical Production Association (VSMPO) to east of the Ural Mountains. Aluminum had arrived as a strategic material.

Nearly 300,000 US military aircraft were produced during World War II, which led to a capacity hangover in peacetime. The US government forced Alcoa to sell numerous facilities and license its technology. This opened the

---

[1]"Alcoa's 125 Years," Alcoa Incorporated, https://www.alcoa.com/usa/en/alcoa_usa/history.asp [retrieved 5 Dec. 2015].

Credit: Arconic

Fig. 10.1 Alcoa Davenport Works sprawls over 130 acres; the company was rebranded Arconic in 2015.

door for US business magnate Henry Kaiser to enter the business; he founded Kaiser Aluminum in 1946.

The jet age spurred the use of another breakthrough material: titanium. With the strength of steel, yet half the weight and excellent corrosion resistance, titanium was ideal for aircraft design. The advent of the Kroll process in 1946, a technique that enabled titanium production in large quantities and at lower cost, set the stage for more aerospace usage. Like aluminum, the Soviet and US governments viewed it as "strategic" and actively supported its development. Eventually, a private sector developed. Titanium Metals Corporation of America (TIMET) and RTI International Metals were founded in 1950 and 1951, respectively. By 1952, TIMET had produced the world's first titanium ingots.[2]

In 1952, the Douglas X-3 Stiletto became the first aircraft to use titanium for structural components. Meanwhile, Soviet aluminum producer VSMPO began using titanium and produced its first large ingots in 1957 for use in aircraft and submarines. Ironically, one of the major incidents of the Cold War—the 1960 downing of an American U-2 spy plane over Soviet airspace—would add impetus to titanium's ascent. Avoiding Soviet surface-to-air missiles meant that the United States needed to design an aircraft that flew exceptionally high and fast, and the job fell to the legendary aircraft designer Clarence "Kelly" Johnson and the Lockheed Skunkworks. Johnson led the development of a new Mach 3 reconnaissance aircraft called the A-12. A vexing engineering challenge was how to design aerostructures that could

[2]"Timeline of Milestones," Timet, http://www.timet.com/about-timet/timeline [retrieved 5 Dec. 2015].

withstand the ram-air temperatures of 800°F over the body of the aircraft. This was above aluminum's temperature range, and left stainless steel alloys and titanium as candidate materials.

Kelly made a bold bet on titanium—a material that lacked a basic supply chain [3]. This required high levels of creativity and problem solving. The first issue was lack of mill capacity. The US Central Intelligence Agency conducted a worldwide search through third parties and managed to purchase the base metal from, ironically, the Soviet Union. With supply secured, the next issue was how to make advanced titanium structures. Lockheed had no idea how to extrude it, push it through various shapes, weld it, or rivet it. Drilling bits used for aluminum simply broke into pieces when used on titanium [4]. Lockheed powered through the challenges and transitioned the A-12 design to the iconic SR-71 Blackbird in 1964. The aerospace titanium supply chain had arrived (Fig. 10.2).

The emergence of jetliners drove the development of other materials. Unlike military aircraft, which favored performance over cost, jetliners required competitive costs as well as durability and corrosion resistance. Early jetliners, such as the Comet and 707, were made primarily of aluminum with steel used for components requiring high strength including landing gear and jet engine disks. The Comet's aluminum fatigue issues, outlined in Chapter 2, led to more robust and durable aerostructure designs and materials. One byproduct of the emergence of jetliners was spot shortages of aluminum in the mid-1950s as suppliers struggled with growing jetliner demand and

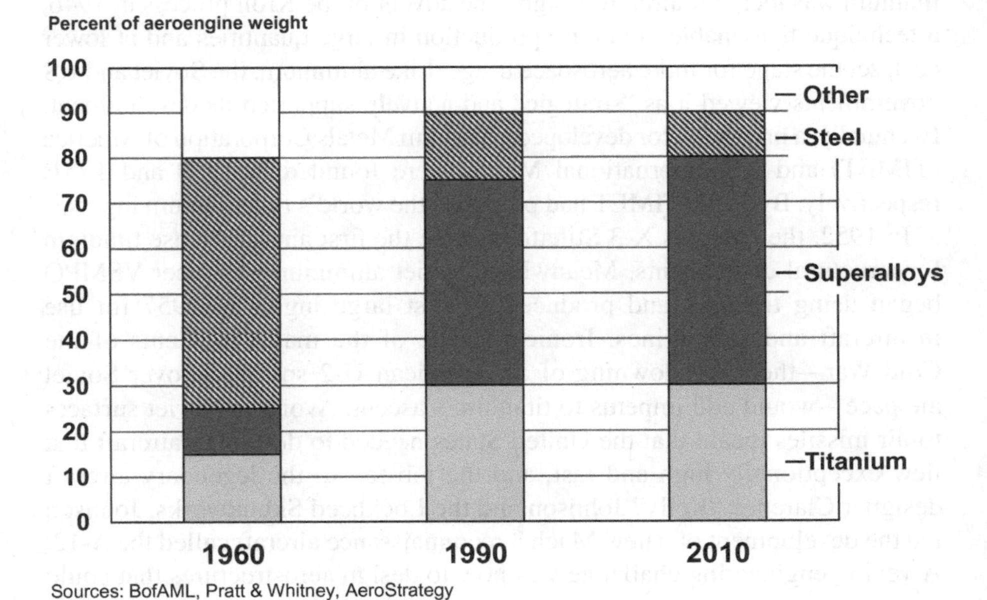

Fig. 10.2 Evolution of aeroengine materials.

surging Cold War military spending. US aluminum demand was projected to double by 1965, which led Kaiser to break ground on an enormous new aluminum mill in Ravenswood, West Virginia, and an alumina (the feedstock for aluminum metals) facility in Louisiana for a staggering $280 million [5]. Kaiser's projections proved to be correct as aircraft production grew in the 1960s. The 727, DC-9, and other popular jetliners were mostly aluminum.

Aeroengine development also spurred new materials. A major design challenge for early gas turbines was to create materials that could stand up to temperatures higher than 1200°F. In the United Kingdom, Wiggin Alloys invented the Nimonic family of alloys (nickel-chromium) for turbine blades in Whittle jet engines. In Germany, a high temperature alloy called Tinidur was developed to support its jet engine programs. Major advances in high-temperature materials, dubbed *superalloys*, would follow. These comprised mixtures of nickel, iron, or cobalt and were used in applications involving the hottest temperatures in the gas turbine, including turbine blades and vanes, discs, and combustion chamber components.

Processing technology also improved. A key breakthrough was application of vacuum melting to superalloy production in the 1950s. While vacuum melting had been utilized for high performance steels for four decades, it was ideally suited for the demanding requirements of superalloys. "Vacuum melting enabled alloys with tightly controlled chemistries free of non-metallic impurities," said raw materials executive Art Kracke." This breakthrough, combined with improved alloy compositions and more demanding jet engine requirements drove superalloy performance to another level."[3] One of the pioneers of vacuum induction melting (VIM) was James Nisbet, a former metallurgist at General Electric. In 1957, he formed Allvac Metals Company (short for "All Vacuum Melted.") to focus on the opportunity. His new company enjoyed early success and in would soon offer double VIM processing as a standard for superalloy production. Allvac's primary competitor was Special Metals, and the two companies would push each other to higher levels of performance. By the early 1960s, VIM became the standard for superalloy production, and also began to influence the aerospace specialty steel segment when major suppliers, including Firth Sterling and Carpenter Steel, installed their own VIM furnaces.

As aeroengine designs became more sophisticated, engineers struggled with how to make large, relatively high-bypass-ratio turbofans like the CF6 and JT9D without adding too much weight to the aircraft. The large "cold section" of these engine designs was a challenge. The fan section, for example, was more than 20% of an engine's weight. Enter titanium, and its superior strength-to-weight ratio compared to steel and aluminum. Titanium was applied to fan blades, engine cases, and other components. Aeroengine OEMs

[3]Art Kracke (former Vice President—R&D & Business Development, ATI Specialty Materials), interview with author, 29 May 2018.

set up their own materials research and development (R&D) organizations to push the state of the art in materials and manufacturing processes.

Turbine blades, which experienced the hottest temperatures, were also a focus area. Higher-temperature superalloys, new design approaches (e.g., internal air cooling passages), and sophisticated manufacturing processes such as directionally solidified casting enabled higher turbine inlet temperatures. The JT9D, for example, had a turbine inlet temperature that was 400°F higher than the JT3D [6]. By the end of the 1960s, aeroengines were composed of near equal portions of titanium, superalloys, and steel.

In the first decades of the jet age, aircraft and aeroengine OEMs were vertically integrated. In-house production of aerostructures, aircraft systems, parts, and interiors fed their final assembly lines. Third-party suppliers were small and fragmented.

Enter Composites

The 1970s brought significant changes including higher fuel prices and massive new twin-aisle aircraft. Yet airframe materials remained metallic. The structure of the 747, for example, was more than 80% aluminum [7].

At the time, Alcoa was the world's largest aerospace aluminum supplier. Other major suppliers included Kaiser and Reynolds in the United States, Pechiney in France, and Alcan Booth in the United Kingdom [8]. Titanium gained popularity, but supply didn't keep pace. The Soviet Union, boasting significant mill capacity to meet Cold War demand, closed the gap. Its titanium hull submarines used 3 million lb of titanium and, despite its strategic significance, the USSR exported significant quantities of titanium to the West [9].

Advances in composites technology, which combined glass, carbon, or boron fibers with epoxy resin, captured the attention of aircraft designers with their weight savings potential. Composites were stronger than steel, yet lighter than aluminum. At first, military aircraft began to experiment with these new materials, and soon jetliner OEMs followed suit. NASA funded a study with Lockheed Martin to integrate carbon fiber composites in the L-1011's vertical stabilizer. One study estimated that every 100 lb in weight removed from a twin-aisle could result in 14,000 lb in annual fuel savings [10]. Soon, carbon fiber reinforced plastics (CFRPs) became the combination of choice for aerospace designers.

Ultimately, Airbus pioneered composites on the A300's secondary structures such as spoilers, airbrakes, and rudder in 1983. Two years later, Airbus went one step further in integrating an all-composite fin on the A310-300 (Fig. 10.3). This was the first use of composites on a commercial jetliner's primary structure. Airbus would use even more composites in the empennage of the A320 later in the decade, driving penetration to 10% of its structural

Credit: RHL Images from England

Fig. 10.3 The A310-300 was the first jetliner with composite primary structures.

weight—a record for jetliners. Still, the unproven A320 was an exception relative to competing aircraft introduced in the 1980s. The MD-80, 737, and 747-400 were composed of 1–2% composites.

Military aircraft designed for the Cold War were even more aggressive in adopting composites. In the early 1980s, the McDonnell Douglas AV-8B Harrier became the first combat aircraft to integrate significant use of composites, which comprised about one-quarter of the aircraft's structural weight. The Northrop Grumman B-2 Spirit stealth bomber was built primarily from carbon fiber composites, but also took advantage of the material's radar-absorbing properties to improve its stealth qualities. Still, these early composites were expensive. The manufacturing cost of composite panel-like structures was $400–$700 per lb in the mid-1980s.[4]

Aerostructures in the 1980s were mostly produced in-house by aircraft OEMs, but new patterns of partnering and outsourcing emerged. Boeing built much of the 767's aerostructures in its own facilities but turned to a consortium of five Japanese companies for a 15% workshare, including fuselage panels, wing ribs, and wing/fuselage fairings. The consortium, which became known as Japan Aircraft Development Corporation, included industrial heavyweights Fuji Heavy Industries, Kawasaki Heavy Industries, Mitsubishi Heavy Industries, Nippi, and ShinMaywa Industries. Sourcing from Japanese suppliers not only reduced Boeing's risk, but also ensured access to the lucrative Japanese market and close relations with Japan Airlines and All Nippon Airways. Boeing sourced wing and tail control surfaces from Aeritalia in Italy.

[4]Chris Red (Managing Director—Composite Market Reports), interview with author, 25 March 2016.

Airbus's four equity partners produced most of its aerostructures, but it did engage in targeted outsourcing. One supplier was Fisher Advanced Composites Corporation (FACC), a business unit of the famous Austrian ski company, which produced parts for the A300 and A310 programs. Another was Rohr, the US supplier of engine nacelles and thrust reversers, which signed its first Airbus contract in 1972 and soon became a key supplier for all of Airbus's programs. By 1988, it would deliver its 1000th nacelle to the European OEM from its French facility.

McDonnell Douglas took aerostructures outsourcing to a whole new level. This was driven, in part, by the company's unwillingness to make significant investments in manufacturing capability. Italian aerostructures supplier Aeritalia started building fuselages in the 1960s as a *quid pro quo* for aircraft purchases by state-owned Alitalia. US aerospace giant General Dynamics supplied DC-10 and MD-11 fuselages. "McDonnell Douglas was capital constrained and outsourced large packages on the MD-11 and later the MD-95," said former FACC CEO Walter Stephan. "They were a pioneer in aerostructures outsourcing."[5] By 1990, jetliner OEMs outsourced 10–15% of aerostructures to outside suppliers, and the trend was accelerating.

In aeroengines, the development of improved superalloys continued. Inconel 718, a mixture of nickel, chromium, molybdenum, and other elements, became a standard element in the hot sections of jets with its 1200°F temperature capability. Aeroengine OEMs developed their own superalloys, including GE's Rene 220 and Pratt & Whitney's waspalloy. Superalloys replaced steel as the temperature requirements of aeroengines continued to grow. Steel did remain for components requiring high strength including aeroengine shafts, bearings, and gears. By the late 1980s, a typical aeroengine was composed of 45% superalloys, 30% titanium, 15% steel, and 10% aluminum and other materials.

Materials processing also improved. One breakthrough was the creation of a *single crystal* turbine blade by Pratt & Whitney in the 1970s. At the time, turbine blades were made of a metallic crystalline structure with grain boundaries, which are areas of weakness that made them prone to fracture. Pratt & Whitney pioneered a single crystal blade that eliminated these crystalline boundaries entirely, which led to improved resistance to fracture, better corrosion resistance, and creep performance. The first application of this invention was in 1980 on the JT9D-7R4 jet engine powering Boeing 747, McDonnell Douglas DC-10, and Airbus A300 aircraft. Single crystal blades soon became an industry standard.

Another significant breakthrough was due to a tragedy. In July 1989, United Flight 232, a DC-10, suffered a catastrophic rupture of the fan disk in its middle

[5]Walter Stephan (CEO—Fisher Advanced Composites), interview with author, 15 Nov. 2015.

engine. Flying debris severed all three hydraulic lines of the triple-redundant flight control system. Pilots were left to control the aircraft solely by thrust settings of the two functioning engines. Miraculously, the crew crash-landed the aircraft, but 111 of 296 passengers died. The cause was determined to be an undetected fatigue crack located in a critical area of the stage 1 titanium fan disk, which had been manufactured by General Electric. It was caused by a *hard alpha inclusion*, a brittle part of the titanium that cracked during forging. A tiny cavity formed in the disk, which grew slightly each time the engine was operated. Eventually it reached a size to cause the structural failure. To prevent this, GE would shift from a *double-vacuum* process, where raw materials were melted together in a vacuum and allowed to cool and solidify before melting them again, to a *triple-vacuum* process. This additional processing step, while expensive, improved control over the material's microstructure to reduce the probability catastrophic failures. Additional improvements would follow, including a *cold hearth* melting process. The aeroengine supply chain would eventually eliminate these types of catastrophic defects.

Material Innovation Golden Age

The 1990s would usher in a period of material innovation on par with the growth of aluminum aircraft in the 1920s. An important catalyst was the Boeing 777, which brought the use of composites to a higher level. The groundwork for this sea change was established in the 1980s. "The early work on the F-22 fighter laid the groundwork for composites usage in aircraft," said aerospace materials expert Chris Red, "and then a slump in military aircraft demand following the Cold War motivated materials suppliers to focus on the 777."[6] The 777's entire empennage would be made of CFRP. Although Airbus pioneered the use of CFRP for jetliner primary structures, Boeing's approach was different. Airbus used a hybrid approach, in which carbon skins were fastened to an aluminum structure. The 777 used composites for the entire empennage structure. And as the world's first jetliner digital design, engineers tailored the design around CFRP's unique properties. Red added, "The 777 took composites to 11–12% of structural weight, were 20% lighter than aluminum designs, and made us realize that composites were viable for primary structures."[7]

New use of composites wasn't limited to the aircraft; the GE90 aeroengine on the 777 included a radical innovation—a composite fan. As highlighted in Chapter 5, some 20 years after the catastrophic failures of Rolls-Royce composite fans on the RB-211, GE bet that technology had advanced enough to make this most flight critical component safe for jetliner operation.

---

[6]Red interview, 25 March 2016.
[7]Ibid.

It wasn't an easy road. GE first gained experience with composites in the 1970s, but manufacturing technologies, inadequate materials, and lack of computing power held back the use of composites in aeroengines. In the mid-1980s, improved epoxy resins became commercially available, and 3D aero computing techniques made it possible to produce durable, wide-chord composite blades. GE made the large external fan blades for the GE36 unducted fan demonstrator from composites, but the program was cancelled. For the GE90, GE leveraged 3D computational fluid dynamics (CFD) to create a unique and beautiful fan design with 22 wide-chord blades with an airfoil geometry optimized for a high-flow, swept configuration (Fig. 10.4). The 4-ft-long solid composite fan blades, GE reckoned, would be significantly stronger yet 10% lighter than the hollow titanium blade designs favored by Rolls-Royce and Pratt & Whitney. GE then had to convince the Federal Aviation Administration (FAA) that the blades were safe, durable, and could withstand bird strikes. The FAA turned to its own experience in certifying composite helicopter blades and gave the GE90 blades a 30,000-cycle life—equivalent to 30 years of operation. GE and CFM partner Safran set up a joint venture in 1993 to produce the radical new fan blades in San Marcos, Texas. They would go on to establish an incredible record of reliability. Just *three* fan blades would be removed for maintenance globally in the engine's first decade of operation [11].

Following the 777's entry into service, Boeing hit the pause button on materials innovation in the late 1990s as it chose to update its venerable 737 with new engines and leave the fuselage and systems largely intact. Airbus, however, had different intentions.

After introducing the A330/340, which used the A300/310's aluminum fuselage, a larger mostly aluminum wing, and a hybrid aluminum-composite empennage, Airbus turned its attention to the A380. By necessity, the new superjumbo needed to make significant advances in the use of advanced materials because Airbus realized that standard alloys had reached maturity.

Credit: GE

Fig. 10.4 GE90 composite fan blade.

The A380 became the first jetliner ever to boast a CFRP composite central wing box, saving up to 1.5 tons. This was very significant, because the wing box—the section of the fuselage that provides support and rigidity for an aircraft's wings—is a key structural component and one of the most vexing aircraft design challenges. It also incorporated a monolithic CFRP design for the fin box and rudder, as well as the horizontal stabilizer and elevators used on previous programs. The scale of parts was a challenge, because the size of the A380's horizontal tail plane was close to the size of the A320's wings [12]. Composites also featured in its rear fuselage section and in its ceiling beams.

One of the A380's most significant innovations came from the Netherlands: glass laminate aluminum reinforced epoxy, or *GLARE*. Originally patented by Akzo Nobel in 1987, GLARE is a hybrid material with very thin layers of metal interspersed with layers of glass fiber and bonded together in an epoxy matrix. GLARE's density was about 75% of aluminum and it had superior fatigue and damage resistance properties. Best of all, it could be used in conventional aerostructures fabrication techniques. Although Boeing, Alcoa, NASA, and others all participated in GLARE's development, the research was kept alive by researchers at Delft University and Fokker. GLARE was chosen for the upper fuselage in the forward and aft sections because of its fatigue performance and damage tolerance (Fig. 10.5). Aluminum was chosen for the center section because of higher static loads due to the presence of the wing. In all, composites would comprise 25% of the A380's structural weight (22% CFRP and 3% GLARE), aluminum was responsible for 61%, and the remaining 14% was made up of steel, titanium, and other materials. Another weight-saving decision was to make the A380's landing gear from titanium rather than steel—another first in jetliner design.

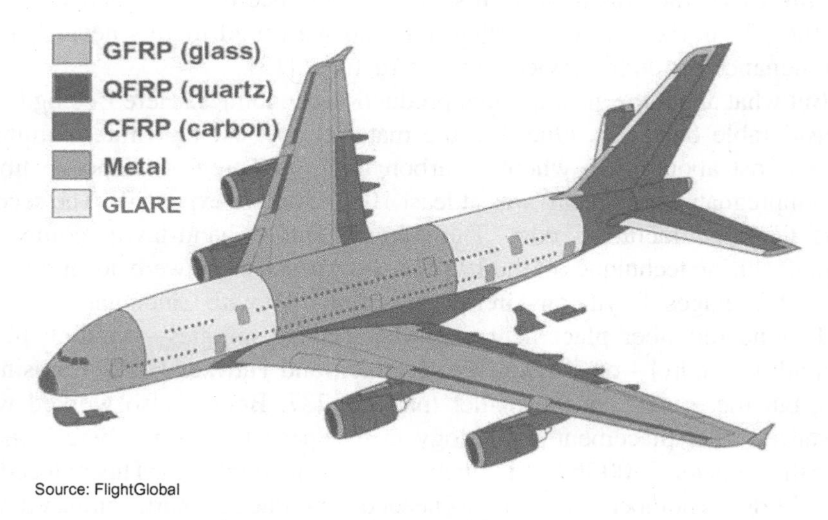

Fig. 10.5 A380's advanced materials.

With Airbus's aggressive use of advanced materials in the A380, Boeing faced important material selection choices on the 7X7 (forerunner to the 787). To what extent would it take risks with its usage of composites and advanced materials? Boeing was historically conservative with materials selection; the structure of its highest technology jetliner, the 777, was composed of just 10% composites.

The 787's design criteria were extremely challenging. Boeing not only pursued the range and fuel efficiency goals outlined in Chapter 3, but also desired significant improvements in maintenance costs and passenger comfort.

Corrosion is a critical maintenance issue for airlines with aluminum aircraft. Over time, moisture corrodes aluminum in hard-to-access locations, such as under galleys and lavatories, and on the inner skin of the fuselage. The insides of aluminum fuselages, much like cold beverage containers on warm summer days, wick moisture from the inside of the cabin, and this moisture, in turn, is trapped by insulation blankets. This phenomenon contributes to the need for significant maintenance inspections, known as C checks, every 18–24 months and major structural inspections (D checks) every 8–10 years. The latter can take aircraft out of service for four weeks or more, with costs of several million dollars plus significant lost revenue. To combat corrosion, airlines purposely reduce humidity in the aircraft cabin to ~5%, which has a negative impact on passenger comfort. Another design tradeoff that works against passengers is the pressurization levels of the fuselage during flight. Aluminum fuselages were designed to withstand a pressure equivalent of 8000 ft altitude in flight to minimize the required thickness of the fuselage wall.

These design considerations pointed to a high-composite design. Composites don't corrode, and the operational history of the 777 and Airbus jetliners backed this up. As 787 Vice President Mike Bair concurred, "One of the things we did was look at in-service experience of maintenance history on the 777 horizontal and vertical tail and we found there's never been a maintenance action in-service on that structure" [13].

But what about the all-important production economics? Here Boeing faced considerable obstacles. One was the material cost. At the time, aluminum alloys cost about $3/lb whereas carbon fiber "pre-preg" (composite fibers pre-impregnated with resin) was at least 10 times more expensive. The second was the manufacturing cost. The labor-intensive hand-layup composite manufacturing techniques of the 1980s were expensive and were not applicable to civil fuselages. Raytheon Aircraft, in conjunction with Cincinnati Milicron, had pioneered fiber placement manufacturing techniques—winding fibers around a mandrel—on its Starship, Premier, and Hawker Horizon business jets, but these were much smaller than the 787. Boeing also worked with advanced fiber placement technology in its Space Division, where it made 13.5-ft-diameter, 3800-lb Sea Launch rocket payload fairings. This reduced by two-thirds the number of labor hours needed to produce the part, compared with hand layup [14]. A third obstacle was the regulators. Here, Hawker Beechcraft had blazed the trail with the FAA, sometimes with considerable pain given the lack of consistency and material databases to satisfy regulators. The composites material supply chain also had matured. "This gave Boeing engineers the ability to maximize performance without impacting safety," according to Chris Red.[8]

THE 787 SHOCKER

In 2003, Boeing shocked the aerospace supply chain in announcing that the 787 would be the industry's first primarily composite jetliner. Like the 777, the empennage would be largely composites. The wings would also be mostly composite—Boeing's first composite jetliner wings. The biggest surprise was an all-composite fuselage: Boeing would construct eight huge, single-piece CFRP fuselage sections and join them in final assembly operations. The cockpit piece would be made in its Wichita operations and the rest by Japanese producers and Italian supplier Alenia. The CFRP pre-preg would be supplied by Toray, another Japanese supplier. In aggregate, the 787's structure was 50% composites by weight—twice the level of the A380 (Fig. 10.6).

Boeing didn't stop there. It also took a step-function advance in the use of titanium in the aircraft's structure. Like the A380, it adopted titanium for

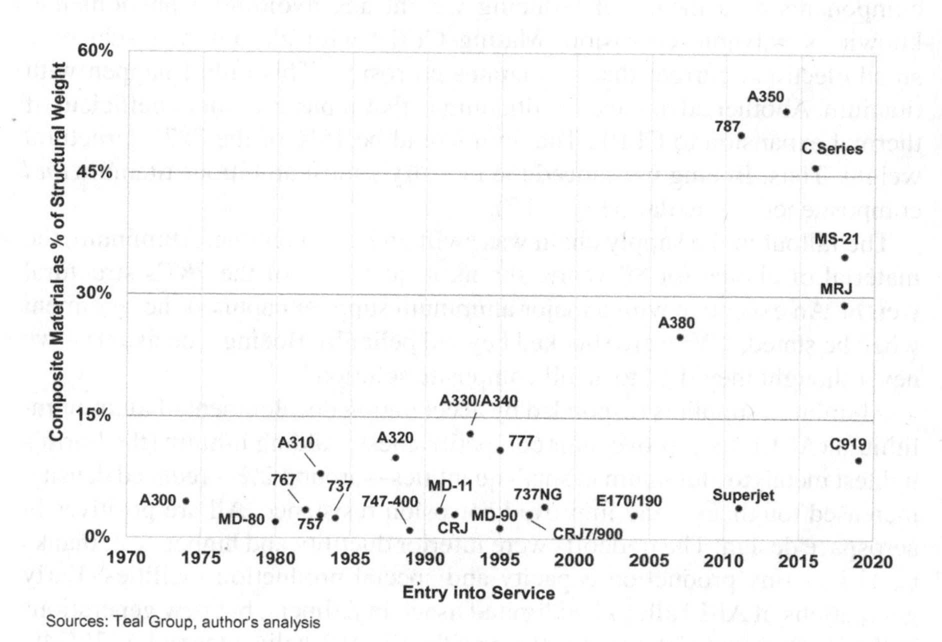

Fig. 10.6 Composites penetration: structures (% weight).

---

[8]Ibid.

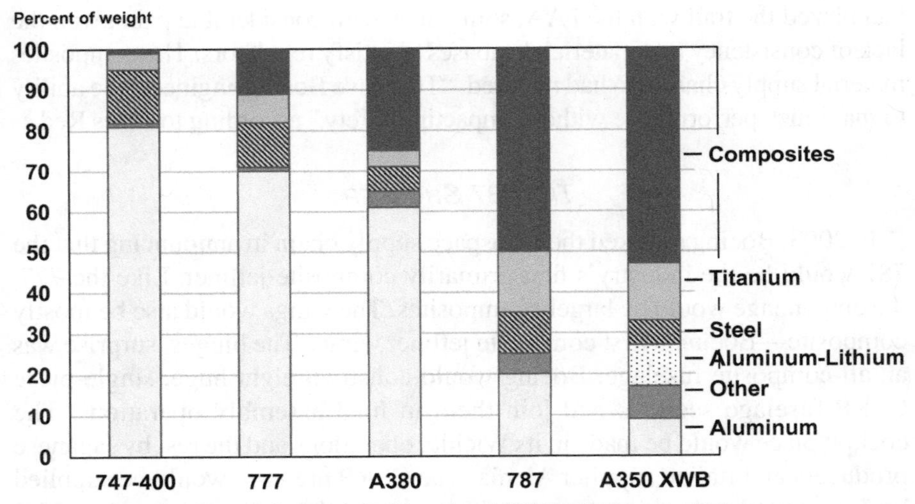

Fig. 10.7 Aircraft structural material composition.

the aircraft's landing gear. It also adopted titanium for numerous structural components as a means of reducing weight and avoiding a phenomenon known as galvanic corrosion. Mating CFRP with aluminum produces a small electrical current that accelerates corrosion. This didn't happen with titanium. Another advantage for titanium is that it has a similar coefficient of thermal expansion to CFRP. Titanium would be 15% of the 787's structural weight. Thus, Boeing introduced the industry's most ambitious titanium *and* composite jetliner to date (Fig. 10.7).

The fallout in the supply chain was swift and considerable. Aluminum, the material of choice for 80 years, shrank to just 20% of the 787's structural weight. An executive with a major aluminum supplier captured the sentiment when he stated, "We were shocked beyond belief by Boeing's decisions—we never thought they'd go to an all composite solution."[9]

Aluminum suppliers responded by accelerating development of aluminum-lithium (Al-Li) to improve their competitiveness. Adding lithium (the Earth's lightest metal) to aluminum in small quantities—around 2%—reduced density, increased toughness, and improved corrosion resistance. All are positives in aerospace design. The tradeoffs were inferior ductility and higher cost, thanks to Al-Li's tiny production capacity and special production facilities. Early generations of Al-Li alloys had limited usage in jetliners, but new generations in the 1980s found niche applications on the EH-101 helicopter and A330/340. Boeing had hoped to make widespread use of the alloy on the 777, but rejected it due to cracking and machinability [15].

[9]Anonymous aerospace aluminum executive, interview with author, Jan. 2016.

Bombardier would ultimately become the Al-Li pioneer. After the first version of the C Series was killed in 2004, Bombardier revisited its material choices when it created a second version with geared turbofan engines. C Series program leader Rob Dewar conducted a trade study and concluded that Al-Li was the best solution for the fuselage of the C Series. Bombardier Commercial Aircraft President Gary Scott, after hearing Dewar's briefing, stated, "Rob, I appreciate your technical expertise, but the A350 (first version) was just killed because it was primarily aluminum, and the 787 is CFRP." Dewar then made an interesting proposal to Scott. "Let's do a workshop with potential customers and I'll share all the Al-Li data with them. I'll embrace whichever way they vote." At the start of the workshop, Dewar took a vote of the 10 participants—all potential C Series customers—before sharing the data. The vote was eight for CFRP and two undecided. After Bombardier shared its results, the tally shifted to seven for Al-Li, two for CFRP, and one undecided. The C Series would thus become the first jetliner to feature an aluminum-lithium fuselage.[10] This material choice would save 600 lb and provide 2.4 times the corrosion resistance versus conventional aluminum. This enabled Bombardier to offer a 15-year service life corrosion guarantee and expend the heavy maintenance intervals on the aircraft. The tradeoffs were that Al-Li was more expensive than conventional aluminum alloys, required special production facilities, and was difficult to recycle because it required segregation from other aluminum alloys.[11]

Airbus was also aggressive in adopting advanced materials including Al-Li on the second version of the A350, the A350 XWB. Like the 787, it would adopt a composite fuselage and wing, although the fuselage design would feature composite panels attached to a metallic structure with heavy usage of Al-Li and titanium. Overall, the A350 XWB would be 53% composites, 19% Al-Li and aluminum alloys, 14% titanium, 6% steel, and 8% other materials [16]. This made the A350 XWB the most advanced jetliner to date in terms of advanced materials usage.

Advances in materials and processing continued in aeroengines. Increasing bypass ratios meant that fan sections became larger and larger. This drove the need for lightweight materials. GE's solution was the CFRP fan blades pioneered on the GE90, and it expanded the use of CFRP to include fan cases on its GEnx for the 787. The LEAP would build on these advances. Rolls-Royce continued to use hollow titanium fan blades, using its novel super-plastic diffusion bonding (SPDB) process, where three layers of titanium are laid upon each other and heated to the point where they can be formed, then inflated with inert gas into the required shape, with the center layer

[10]Rob Dewar (Program Manager—C Series, Bombardier Aerospace), interview with author, 28 June 2016.

[11]Dewar interview, 28 June 2016.

becoming an internal reinforcement. And Pratt & Whitney, as described in Chapter 6, shifted from titanium to hollow aluminum-lithium blades for its geared turbofan.

The fiery hot section of aeroengines also saw new developments as pressures and temperatures increased. Single crystal and directionally solidified casting manufacturing processes continued to improve. The 1990s brought the application of thermal barrier coatings to turbine blades and vanes, and added several hundred degrees to their temperature limits. The temperature capability and component life of hot section disks and drums also improved thanks to hot isostatic processing techniques using metal powders. These developments facilitated much higher turbine inlet temperatures—as high as 2800°F or more.

Aeroengine OEMs and specialist material suppliers competed to develop unique superalloy formulations to gain advantage. Rolls-Royce, for example, developed an RR1000 powder metallurgy superalloy for the last two stages of the high pressure (HP) compressor drum and the HP turbine discs for its Trent 1000 engine. Pratt & Whitney also expanded its powder metallurgy capability after it purchased its leading supplier, Homogeneous Metals Inc., in 1980. The subsidiary went on to produce powder metals such as Inconel 100 and ME16 for civil and military aeroengines. General Electric had its own powder metals, including René 88 and René 104. The search for the best elements to add to nickel and cobalt-based superalloys was never-ending, with exotic elements like tantalum, hafnium, and rhenium used to improve strength. Rhenium is one of the scarcest substances on Earth; it forms only one part per billion of the Earth's crust. In the mid-2010s, the entire annual worldwide production of rhenium was just 40 tons, with more than 75% of it used in strengthening superalloys.[12] High demand coupled with very limited supply led to rhenium prices that sometimes exceeded $1000/lb.

There were also material advances in the lower temperature aft section of the engine. In nongeared (direct-drive) architectures, low-pressure turbine blades and vanes got progressively larger and heavier as bypass ratios grew. GE's solution was to introduce titanium aluminide alloys for these crucial components.

The combination of better (and often more expensive) materials coupled with escalating jetliner production rates led to surging spending on aerospace raw materials. Jetliners that were once made mostly of aluminum, which might cost $3–4/lb, were increasingly dependent on $15–20/lb titanium and $40/lb CFRP pre-preg. By 2015, total aerospace raw material demand reached 1.56 billion lb for civil and military aircraft and aeroengines, according to

[12] Miguel-Descalso, A. R., "Nickel Superalloys are the Superheroes of Commercial Flight," Born2Invest.com, 3 Nov. 2016, https://born2invest.com/articles/nickel-superalloys-commercial-flight [retrieved 9 Feb. 2018].

consultancy ICF International. Aluminum alloys were the largest category (47%), followed by steel alloys (21%), titanium (11%), superalloys (10%), and composites (5%) [17] (see Fig. 10.8). The value of this material was $10–15 billion. Jetliners comprised 75% of this total, with the balance for military and business aircraft.

The irony of the huge spending on aerospace raw material was that most of it was wasted. Some 80% of total raw material consumption ended up on shop floors as scrap material, or "revert," rather than in finished aircraft and aeroengines. Thick aluminum plate, for example, might be machined into a much smaller component. Across all part types, 5 lb of raw material were purchased for every 1 lb of material in an aircraft rolling off the assembly line. For some components, it was more than 20 lb. Some suppliers began to search for ways to reduce waste and efficiently recycle the huge amounts of revert.

EMERGENCE OF MAJOR TIER 1 AEROSTRUCTURES SUPPLIERS

While aircraft and engine OEMs adopted ever more advanced materials, another important change unfolded in the skin and bones of jetliners: the emergence of major Tier 1 aerostructures suppliers. Because of OEM outsourcing and the need for risk-sharing partners, a cadre of sophisticated aerostructures suppliers took their place as critical elements of the aerospace ecosystem. As outlined in Chapter 8, the largest suppliers, called *Tier 1s*, supplied directly to the aircraft OEMs, whereas smaller suppliers, known as *Tier 2s*, focused on subassemblies and parts.

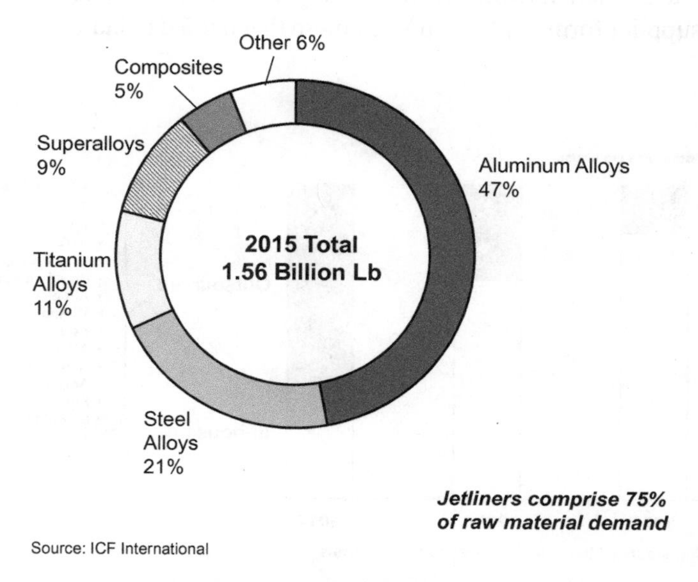

Fig. 10.8 2015 Aerospace raw materials demand (buy weight).

A major shift of aerostructures work to Tier 1 suppliers took place for several reasons. First, Boeing and Airbus spun out critical elements of their aerostructures capabilities as they redefined their core and pursued better financial returns. Thus, Boeing's massive Wichita aerostructures facility became Spirit AeroSystems (2005). Airbus sold its Filton, United Kingdom wing facility to GKN in 2008. And in 2009, Airbus created two companies—Aerolia and Premium Aerotec—from five of its French and German aerostructures facilities.

Second, jetliner OEMs needed risk-sharing partners as aircraft development costs escalated. The Japan Aircraft Development Corporation, for example, increased its share of Boeing's aerostructures from 15% on the 767 to 21% on the 777 and 35% on the 787—including the aircraft's composite wings. Airbus and regional jet OEMs also expanded aerostructures outsourcing in the 2000s.

Third, this outsourcing enabled aircraft OEMs to reduce their internal engineering requirements. Many of the large aerostructures contracts were "design-build," where the supplier was responsible for design, structural analysis, and production.

By 2012, industry consultant Counterpoint Market Intelligence estimated that jetliner OEMs retained just 36% of aerostructures production (Fig. 10.9). Some $20 billion worth of jetliner aerostructures was sourced from outside suppliers. Approximately 60 Tier 1 suppliers captured most of this business [18].

Aerostructures suppliers were distributed globally. Major European suppliers included Airbus spin-outs Premium Aerotech and Aerolia; GKN aerospace, a UK-based, high-technology wings and structures supplier; Aircelle, a French nacelle systems supplier; Aernnova, an independent Spanish supplier formerly known as Gamesa that helped to launch the Embraer

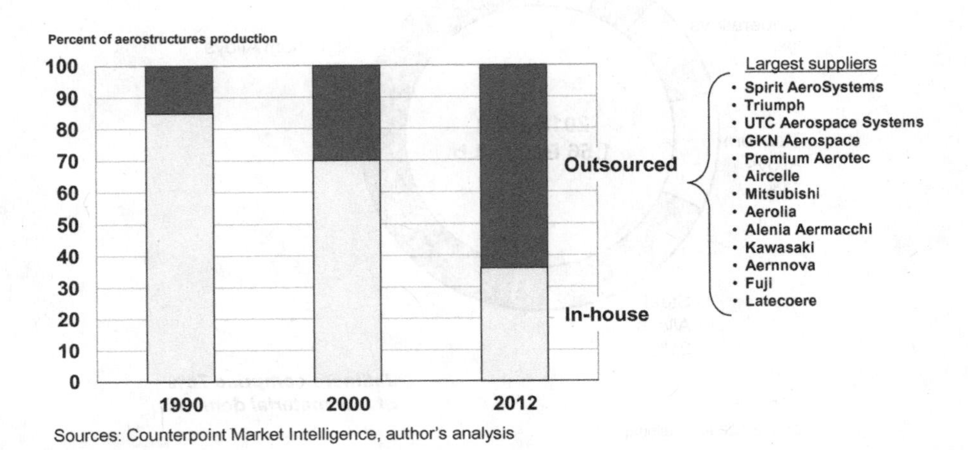

Fig. 10.9 Aerostructures outsourcing (%).

Regional Jet; and Latecoere, a Toulouse-based supplier with significant Airbus content.

In North America, the largest supplier was Spirit AeroSystems, which added operations in the United Kingdom and Malaysia after being spun out by Boeing and reaching $5.2 billion in revenue; Triumph, which acquired Vought Aircraft in 2010 and was a major Boeing fuselage and wing supplier; and UTC Aerospace Systems, a major nacelle supplier, formerly known as Goodrich Aerostructures and Rohr.

Finally, in Asia, aerostructures capability was dominated by the three Japanese "heavies"—Mitsubishi Heavy Industries, Kawasaki Heavy Industries, and Fuji Heavy Industries.

MATERIAL SUPPLIER COMPETITION

While OEMs made bold material selection decisions and large Tier 1 aerostructures suppliers emerged, the suppliers of raw materials engaged in intense competition as they rode a roller coaster of demand.

The emergence of aircraft with high advanced materials content, coupled with surging orders in the mid-2000s, led to projections for skyrocketing titanium demand. Titanium was long a fixture of aeroengines, but significant amounts in the aircraft's structure took demand to another level. The 787, for example, would use 250,000 lb of titanium. Moreover, the anticipated ramp up in production of the 787, A380, and A350 XWB would overlap with another high titanium aircraft—the F-35 Joint Strike Fighter. One forecast predicted that global titanium demand would reach 360 million lb by 2011—triple the level in 2006. Titanium prices surged, and the industry's four major titanium mill suppliers scrambled to add capacity.

US-based Allegheny Technologies announced plans in 2008 for a $1.2-billion capacity expansion including the world's first new titanium sponge plant in 30 years [19]. Titanium sponge is a porous form of titanium that is created during the first stage of processing. RTI International Metals (United States) unveiled plans for a $300-million expansion in Mississippi to support surging demand from Airbus and for the F-35 Joint Strike Fighter [20]. The third major supplier, TIMET, unveiled plans to expand its aeroengine titanium capacity in Morgantown, Pennsylvania. However, it was Russian company VSMPO-AVISMA—the industry's largest—that had the most attention-grabbing actions.

Verkhnaya Salda Metallurgical Production Association (VSMPO) grew substantially during the Cold War with booming Soviet titanium demand for submarines and aircraft. In 1992, following the collapse of the Soviet Union, VSMPO leader Vladislav Tetyukin developed a bold strategy to integrate it into the global economy. Slowly it gained foreign customers and certifications. A breakthrough was developing Boeing as a major customer.

Credit: VSMPO-AVISMA

Fig. 10.10 VSMPO titanium facility: Salda, Russia.

By 2002, it exported more than 10,000 tons of titanium—more than all US suppliers combined. Soon, a saying developed around VSMPO that "the Boeing runway starts in Verkhnaya Salda"[13] (see Fig. 10.10). It was a rare example of a globally competitive Russian company outside of the oil and gas sector. In 2005, VSMPO merged with another Russian firm, AVISMA, the world's largest producer of titanium sponge. This positioned the new company, VSMPO-AVISMA, as a fully integrated and competitive aerospace titanium supplier. Two years later, it formed a joint venture (JV) with Boeing to produce titanium forgings for the 787's landing gear. VSMPO would eventually provide up to 35% of all the needed titanium for Boeing, 65% for Airbus, and 100% for Embraer.[14] In 2016, Boeing and VSMPO announced plans for a second JV for titanium parts called Ural Boeing Manufacturing.

Collectively, the actions of ATI, Timet, RTI, and VSMPO indicated that the sky was the limit for aerospace titanium. Then the world changed. The entry into service dates for the A380 and 787 slid by several years. A new version of the A350, the 350 XWB, meant a later introduction date than anticipated. The military F-35 program struggled, and then the Great Recession slammed the global economy. Suddenly what appeared to be a shortage of titanium capacity became a glut. Predictably titanium prices plunged, and in 2010 RTI cancelled plans for a new titanium sponge plant in Mississippi. Meanwhile, titanium stocks built up in warehouses of aircraft OEMs that had entered "take or pay" contracts during the boom. ATI would also regret building its major titanium sponge plant in Utah. It idled the facility in 2016 when global titanium prices fell below its cost of production. ATI CEO Rich Harshman summed up the dilemma of making long-term investments in what is sometimes a boom–bust

[13]"In The World Market," VSMPO Avisma, http://www.vsmpo.ru/en/pages/V_mirovoj_rinok [retrieved 29 Jan. 2017].

[14]"Our Partners," VSMPO Avisma, http://www.vsmpo.ru/en/pages/Partneri [retrieved 29 Jan. 2017].

market: "Raw materials are risky—sometimes the bets pay off; sometimes they don't, and the world changes in ways beyond your control. In business, there is no Mulligan."[15]

While titanium grappled with oversupply, the aluminum sector received encouraging news when Airbus and Boeing chose to reengine the A320 and 737 rather than develop a white-sheet aircraft. This avoided the possibility of composite single-aisles and locked aluminum into the aerostructures of the world's two most popular jetliners until at least 2030. To put this into perspective, these two aircraft models comprised more than 30% of the 600 million lb in total aerospace aluminum demand (civil and military) in 2011 [21].

There were four major producers of aluminum in the early 2010s that supplied more than 90% of aerospace demand: Alcoa, Constellium, Aleris, and Kaiser (Fig. 10.11). Alcoa was the largest and best-known supplier with major aerospace aluminum mills in the United States, the United Kingdom, and Russia. Its largest mill in Davenport, Iowa, was one of the largest industrial complexes in the United States and covered a staggering 130 acres under roof. Alcoa positioned itself to be at the cutting edge of aluminum metallurgy and pursued higher yield components like wing and fuselage skins. In 1965, it founded the world's largest light metals research center, the Alcoa Technical Center, outside of Pittsburgh. As aluminum-lithium gained popularity following the C Series decision, it opened the world's largest dedicated Al-Li mill in Lafayette, Indiana.

European supplier Constellium was also a technology leader in aerospace alloys. Formerly known as Alcan, it changed its brand to Constellium in 2011. With seven aerospace facilities in Europe and the United States, it was

Aluminum Alloys	Steel Alloys	Titanium Alloys	Superalloys	Composites (pre-preg)
•Arconic**** •Aleris •Constellium •Kaiser	•Allegheny Technologies •Aubert & Duvall •Bohler •Carpenter Technology* •Universal	•Allegheny Technologies •Arconic** •Timet •VSMPO-AVISMA	•Allegheny Technologies •Aubert & Duval •Carpenter Technology •Firth Rixon •Haynes International •Special Metals •OEM internal production	•Hexcel •Solvay*** •Toray

Source: Author's analysis
* Acquired Latrobe Specialty Metals in 2012
**Acquired RTI International in 2015
***Acquired Cytec in 2015
****Rebranded in 2016; formerly Alcoa

Fig. 10.11 Leading aerospace raw material suppliers: 2016.

[15]Rich Harshman (CEO—Allegheny Technologies), interview with author, 14 July 2016.

a key supplier to Airbus and a pioneer in aerospace Al-Li. It also brought marketing to the staid aluminum industry, branding its lightweight solutions as "Airware."

In contrast, Kaiser was positioned as a technology follower. After several near-death experiences and restructurings, it emerged from bankruptcy in 2006 as a lean and focused supplier. Its major aerospace mill was in Spokane, Washington, leveraging cheap electricity in the region.

The fourth major aerospace aluminum supplier was Aleris, formed in 2004 through the merger of Commonwealth Industries, Inc. and IMCO Recycling. In 2006, it acquired the Corus Group for $900 million, which doubled its size and significantly expanded its presence in Europe and China. Aleris was the first western supplier in China, opening an aluminum mill in Zhenjiang, China in 2013. Three years later, China Zhongwang Holdings, the world's second largest aluminum parts maker, made a $2.3-billion bid for Aleris. The deal, however, was blocked by the US government in late 2017 due to economic and national security concerns.

Aluminum's nemesis, composites, had a very consolidated supply base with three major pre-preg suppliers. Pre-preg is composite fabric or tape that has been pre-impregnated with a resin system (typically epoxy) and a curing agent. As a result, the pre-preg is ready to lay into the mold without the addition of any more resin. The composites supplier base wasn't always consolidated. In the 1980s and 1990s, there were dozens of fiber and resin suppliers as aerospace standards were yet to emerge. Cytec Industries, a company formed in 1993 by a spinout from American Cyanamid, began an acquisition spree of resin, chemicals, and fiber suppliers that would position it as a top aerospace composites supplier. It was particularly strong in resins and high-performance military applications. In 2015, Belgium's Solvay bought Cytec for $5.5 billion.

The second major pre-preg supplier, Hexcel, had a much longer history stretching back to 1946, when it was founded in California by Roger Steele and Roscoe Hughes. The new company would acquire a British composites pioneer, Aero Research Ltd, in 1947 and go on to build a leading composites supplier known for its pre-preg and it famous honeycomb structural material. By 2016, its revenue was nearly $2 billion, with most coming from aerospace. The third major pre-preg supplier, Japan-based Toray, made a major splash when it won huge contracts worth billions to supply CFRP pre-preg to Boeing for the 787. Beyond the three major pre-preg suppliers were smaller resin, chemical, and fiber suppliers with niche positions on the aircraft, including interiors.

Superalloys had their own set of competitors, with several major specialists. The largest were US-based Allegheny Technologies and Special Metals. Allegheny Technologies, also a major titanium and steel supplier, made a breakthrough when it developed 718Plus, an alloy with a 100°F operating temperature advantage over Inconel 718, the industry's workhorse alloy for aeroengines. Other major suppliers were Aubert & Duval, Carpenter Technology, Firth Rixon, and Haynes International.

What about steel? The demand for lighter aircraft and the growth of strong and lightweight titanium reduced steel content in some aircraft. Landing gear, for example, were once one of the major consumers of steel. Titanium replaced steel as the primary material for the landing gear on the 787, A380, and A350 XWB. In aeroengines, higher operating temperatures and the imperative for light weight meant steel was replaced by superalloys and titanium for most parts except the shafts, gears, and bearings. This pinched leading aerospace steel suppliers; in the United States these were Allegheny Technologies, Latrobe Specialty Metals, and Universal; in Europe they were Aubert & Duvall and Bohler.

Revolution from Below

The massive losses on the A380 and 787 and supply chain initiatives highlighted in Chapters 7 and 8 put the aerostructures supply chain in the bulls-eye for aircraft cost reduction in the early 2010s. One target was reducing the massive amount of scrap from aerospace manufacturing by designing more "near net shape" parts. OEMs also redesigned parts to reduce the use of expensive materials, like titanium, which cost $20/lb or more. On the 787, Boeing redesigned hundreds of parts to shift from titanium to lower cost materials. An added benefit of the shift was reduced tooling costs. Titanium is a notoriously difficult-to-machine material, with cutting tools sometime lasting less than one hour before needing to be replaced.

Another opportunity for cost savings was reducing the complexity of the supply chain. In some instances, parts were shipped from region to region—logging tens of thousands of miles in travel—before being delivered to the production line. An Al-Li billet, for example, might be shipped to from Europe to Asia for machining, then to North America for specialty processes, before returning to Europe for final assembly.

The aerostructures supply chain was not only global, but also spread across the four supplier tiers described in Chapter 8. Supplying the 60 or so Tier 1 aerostructures suppliers were hundreds of Tier 2 manufacturers of subassemblies and parts. Next were thousands of Tier 3 make-to-print manufacturers. These suppliers, also known as *machine shops*, were typically small and often located near final assembly facilities or Tier 1s. Finally, there were Tier 4s, which included raw material mills suppliers of forgings, castings, and extrusions. Due to the large capital requirements to build an aerospace mill (billions of dollars) or a major forcing facility (hundreds of millions of dollars), Tier 4 was much more concentrated, and included less than 150 suppliers (see Fig. 10.12).

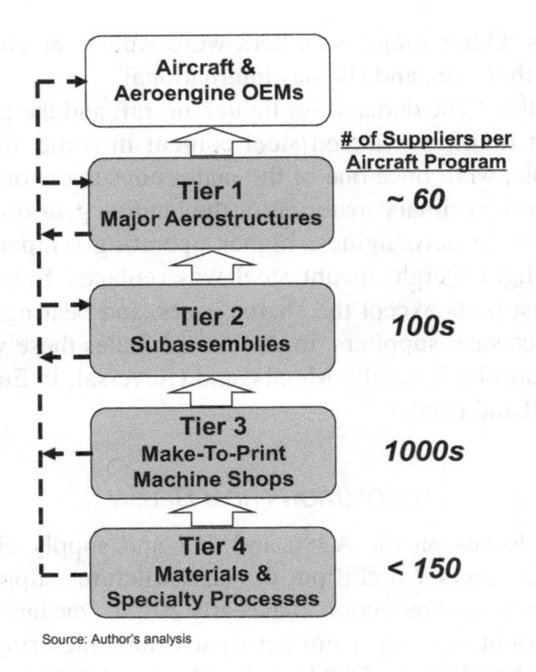

Fig. 10.12 Aerostructures supply chain.

For some raw materials, there might be just two or three qualified suppliers on an aircraft program to leverage economies of scale and ensure that there is adequate volume to meet "mill minimum" melt conditions. This relative concentration created a dilemma. How could small and fragmented Tier 2 and Tier 3 suppliers receive competitive raw material prices when negotiating with large, concentrated raw material mills? Boeing addressed the issue by striking a 10-year deal with supply chain services company TMX Aerospace in 1998 to manage aluminum and titanium for its extended supply chain. Boeing negotiated a price with raw material mills that its entire supply chain could leverage, and TMX delivered the material, often cut to order or kitted, to its suppliers. It renewed the agreement for another 10 years in 2008.

While OEMs searched for ways to improve aerostructures supply chain efficiency, subtier suppliers pursued their own agenda. The most prominent was Portland, Oregon—based Precision Castparts Corporation (PCC), which would write one of the great value creation stories in aerospace history.

Founded in 1953, PCC was a niche manufacturer of investment castings and components for aerospace and industrial gas turbine applications. In 2001, it named Mark Donegan as president and COO (Fig. 10.13). The former GE executive had joined Precision Castparts in 1985 as a manufacturing

Credit: Bloomberg / Getty Images

Fig. 10.13 Precision Castpart's CEO Mark Donegan (left) built a giant and sold it to Berkshire Hathaway for $37 billion.

supervisor and helped turn around its troubled Wyman-Gordon forging business after its acquisition in 1999. Amid the aerospace downturn in the early 2000s, Donegan was confident in PCC's ability to reduce cost structures in the future and offered major aerospace customers deflationary pricing in exchange for longer term contracts and market share gains. This approach proved so successful that, when the new contracts kicked in at the beginning of 2003, the company had the largest number of parts under development in its history. As these components moved into production and the cycle ramped back up, Precision Castparts achieved unprecedented sales levels.[16]

Obsessed with driving improved operations from his businesses, Donegan, now CEO, was confident that he could create value by applying PCC's processes to acquisitions. Thus began one of the great acquisition sprees in aerospace history (see Fig. 10.14). In 2003 he acquired SPS Technologies, moving the company into fasteners—a new segment. Cannon-Muskegon, a manufacturer of nickel-based alloys for the casting industry, was part of the SPS deal. This moved PCC upstream into Tier 4 raw materials. He added to this vertical integration play with the 2006 acquisition of Special Metals, one of the largest superalloy suppliers. Previously, Wyman-Gordon was buying all its nickel billet from outside suppliers; now it procured more than 70%

[16]"PCC History," Precision Castparts Corporation, http://www.precast.com/overview/history [retrieved 20 May 2017].

Tier 4 Raw Material & Processes	Tier 2 & Tier 3 Machining, Parts, & Subassemblies	Fasteners
Wyman Gordon (1999) Cannon-Muskegon (2003) Specialty Metals Corp (2005) Caledonian Alloys (2007) Carlton Forge Works (2009) Titanium Metals - TIMET (2012) SOS Metals (2014) Schultz Steel (2016)	Primus International (2011) Tru-Form (2011) Klune Aerospace (2012) Centra Industries (2012) Héroux-Devtek-aerostructures (2012) Synchronous Aerospace (2012) Aerospace Dynamics (2014) Noranco (2015)	SPS Technologies (2003) Cherry Aerospace (2007) PB Fasteners (2012) Permaswage (2013)

Fig. 10.14 Precision Castparts' major acquisitions: 1999–2016.

of its billet from the new acquisition.[17] It added Carleton Forge Works, the leading supplier of seamless rolled rings, to its portfolio in 2009 to strengthen its aeroengine content.

Why the push for vertical integration? As PCC made productivity improvements of individual manufacturing facilities, it grasped a new efficiency frontier: to create a closed-loop ecosystem where the massive amount of revert from aerospace manufacturing could be captured and then sent to PCC's mills for recycling. PCC would attack the industry's 5:1 buy-to-fly ratio as a productivity opportunity. In 2007, it bought Caledonian Alloys, a materials scrap dealer, to facilitate the initiative.

In subsequent years, PCC would build out its fastener capability with several acquisitions and expand into aircraft structures by buying numerous Tier 2 and 3 suppliers. It applied a similar playbook to each acquisition. A PCC transition team descended on the acquisition and spent months transferring its processes, standards, and culture. CEO Donegan would then visit every three months for management reviews that were challenging and aggressive. Executives that didn't embrace continuous improvement or make the numbers didn't last long. PCC's laser-like focus on productivity not only drove higher margins, but also enabled it to outbid competitors for acquisitions.

PCC made another blockbuster acquisition in 2012 when it acquired Titanium Metals, a major titanium supplier with a strong position in aeroengines. Combined with its prior acquisition of Special Metals, PCC now controlled the two largest aeroengine raw material suppliers and was also the largest supplier of forgings and castings. It would go on to make a dozen acquisitions from 2013 to 2016 and build a portfolio that stretched from Tier 4 through Tier 2 components. It became one of the top two suppliers of nickel alloy, rotating-grade titanium, investment castings, forgings, fasteners, and large structural castings. On the 787, its shipset revenue was an eye-popping $10 million; for the A380 and 777-300ER it was $5 million. By 2015, its revenues pushed through $10 billion with an

[17]Ibid.

earnings before income and taxes (EBIT) margin of more than 25%. Mark Donegan and PCC were Wall Street darlings. Putting its performance into context, PCC's value (market capitalization) at the end of its fiscal year in March 2014 was $36.7 billion. Airbus ended its fiscal year in December 2014 with a value of $38.7 billion.

PCC's stellar performance and predictable earnings caught the eye of Berkshire Hathaway CEO Warren Buffett, arguably the world's greatest investor. His firm was already the largest shareholder of PCC stock, and in August 2015, Berkshire Hathaway announced a blockbuster deal to buy Precision Castparts for $37 billion. It was the largest acquisition in aerospace history.

PCC wasn't the only supplier to consolidate the subtiers. Titanium supplier RMI Titanium (later renamed RTI International Metals) made several major acquisitions, but for different reasons than PCC. "When VSMPO-AVISMA entered the titanium market, they were larger than all three incumbent titanium suppliers combined…we knew that we were vulnerable and had to pivot," explained former CEO Dawne Hickton (Fig. 10.15). "We chose to expand horizontally and vertically."[18] First, RTI diversified by supplying titanium to the energy sector in the late 1990s. It then began a vertical integration campaign.

Credit: Dawne Hickton

Fig. 10.15 Dawne Hickton was RTI International Metals CEO from 2007–2015.

[18]Dawne Hickton (former CEO—RTI International Metals), Interview with author, 29 May 2018.

"We moved into Tier 3 by doing extrusion and hot-forming work," recalls Hickton, "and the next step was Tier 2 when in 2004 we bought fabrication facilities in Montreal. We later added machining and aluminum capabilities when we bought Remelle Engineering in 2012."[19] Like PCC, RTI proved that it could create value by moving downstream into advanced fabrication.

There was more consolidation in the supply chain subtiers. Superalloy and steel supplier Allegheny Technology acquired Ladish, a leading forging and casting supplier. ATI competitor Carpenter Technology bought specialty steel supplier Latrobe. Meanwhile, Alcoa, arguably PCC's biggest competitor, made several key acquisitions. In 2000 it purchased Howmet, a major aeroengine investment casting supplier. It followed with a string of fastener acquisitions that would position it as PCC's principal competitor in this segment. In 2014 it paid $2.85 billion to buy Firth Rixson, a major aeroengine component manufacturer. Then it pushed into titanium by purchasing RTI International Metals and German casting supplier Tital in 2015. The "Aluminum Company of America" was no longer just about aluminum. It then announced that it would separate into two independent companies. Aerospace and transportation specialty components and materials would become Arconic, while the other capabilities and the upstream mines would remain Alcoa.

New Frontiers

By the mid-2010s, Tier 4 was effectively consolidated, and attention shifted to a disruptive technology that was both a threat and an opportunity: additive manufacturing. GE's aggressive adoption of this technology on the LEAP engine (detailed in Chapter 6) was a catalyst. Soon most OEMs pursued their own additive manufacturing strategies. Pratt & Whitney announced that it would print a stator vane section of its geared turbofan engine. Still, progress was frustratingly slow. A major barrier was the jetliner sector's extremely demanding certification standards. The properties of parts made from powder metal melted by a laser or electron beam could differ by process and even when made by different machines. Much like the introduction of composites four decades before, material databases and standards would be required for additive manufacturing to assume center stage. There were also economic barriers, including slow production speeds and the high cost of powder metal. Perhaps the strongest barrier to greater use of additive manufacturing was the "mental models" carried in the heads of design engineers themselves. All had grown up designing aircraft and parts for subtractive manufacturing methods. Designing for additive manufacturing was a different endeavor altogether, and many companies therefore assigned their young engineers to tackle the challenge.

[19]Ibid.

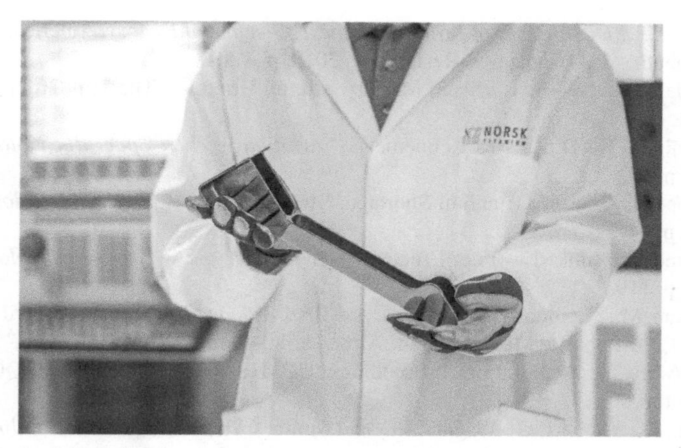

Credit: Norsk Titanium

Fig. 10.16 Norsk Titanium's additive-manufactured 787 part.

General Electric doubled down on its commitment to additive manufacturing in 2016 when it announced the creation of GE Additive, a business dedicated to producing 3D-printed parts for internal and external customers in aerospace and other industrial applications. GE purchased stakes in Concept Laser and Arcam, two leading European additive manufacturing equipment suppliers, and established a revenue goal for its new business of $1 billion by 2020.

Several breakthroughs came in 2017. Boeing announced that it would integrate titanium structural components on the 787. The parts would be made by Norwegian newcomer Norsk Titanium, which had pioneered the rapid plasma deposition (RPD) process that melted titanium wires to produce large structural components (Fig. 10.16). GE announced that 35% of the parts in its new turboprop engine, the Advanced Turboprop, would be printed, reducing its serialized part count by more than 800. And aircraft OEMs led by Airbus contemplated "bionic" aerostructures design that would combine proven design concepts from nature with additive manufacturing.

With the jetliner business in a "more for less" era, and more than 1 billion lb of scrap metal created annually, the future of additive manufacturing appeared strong.

REFERENCES

[1] Sabbagh, K., *21st Century Jet: The Building of the 777*, Scribner, New York, 1996, p. 97.

[2] Sayers, W. H., "Steel Aircraft Construction in Great Britain," *Aviation*, 15 June 1929, p. 94.

[3] Johnson, C. L., *Kelly: More Than My Share of It All*, Smithsonian Institution, Washington, D.C., 1985, p. 139.

[4] Rich, B. R., *Skunk Works*, Back Bay Books, New York, 1994, p. 202.

[5] "Aluminum Expansion," *Aviation Week & Space Technology*, 19 Dec. 1955, p. 33.

[6] Connors, J., *The Engines of Pratt & Whitney: A Technical History*, American Institute of Aeronautics & Astronautics, Reston, VA, 2010, p. 411.

[7] Aboulafia, R., "World Civil and Military Aircraft Briefing," The Teal Group, July 2015, p. 451.

[8] "Aluminum R&D Stresses Refinements," *Aviation Week & Space Technology*, 26 Jan. 1976, p. 85.

[9] "Europeans Gear for Titanium Shortage," *Aviation Week & Space Technology*, 10 Dec. 1979, p. 64.

[10] "Composites Linked to Fuel Efficiency," *Aviation Week & Space Technology*, 25 Jan. 1976, p. 119.

[11] Mecham, M., "Composite Power," *Aviation Week & Space Technology*, 17 April 2006, p. 51.

[12] Pora, J., "Composite Materials in the Airbus A380—From History to Future," 2001, p. 13. http://www.iccm-central.org/Proceedings/ICCM13proceedings/SITE/PROCEEDING/PRO.htm

[13] Norris, G., Thomas, G., Wagner, M., and Smith, C. F., *Boeing 787 Dreamliner—Flying Redefined*, Aerospace Technical Publications International, Perth, Australia, 2005, p. 51.

[14] Brown, S., "The Stuff in Stealth Bombers Goes Civilian as Clever Machines Shrink the Cost of Making Big Parts from Carbon Composites, They're Going into Business Jets and Commercial Rockets," *Fortune*, 6 Dec. 1999, http://archive.fortune.com/magazines/fortune/fortune_archive/1999/12/06/270013/index.htm. [retrieved 2 February 2016].

[15] Sabbagh, pp. 101–105.

[16] Aboulafia, p. 397.

[17] Zimm, P., "Aerospace Raw Materials," SpeedNews Raw Material Conference, Beverly Hills, CA, 7 March 2016.

[18] Counterpoint Market Intelligence Ltd., *Aerostructures 2013*, Counterpoint Market Intelligence, Wiltshire, UK, 2013, pp. 16–28.

[19] Anselmo, J., "Titanium Titan," *Aviation Week & Space Technology*, 26 May 2008, p. 70.

[20] Ladd, D., "RTI to Build Titanium Sponge Plant in Mississippi," *Jackson News*, 17 Sept. 2007, http://www.jacksonfreepress.com/news/2007/sep/17/breaking-rti-to-build-titanium-sponge-plant-in. [retrieved 26 Jan. 2017].

[21] Michaels, K., "Aerospace Raw Materials Outlook," AMM Aerospace Raw Materials Conference, Pittsburgh, PA, 24 April 2012.

Chapter 11

INTERIOR DESIGNERS

If the modern jet age is characterized by equipment standardization, the interior of the aircraft is about customization. It is core to airline branding and service differentiation. The drive to customize, coupled with stringent, constantly evolving regulations and fast-changing consumer electronics technology, makes it one of the most complex parts of the aircraft, despite appearances.

The interior of a jetliner can be split into six main segments: seats; panels, bins, floor coverings, and lighting; lavatories; inflight entertainment (IFE); galleys; and cargo handling systems (see Fig. 11.1).

What began as a fragmented market made up of a patchwork of players has turned into a heavily consolidated market with only a few major suppliers. By 2017, the value of interiors for production aircraft had reached $7.6 billion.

JETLINER INTERIORS: EARLY HISTORY

The onset of the jet age, with larger cabins and long-range flights, caused airlines to think deeply about the aircraft's cabin. What should it look like? What functionality and creature comforts should it offer? There were several thought-leaders in the late 1950s. Henry Dreyfuss, designer of the Lockheed Electra interior, believed that passengers were acclimatized to rooms, not long tubes, and advocated partitions to create separate cabins in the aircraft's interior. Walter Dorwin Teague Associates, designer of the Boeing 707 interior, emphasized "confidence and security," and eliminating wood and "the textile look." Most of the cabin interior—walls, ceiling, and baggage rack—were made of plastic or plastic laminated on metal. The firm also believed that any item in the cabin equipment should be capable of being replaced in a matter of minutes. Raymond Loewy Associates, designer for United Airlines, believed in the use of "light and airy colors," and functionality. It believed

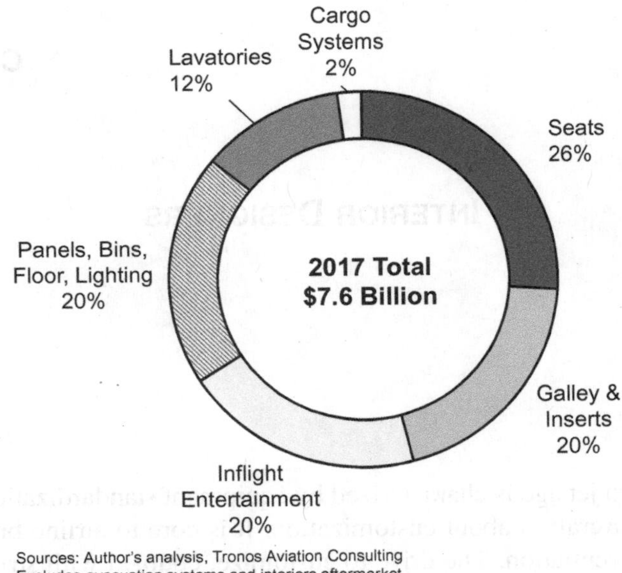

Fig. 11.1 Aircraft interiors segment: 2017.

that all passenger services should be "featured rather than hidden" and should be functionally grouped so that the passenger could find the right button or switch [1]. The buzz surrounding new interior designs was palpable. Teague and Boeing set up a studio in Manhattan in 1956 with a full 707 interior mock-up, and invited the world's leading airlines to visit and imagine the possibilities for their own airlines. It was an experimental sales technique, and it was successful. The cost of a 707 interior was about $500,000 at the time [2]. Airlines experimented with interiors for early jetliners including the 707 and DC-9—hardly any airline was fitted with the same interior. It wasn't until the 727 that some standardization took hold [3].

Airlines experimented with inflight entertainment (IFE) in the 1960s. Trans World Airlines spent $750,000 per year to show movies on its transatlantic flights. American Airlines equipped all 47 of its 707s with an "Astrovision" IFE system in 1964 at a cost of $4 million. Pakistan International and United Airlines were also early adopters. Other airlines, including Air France and Alitalia, remained on the sidelines and waited to see if industry leader Pan Am would adopt IFE before making their own decisions. Swissair was opposed to IFE given its high percentage of business travelers who would prefer to work during a trip rather than watch a movie. "Movies will do nothing to broaden the air travel market base," a Swissair spokesman said [4].

The introduction of the 747 brought more change to interiors; the interior possibilities were endless. Boeing contemplated a layout with five cabins—luxury, first, standard, economy, and coach—and the possibilities of a nonsmoking section or a theater. With the advent of the 747, airlines realized they had additional passenger volume and investigated whether it was possible to differentiate these customers based on willingness to pay different fares. Layouts diverged considerably. American Airlines had 340 seats on the 747's main deck, whereas Pan Am, pursuing a 30% unit cost reduction, had 373–380 [5]. Airlines sought to differentiate other twin-aisles from the 747. Trans World Airlines's L-1011s, for example, had high ceilings, smaller cabins, and wider seats than the 747; it featured eight-abreast seating rather than the 747's nine-abreast configuration, and large overhead storage bins [6].

The jetliner interiors evolution was fast-tracked by 1978 US deregulation, allowing airlines to charge what the market demanded. A previous two-class configuration had existed up until this point, consisting of first class and economy. This configuration was a bit of a misnomer, however, because economy consisted of passengers paying full fare, those on business travel, and leisure passengers paying a discounted fare. Many companies refused to pay for first class travel for their employees, thus forcing them to sit with discounted passengers, causing discontent. At this point, the seeds of the modern-day business class were planted.

Initially, some airlines such as Japan Airlines and KLM created a new section for these business travelers, with upgraded amenities, but the creation of the new business class is generally credited to Pan Am, who rolled out its Clipper Class in 1978. The *C* used to denote Clipper Class is still used as the modern-day notation for business class. In 1979, Qantas began altering its cabin by offering wider business class seats; Pan Am and TWA following in 1981 with seats with extendable leg rests. Pan Am and TWA also started to reconfigure their business classes with wider seats, featuring as few as 6 abreast compared to the previous 10 abreast. Pan Am also introduced "sleeperette" seats on 747s in 1979, which greatly increased passenger comfort.

While business class cabins were getting less seats, economy class was experiencing "densification," or the addition of seats. Newly designed, thin seats replaced the older plush seats and had reduced pitch—the distance between one point on a seat and the same point of the seat in front or behind it. Traditionally, economy class pitch had been approximately 35 in., but these newer seats installed at 32 in. of pitch provided nearly the exact same legroom, a bonus for passengers. Thinner seats allowed for more economy passengers to be packed in while also weighing less than older seats. Additionally, Lockheed, Boeing, and McDonnell all began expanding the size of overhead bins as passengers checked fewer bags and carried them onboard more often. Thus, the arms race was on as airlines and suppliers began experimenting with new seat designs and configurations to provide the most comfort at the lowest cost [7].

Safety standards for aircraft seats also changed. Following the lead of the automotive industry, the US Federal Aviation Administration (FAA) required aircraft seats to meet a 16-*g* static crash test, up from a 9-*g* standard, to demonstrate the structural strength of the seat and restraint system. At the time, the FAA estimated the rule would add $36 in additional manufacturing costs per seat and that improved standards would result in a 3–15% reduction in fatalities and a 2–9% drop in the number of serious injuries during a given accident [8]. Soon, the FAA's ruling became the international standard. Seats on the MD90, 737NG, A340, and 777 were certified to the new standard. In 1990, the FAA also released a new rule requiring better flame and smoke resistance of interior panels. This required all new manufactured aircraft and older aircraft undergoing major interior reconfigurations to have significantly improved fire protection properties in their interiors.

The Unlikely Consolidators

As passenger cabins were getting reconfigured, so, too, was the aircraft interiors market, as an industry outsider began his ascent to the top. Amin Khoury did not intend to get involved in the aircraft industry. A chemist by trade, he formerly worked in the medical field, and his investment group was looking for opportunities in niche markets; the interiors market was attractive due to its fragmented nature. Bach Engineering, known for making the passenger control units (PCUs) for airline seats, fit the bill with its strong portfolio. In 1987, Amin and his brother Robert (Fig. 11.2) entered the aircraft interiors segment with the acquisition of Bach, followed shortly thereafter by the acquisition of Bach's main competitor, EECO Avionics Division, a leader in audio systems. The new entity was named B/E Avionics (Fig. 11.4). B/E achieved a 70% share of the PCU market within two years of its inception and continued to diversify its interior offerings [9].

Credit: B/E Aerospace, Inc.

Credit: Unitech Aerospace

Fig. 11.2 B/E Aerospace cofounders Amin and Robert Khoury.

The year 1992 saw the acquisition of two more targets as B/E expanded its product line into seats and galley equipment with acquisitions of Flight Equipment & Engineering Ltd. and two units of the Pullman Corp. The newly christened B/E Aerospace was now a major player with sales growing from $3 million to $180 million in just six years. It also began to become involved in the inflight entertainment. The Khoury brothers believed that B/E's integrated seat and video capability was a unique value proposition. In 1988, Northwest Airlines became the first airline to introduce in-seat audio and video-on-demand (Fig. 11.3), and Virgin Atlantic became the first airline to introduce integrated seat-back entertainment in 1991 in all classes.

By 1995, B/E Aerospace retained dominance of the PCU market while controlling 85% of aircraft ovens and refrigeration systems and boasting the most in-seat video units. One of B/E Aerospace's primary competitors, France-based Zodiac Aerospace, was following the same growth strategy as it consolidated the interiors sector with acquisitions of Air Cruisers (evacuation systems), Weber Aircraft (seats), Monogram Systems (water and waste), and Sicma Aero Seat. Like B/E Aerospace, this nonorganic growth strategy allowed the company to quickly expand into aircraft safety systems, seats, cabin interior, and aircraft systems (Fig. 11.3).

As the decade progressed, B/E Aerospace and its competitors focused on the emerging IFE market, believed to be worth more than $1 billion. B/E Aerospace pursued live TV with negative results and also struggled with IFE reliability throughout the 1990s. The IFE industry took a hit in 1998 with the

Credit: Delta Air Lines

Fig. 11.3 The first seat-back inflight entertainment system: Northwest Airlines, 1988.

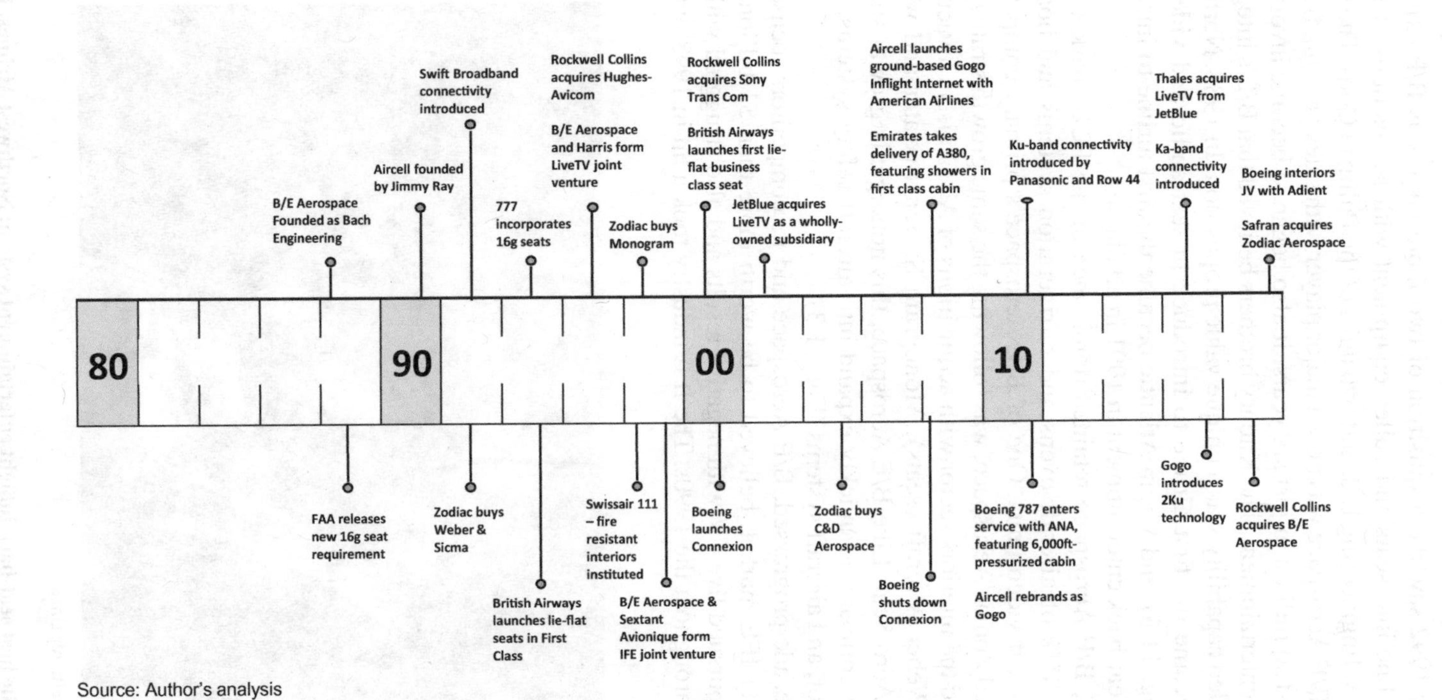

Fig. 11.4 Aircraft interior milestones.

crash of Swissair Flight 111, which took the lives of 229 people due to a fire that appeared to have started within the plane's IFE system. This incident played a large role in the FAA revamping its material flammability requirements.

The growing IFE industry drew the attention of major avionics suppliers, who wanted a piece of the action. In 1997, Rockwell Collins purchased IFE supplier Hughes-Avicom, the second largest supplier, for $120 million. Rockwell believed it could help solve the IFE industry's reliability issues by leveraging its highly regarded engineering and customer support functions. Hughes, which had already been the first entrant in interactive IFE, hoped Rockwell could also provide help with its video-on-demand and live TV streaming development.

Rockwell Collins's competitor, Honeywell, also struck up a strategic alliance with Matsushita as a response to Rockwell Collins's move. Honeywell believed its avionics and communications expertise could help strengthen Matsushita's IFE offerings. Delta Air Lines would become the first airline to demonstrate live broadcast television programing in 1996 on a 767 dubbed *The Spirit of Delta*.

In the late 1990s, B/E completed more acquisitions that further diversified its business, breaking into the oxygen delivery segment via the acquisition of Puritan-Bennett Aero Systems, along with business and passenger jet cabin interiors. Its own growth, however, led to ballooning debt. External macroeconomic factors such as the Asian financial crisis and Boeing's decision to cut widebody production forced B/E to sell 51% of its IFE business to Sextant Avionique, an avionics competitor of Rockwell Collins, who hoped to follow Rockwell Collins's lead by using its proven capabilities to increase the reliability of IFE systems. Unfortunately, this alliance was short lived as B/E Aerospace made agreements in 1999 to sell the remaining 49% to Sextant and exit the IFE business to focus on its core business. This move allowed B/E to narrow its product line and focus on its core competencies, which would serve it well into the next decade. Sextant would rebrand itself into Thales in 2000.

As the major suppliers were consolidating, they also began dealing with new FAA safety regulations and requests from airlines. In addition to new fire and smoke-resistance standards for ceilings, walls, partitions, and overhead bins, the influence of the 16-*g* standard grew on new production aircraft including the 737NG and 777. In contrast, legacy aircraft such as the 767 and 757 were free to install the less-stringent 9-*g* seats.

The early 1990s saw airlines dealing with rising oil prices and decreased air travel demand. In a bid to reduce operating costs they urged interior suppliers to design not only for passenger comfort, but also for reduced maintenance and increased reliability [10]. Equipment that is lighter and more durable would require less money to maintain and operate. For example, vacuum toilet assemblies that needed to be serviced after 9000 operating hours in 1991 required maintenance only every 50,000 hours three years later. Another key to cost reduction was the rollout of self-diagnostic equipment. B/E Aerospace introduced self-diagnosing galleys in 1993. This type of equipment made it easier to identify and address issues, reducing lengthy downtime and allowing airplanes to be in the sky more often [11]. At the same time, first and business class seats became more complex. By the mid-1990s, most international first class seats were electrically actuated, and Delta installed the first electrically actuated business class seats in the late 1990s. In 1995, British Airways introduced the lie-flat seat in its first class cabins. It justified its lie-flat decision, which reduced the number of first class seats on its 747s from 18 to 14, by noting that premium traffic was growing at a double-digit pace and that business travelers accounted for just 15% of total passengers but provided more than a third of revenue [12].

The traditional three-class configuration onboard aircraft lasted a few decades and even saw some major improvements. Virgin Atlantic was the first to chip away at this paradigm in the 1980s by offering two-class service on its 747s: upper class, with first class service at business class fares, and economy. Following Virgin's lead, in 1992 Continental Airlines introduced its BusinessFirst class, a blend of first and business class, thus becoming a two-class operation. With 55-in. pitch and reclining capability, its chairs provided all the comfort of a first class operation without the sticker-shock price. Its competitors soon followed as KLM and Northwest eliminated their first class in 1993, followed by TWA in 1994 and Delta in 1998. Business class was now the new first class.

Interiors in the New Millennium

By the end of the 1990s, first class cabins had evolved to include both angled and fully lie-flat seats. Business class cabins were not far behind, typically using mechanically actuated reclining seats.

In 2000, British Airways faced increased competition and higher fuel prices. It needed a breakthrough in its Club World product, and presented interior design firm tangerine with a challenge: "Find us the holy grail of airline travel sprinkled with a bit of pixie-dust—and astound us along the way." tangerine's answer was novel. Instead of all passengers facing the front of the cabin, seats were paired in a forward/rearward formation. It became known as the "yin-yang." "No one had ever considered arranging [business] cabin space in this way," said tangerine Chief Creative Officer Matt Round. "It gave BA a fully flat six-foot bed in business class while still keeping eight seats abreast." BA customers loved the new design. The company saw a return on its investment within 12 months, and best of all the design was patented.[1] Six years later,

[1]tangerine "Club World—British Airways," http://tangerine.net/our-work/aircraft-interior-design-british-airways-club-world [retrieved 19 Jan. 2018].

tangerine would develop an improved yin-yang configuration with more privacy and 25% greater width (Fig. 11.5). It was called a "pod" configuration.

As the amenities in business class became more and more like those offered in first class, it became more and more difficult for customers to justify the additional expense of a ticket in first class. First class cabins therefore became increasingly populated by those passengers who were upgraded to the cabin free of charge. As a result, many airlines began to remove seats from their first class cabins, electing to replace them with additional business class seats.

As the business class product moved closer and closer to what first class used to be, and became more and more expensive, corporate customers became less willing to pay for their employees to fly in business class. As a result, airlines started experimenting with their economy class cabins. In the United States, several major airlines increased legroom, typically achieved by removing one or two rows of economy class seats. American Airlines, for example, rolled out increased legroom throughout economy class across its entire fleet in 2000. It did this to differentiate its service, and did not charge more for the additional seat pitch. However, as is typical in the airline industry, passengers spoke with their wallets, almost always choosing lower fares over additional legroom [13]. Other carriers did not make the same mistake. United Airlines, for example, introduced its Economy Plus section by removing one or more rows of seats, but increasing the legroom of far fewer seats. This allowed United to increase legroom in those seats by up to 5 in.—far more than what American could

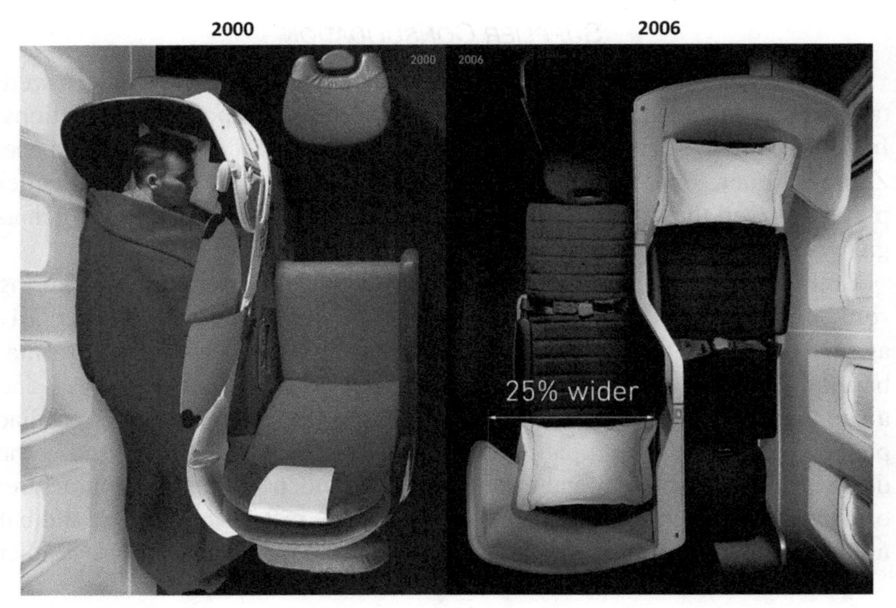

Fig. 11.5 The British Airways yin-yang layout.

offer. Although the seats were initially meant for its most loyal customers and became available to all others at check-in, United quickly realized the value of these seats and began charging access to the seats.

While US carriers focused on providing more legroom in economy class, other carriers began to introduce a new class of service positioned between economy class and business class; it was called *premium economy*. Taiwan's EVA Air and Virgin Atlantic were pioneers, introducing Evergreen Class and Premium Economy, respectively. British Airways responded to Virgin Atlantic's challenge with its World Traveler Plus product in 2000. A "true" premium economy offering became a separate fare class located in a separate cabin onboard the aircraft. The seat was wider than a regular economy class seat, and it provided a 38-in. seat pitch versus 31-in. in economy. It also included seat-back video screens, laptop power, and additional inflight amenities such as premium catering and amenity kits. Tickets cost 20% more than economy fares [14].

At the same time, as first class was being pulled back in favor of business class, some airlines went the other direction with their first class cabins, launching lavish super-premium products that redefined airborne luxury. Emirates was the first to launch a first class suite, where the seat is fully enclosed and includes closable doors (Fig. 11.6). Emirates also installed a shower on the A380 for these premium customers. Other carriers followed suit, with Singapore Airlines introducing its first class suite on its A380 aircraft, which began arriving in 2007.

Supplier Consolidation

While these changes were occurring in the cabin, structural changes were taking place in the interiors supply chain, with several important transactions. In 2000, there were five major seat suppliers: Zodiac-Sicma, B/E Aerospace, Zodiac-Weber, Britax-Rumbold, and Recaro. There were many smaller players as well that focused on other aspects of the aircraft's interior, such as galleys, passenger service units, and overhead bins.

Both Zodiac and B/E Aerospace continued their roll-up strategies, acquiring some of these smaller suppliers during this period. Zodiac was particularly aggressive. In 2004, it completed a $600 million purchase of California-based C&D Aerospace, one of the leading integrators of cabin interiors and a supplier of overhead bins, sidewalls, and passenger service units. Zodiac purchased three other companies in 2007: Driessen, which specialized in the design and manufacture of galley and cargo equipment; Adder, a manufacturer of composite materials supplying stowages, crew rest areas, galleys, and bar units; and TIA, which designed and manufactured insets for galleys.[2] Rather

---

[2]Zodiac Aerospace, "Our Heritage: 120 Years of Entrepreneurial Journey," http://www.zodiacaerospace.com/en/corporate/history/milestones

Credit: Richard Moross / Wikimedia Commons

Credit: Richard Moross / Wikimedia Commons

Credit: JT Genter / The Points Guy

Credit: Travelarz / Wikimedia Commons

Clockwise from top left:
1. Cradle Business Class Seat: British Airways Boeing 747
2. Angle Lie-Flat Business Class Seat: Japan Airlines Boeing 777
3. Lie-Flat Business Class Seat: American Airlines Boeing 767
4. First Class Suite: Etihad Airbus A380

Fig. 11.6 The evolution of premium cabins.

than consolidating its acquisitions, Zodiac left them as stand-alone companies that sometimes competed against each other. B/E Aerospace bought several machined components suppliers for interiors and passenger-to-freighter conversions, and in 2007 bought oxygen systems supplier Draeger Aerospace from Cobham plc.

A third interiors conglomerate, Premium Aircraft Interiors Group (PAIG), emerged during this time and went on to purchase Sell (galleys), Dasell (lavatories), HeathTechna (interior retrofits), and Contour (seats). The IFE segment also consolidated, when in 2000 Rockwell Collins purchased Sony Trans Com, one of the few remaining IFE suppliers not aligned with an avionics OEM.

INTERNET CAFES IN THE SKY?

The dot.com craze swept the aerospace industry in the early 2000s as aircraft OEMs began focusing on how to bring two-way Internet services to passenger cabins. In 2000, Boeing launched Connexion, its own foray into what became known as inflight entertainment and connectivity (IFEC). Connexion was a Ku-band satellite-based system that would leverage advanced phased-array antennas to provide 1 megabyte per second (Mbps) upload speed (passenger to ground). In contrast, Inmarsat's service at the lower frequency L-band satellite had less than half of the upload speed. Download speeds (ground to passenger) would be up to 20 Mbps. Connexion also offered live television, and passengers connected to the new service via an onboard WiFi network—another innovation.

At the 2001 Paris Air Show, Boeing announced that three major US airlines—American, Delta, and United—would be the first customers for this groundbreaking service. Boeing estimated that Connexion could become a $5 billion/year business by 2010 [15]. The euphoria surrounding broadband IFEC took a hit, however, with the 9/11 terrorist attacks and subsequent recession, and the three US launch customers dropped out. However, Lufthansa committed to launching the service, and British Airways and Japan Airlines planned to follow. Lufthansa deputy chairman Wolfgang Mayrhuber called the system's Internet messages "air mails." "Business travelers will value the significantly improved possibilities to communicate while on board," he said in 2002 [16]. Although it took some time for the service to be fully qualified, it was first commercially demonstrated in 2003 on Lufthansa and British Airways during flights from Frankfurt to Washington, D.C. and London to New York, respectively. The launch followed shortly thereafter in May 2004. Other new airlines signed up for the service in 2004 and 2005, including All Nippon, China Airlines, Singapore Airlines, Asiana, Korean, SAS, El Al, and Ethihad. Prices for the Connexion service varied from airline to airline, but were typically $9.95 for one hour of access, up to $26.95 for 24-hour access [17].

Not surprisingly, Airbus responded to Boeing's new service. In June 2001, it bought a 30% share of Seattle-based Tenzing Communications. Rather than high cost broadband, it would offer a lower cost narrowband service. Finnair, Air Canada, and Varig were early customers.

Despite the euphoria surrounding the "Internet café in the sky," the economic case didn't work. Connexion's service was perceived as slow, expensive, and sometimes hard to use, and passenger uptake was well below initial projections. Boeing did what it could to cut costs, including reducing system and antenna installation costs of up to $500,000 by having airlines perform the hardware modification to the aircraft that Boeing had previously done in-house. It also offered a maritime service. It wasn't enough, and Boeing shut Connexion down in 2006. Estimates are that Boeing spent $800 million to $1 billion on the failed initiative [18].

Passengers remained interested in IFEC, but at a lower price point. US company Aircell had a different idea. Why not avoid expensive satellites and leverage terrestrial cellular communications infrastructure instead? In 1991,

Credit: Gogo

Fig. 11.7 Jimmy Ray popularized jetliner Internet service.

Aircell founder Jimmy Ray (Fig. 11.7) sketched an idea for an affordable system on a paper napkin at a barbecue restaurant in Denison, Texas. Ray's "air-to-ground" (ATG) approach used a cellular radio network where jetliners communicated with more than 200 towers in the continental United States, Alaska, and Canada. This was a "line of sight" system where ground-based antennas needed to be visible to aircraft, so the power requirements and installation costs were significantly cheaper than for satellite communication. There was a drawback, however—service wasn't available over the oceans or geographic regions outside of North America. Aircell initially focused on providing telephone service, where it competed with GTE's Airphone and AT&T's Air One. Passenger utilization of the new services, however, was low, with aircraft logging just a few telephone calls per flight. Airlines, sick of carrying around the equipment without corresponding revenue, eventually discontinued the service.

Aircell shifted its strategy to emphasize inflight connectivity instead. In 2006, it was granted an exclusive license from the US Federal Communications Commission for air-to-ground communications in the 3-GHz band for inflight Internet access. Aircell launched the new service with American Airlines in 2008, starting on premium routes from Kennedy Airport in New York City to San Francisco, Los Angeles, and Miami before expanding to its entire domestic route network. Delta, Air Canada, Virgin America, and United followed suit, and soon hundreds of aircraft offered the Gogo service (Fig. 11.8). In contrast to Connexion, Gogo was a narrowband offering and inexpensive—typically less than $10 per flight. It found a sweet spot in the IFEC market and pioneered jetliner inflight Internet connectivity. In 2011, Aircell changed its name to Gogo to reflect the company's primary product offering.

In addition to changes to inflight connectivity, there were changes to inflight entertainment. Personal screens were introduced in the late 1980s, and continued to increase in popularity. Since its inception, JetBlue has been known to have live television available at each of its seats. LiveTV, which

Technology	Year Introduced	Coverage	Relative Cost	Peak Data Speeds	Connectivity Service Providers
SwiftBroadband (L-Band)	1991	Global	$$$$	✓	Inmarsat, Iridium, OnAir, Thales
Early Ku-Band (Connexion)	2003	Global	$$$$	✓	Connexion
Air-To-Ground	2008	North America	$	✓	Gogo
Air-To-Ground 4	2012	North America	$	✓✓	Gogo
Ku-Band Broadband	2010	Global	$$$	✓✓✓	Panasonic, Gogo, Global Eagle
Ka-Band Broadband	2015	Global	$$	✓✓✓✓	Inmarsat, Viasat, Thales, SES, Gogo
2Ku-Band Broadband	2016	Global	$$	✓✓✓✓	Gogo

Source: Author's analysis, Gogo

Fig. 11.8 Overview of air transport passenger data channels.

JetBlue acquired in 2002, was the provider of this service until it was sold to Thales for $399 million in 2014. In addition to JetBlue, Continental Airlines installed LiveTV on some of its Boeing 737 and 757 aircraft, while Delta Air Lines installed it on the 757s of Song, its new low-cost carrier. Other major suppliers of inflight entertainment systems included Honeywell Aerospace, Rockwell Collins, and Panasonic Avionics.

While there is surely money to be made in in-flight entertainment systems, there are also some challenges to profitability, especially for larger firms. IFE tends to be a very customized product, with airlines typically asking for customized systems and user interfaces. This drives up the fixed costs for each installation and limits the ability of suppliers to achieve economies of scale. This equated to low supplier profitability. As a result, Rockwell Collins decided not to offer IFE systems on the aircraft that were launched in the 2000s, such as the Boeing 787 and A380, effectively signaling its pullback from the jetliner IFE market.[3] "The strategy for getting into IFE was sound, but it turned out to be a bad idea," reflected former Rockwell Collins CEO Clay Jones.

> "It was the singularly worst decision that I made during my time at Collins. We won $650 million in our first year but our problems were twofold. First, we were timid about integrating Hughes Avicom into the Collins system…we didn't fix the quality problems fast enough. Second, we underappreciated the fact that IFE is like a narcotic. Once you're on it, it's hard to get off of it. And none of the airlines really want to have it. It adds weight and cost, and airlines didn't want to pay anything for it. We thought if we improved the quality that airlines would pay for it."[4]

[3]Mary Kirby, "Rockwell Collins to re-enter in-seat IFEC market," FlightGlobal, 13 Sept. 2011, https://www.flightglobal.com/news/articles/interiors-rockwell-collins-to-re-enter-in-seat-ifec-361974/. [retrieved 12 Feb. 2018].

[4]Clay Jones (former CEO—Rockwell Collins), interview with author, 5 Nov. 2015.

INTERIORS IN THE 2010S

The introduction of the Boeing 787 into revenue service in 2011 represented a step change in aircraft interiors and passenger comfort, primarily due to its composite construction. As a result, the passenger cabin could be pressurized to the equivalent of 6000 ft of altitude compared to 8000 ft in other jetliners. The 787 also featured a cabin humidifier for premium cabins, increasing passenger comfort and mitigating the effects of jetlag. The windows on the 787 were also about 65% larger than other aircraft, offering a broader outside view for passengers, as well as an electrochromic shading system that replaced traditional window shades. The 787 also served as the debut for LED "mood lighting" on a production aircraft, which offered gentler lighting themes compared to fluorescent lights. Larger, pivoting overhead bins also provided more room for carry-on luggage, as well as a more open cabin.

While Boeing innovated with interior features on the 787, airlines also started to ramp up new interior modifications on existing fleets. With improved airline profitability following rounds of consolidation and the decline in fuel prices after the financial crisis, airlines could spend on discretionary items like the interior and passenger amenities. Many airlines also realized that modifying the interior to look like new from the inside was a cost-effective way of operating an older fleet, while still providing the latest amenities to passengers and avoiding costly new aircraft. All of this was part of the renewed focus on return on invested capital (ROIC) as a key performance indicator of airlines.

From the passenger point of view, one of the most obvious changes in the cabin was the reconfiguration of seats to offer more capacity, also known as "densification" (see Fig. 11.9). By increasing seat capacity, airlines could effectively manage the operating costs of an aircraft, given that the additional seats are filled. For example, Air Canada outfitted its 777-300ER with 450 seats, which previously had configurations with 350 seats that is typical of most operators. By increasing seat capacity, Air Canada experienced a 21% decline in unit operating costs on the aircraft [20].

To squeeze the additional seats into the same fixed amount of space, some airlines introduced "slimline" seats that allowed operators to squeeze more rows into the same area. Others reduced seat width to increase the number of passengers abreast. Air Canada, for example, modified its 777-300ER cabin from a 9-abreast cabin to a 10-abreast configuration. The cabin densification phenomenon was not limited to seats, either, as airlines employed new lavatories and galleys that consumed less cabin real estate.

In North America, airlines also began to embrace true premium economy seats on international flights (Fig. 11.10), like the product their European and Asian counterparts began to install in the 2000s. Air Canada was the first to offer premium economy in 2013. American Airlines introduced premium

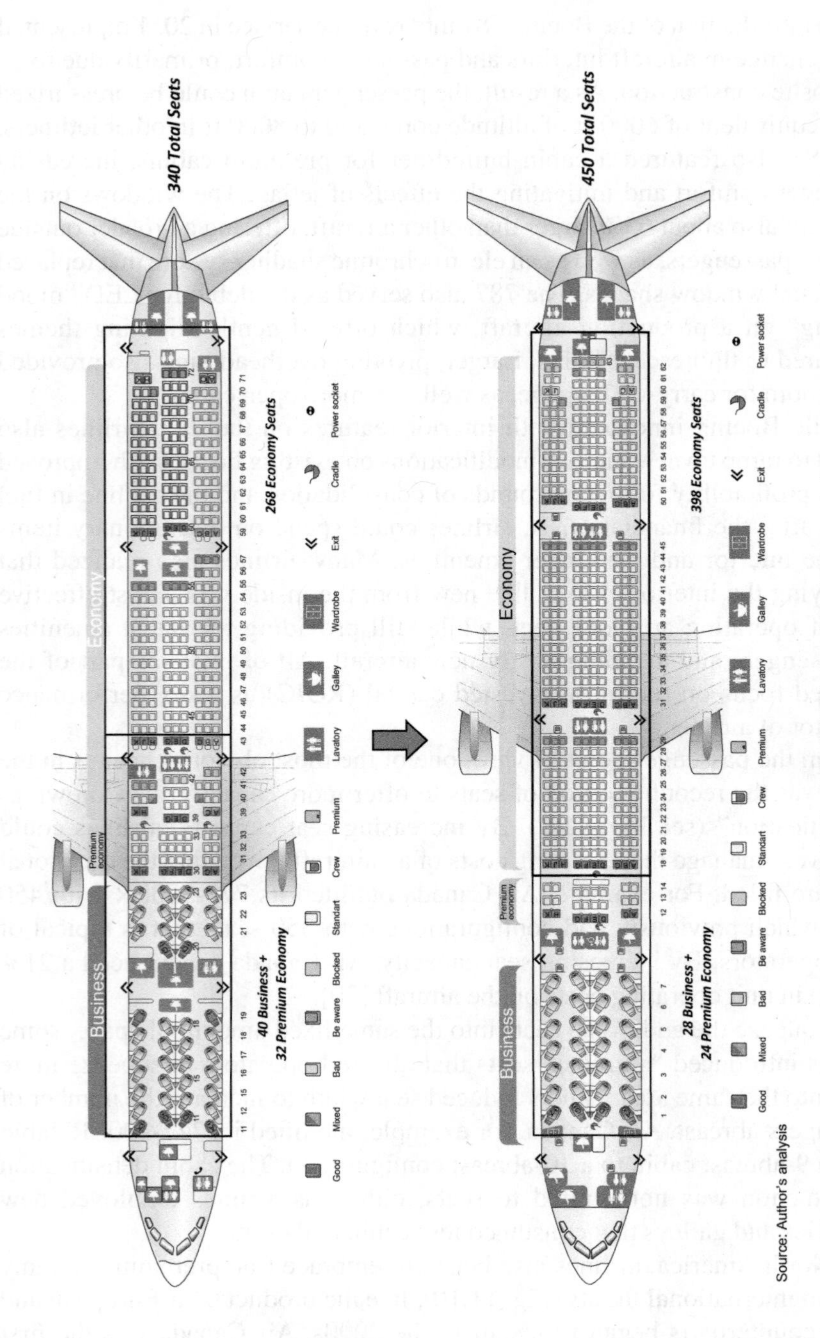

Fig. 11.9 777-300ER densification example.

Credit: Alex S.H. Lin

Fig. 11.10 Premium economy seats.

economy on its Boeing 787-9 fleet in 2016, and Delta Air Lines followed with a new Premium Select product on its A350 XWBs. Soon, North American carriers began offering premium economy on domestic flights.

While premium economy gained traction, the emergence of *ultra-low-cost carriers* in the 2010s created demand for extremely simple, no-recline, slim-line seats. There was also more focus on lightweight seats. New designs and materials, including magnesium, titanium, and composites, helped to reduce the weight of an economy seat by 30% compared to earlier generation seats.[5]

Cabin designs had come a long way in several decades. Which airlines had the biggest impact on interiors? "In my mind, there are three airlines that stand out," said aircraft interiors expert Gary Weissel:

> British Airways came out with the yin-yang seat—the seats that face forwards and backwards. This made it possible for high-density lie-flat seats. The second was Singapore Airlines, which was always known for excellent service and comfortable large seats. They were the first to roll out a shower, and now they have bedrooms in the A380. Finally, there is Emirates Airlines, which is known for glitz, glamor and creative cabin designs. They introduced first class cabins with doors.[6]

Operations Headaches

The second decade of the new millennium brought new challenges to interiors suppliers, as they dealt with higher production rates for single-aisles as well as the introductions of the 787 and A350 XWB. At the same time,

---

[5]Gary Weissel (Managing Director—Tronos Aviation Consulting), interview with author, 18 Jan. 2018.
[6]Gary Weissel, interview with author, 21 June 2015.

there was pent-up demand for interior updates from airlines that deferred investment in the wake of the Great Recession. This spike in demand, coupled with the complexity of the new interior products, caused a breakdown in interiors supply chain performance. Zodiac Aerospace quickly became the poster child for this issue and was publicly blasted by Boeing and Airbus for delaying delivery of their new widebody jets, for which the OEMs were trying to pay down the large development costs.

At the heart of Zodiac's problems was an unwieldy organization resulting from the large number of acquisitions over more than two decades. As production rates increased with a growing product catalog and the bespoke nature of interiors, the weaknesses in its organization were laid bare. The sales function, which Zodiac decided to centralize, signed contracts without sufficiently communicating with its businesses about plant capacity. At the same time, Zodiac pursued supply chain consolidation among its formerly decentralized businesses, which led to inevitable stock-outs and growing pains. The result was numerous delays, missed deadlines, customer complaints, and declining profits. This wasn't dissimilar to B/E Aerospace's experience in consolidating its acquired seating companies in the late 1990s, which led to significant delivery issues at the time with Boeing.[7]

In part to address operational issues at suppliers like Zodiac, Boeing recruited new suppliers. One example was LIFT by EnCore, a new California-based seat manufacturer that would exclusively supply economy seats for the 737. The founders, Jim Downey and Tom McFarland, also ran C&D Aerospace before selling it to Zodiac Aerospace in 2005. For Boeing, the move represented a significant step in supply chain rationalization by standardizing the type of economy seat available for 737 customers.

Boeing would continue working with new suppliers, even from outside the aerospace industry. In March 2017, Boeing and Adient, a supplier of seats for the automotive industry, announced a partnership in which Boeing would provide technical specifications and insight into FAA certification. Again, for Boeing, the move represented the willingness to train up new suppliers who would agree to its revised terms and conditions. Adient, meanwhile, would be able to grow its aerospace business significantly [21].

Despite decades of consolidation already behind it, the interiors industry continued to rationalize into the 2010s. The industry behemoths, B/E Aerospace and Zodiac Aerospace, conducted several transactions in the first half of the decade. Zodiac led the total number of acquisitions, with 10 from 2010 to 2014, including niche suppliers like Sell GmbH, a provider of galley equipment, and Cantwell Cullen & Co., a provider of electrical wiring. Larger Zodiac acquisitions in the decade included Greenpoint Technologies, a US-based provider of VIP cabin interiors

[7]Weissel interview, 18 Jan. 2018.

Timeframe	B/E Aerospace	Zodiac Aerospace
1980s	• Bach Engineering (passenger seat controls) • EECO Avionics (passenger audio systems)	• Air Cruisers (evacuation slides and rafts)
1990s	• Aircraft Products Company (galley equipment) • PTC Aerospace (seats) • Flight Equipment & Engineering Ltd. (seats) • Aircraft Furnishing Ltd. (seats) • Royal Inventum (galley equipment) • Acurex Corporation (onboard refrigerators) • Nordskog Industries (galleys) • Phillips Airvision (aircraft audio and video systems) • Burns Aerospace (seats) • Puritan-Bennett Aero Systems (oxygen systems) • Aircraft Modular Products (business jet interiors) • SMR Aerospace (cabin interior design) • Aerospace Lighting Corp. (business jet lighting) • C.F. Taylor Interiors (cabin interior design)	• Weber Aircraft (seats) • Sicma Aero Seat (seats) • Monagram Systems (water & waste treatment) • Intertechnique (oxygen, fuel, actuation, & electrical)
2000s	• T.L. Windust Machine (machined components) • Alson Industries (machined components) • DMGI (machined components) • Maynard Precision (machined components) • Draeger Aerospace (oxygen systems)	• Icore (electrical harness components) • Avox Systems (oxygen systems) • C&D Aerospace (interior products) • Driessen (galleys) • Adder (interior composites) • TIA (galley inserts)
2010s	• TSI (power and thermal management) • LaSalle Electric (aircraft lamps & lighting products) • Teklam (honeycomb panels) • WASP (lighting, control units, & switches) • EMTEQ (interior lighting & power management) • Fischer (helicopter seating)	• Sell GmbH (galleys and galley inserts) • Cantwell Cullen & Co. (electrical and wiring) • Heath Tecna (interior components) • IMS (inflight entertainment) • Contour Aerospace (cabin reconfiguration) • Threesixty Aerospace (interior products) • TriaGnoSys (inflight entertainment & connectivity) • Pacific Precision Products (oxygen systems) • Greenpoint Technologies (VIP cabin interiors) • Enviro Systems (environmental control systems)

Source: Author's analysis

Fig. 11.11 Zodiac and B/E Aerospace interior acquisitions: 1987–2017.

and modification services, and Contour Aerospace, a provider of aircraft reconfiguration services. B/E Aerospace, on the other hand, conducted six transactions during the first half of the 2010s. Notable acquisitions included EMTEQ in 2014, a manufacturer of aircraft interior lighting solutions and power management systems. See Fig. 11.11 for more detail on these two companies' history of acquisitions.

By the middle of the decade, the interior supplier landscape appeared drastically different from the industry of the late 1980s and early 1990s. Between 1987 and 2014, Zodiac Aerospace and B/E Aerospace conducted a total of 48 acquisitions, rolling up suppliers all across the interiors supply chain.

EVOLVING CONNECTIVITY

The 2010s would bring major changes in aircraft connectivity. There are three broad types of data that need to move on and off a jetliner:

- *Aircraft control and safety*: Air traffic control communications, navigation, weather, and aircraft dispatch
- *Airline operations*: Aircraft health management, flight operations, and electronic flight bag information
- *Passenger services*: Inflight entertainment, Internet, messaging, and voice

Aircraft control and safety was handled by the avionics discussed in Chapter 9—VHF radios for voice and datalinks when over land, and HF radios or satellite communications when over the ocean or in remote areas. These communications pipes needed to be secure, and there were three significant communications services suppliers: ARINC (owned by avionics supplier Rockwell Collins), SITA (a cooperative founded in 1949 and owned by 400 airlines, OEMs, governments, ground handlers, and airports), and Iridium Communications (an operator of a constellation of 66 L-band low-Earth-orbit satellites). Their communications channels were narrow band and expensive, which meant that airlines were selective about the data transmitted over them. The system used for data transmission was called ACARS (aircraft communications addressing and reporting system), which was developed by ARINC in 1978.

The second category of information, airline operations, was not considered as sensitive as aircraft control and safety but transmitted over the same secure and expensive communications channels. The onset of aircraft and aeroengine health management systems in the 1980s and 1990s created demand for transmitting data; however, the expense of the communications services meant that snippets of data were downloaded three or four times per flight. In most instances, airlines manually downloaded aircraft health management data when on the ground or via an airport datalink or mobile telephony to avoid the expense of using ARINC or SITA. Although airlines valued the transmission of operations data, the public didn't give much thought about it until 8 March 2014. On this day, Malaysia Airlines Flight 370 took off from Kuala Lumpur bound for Beijing. The Boeing 777-200ER carried 239 passengers and flight crew. The aircraft disappeared from air traffic controllers' radar screen over the South China Sea less than one hour after takeoff. Military radar continued to track the aircraft as it deviated westwards from its planned flight path and crossed the Malay Peninsula. It disappeared about 200 nm northwest of Penang, Malaysia. The largest and most expensive aviation search in history ensued. The investigation showed that the aircraft sent an ACARS transmission shortly after disappearing from radar to transmit health management data from the aircraft's Rolls-Royce engines. The aircraft's satellite data unit aboard the aircraft responded to hourly status requests for another seven hours until the aircraft was lost. The location where the aircraft went down was somewhere over the Indian Ocean. The search found a few parts from the aircraft in Mozambique and Reunion Island, but the aircraft was never found. A modern jetliner and 239 passengers simply vanished. The public and the victims' families were shocked. How could this happen in the age of connectivity? It turns out that continual connectivity was not required and very expensive. In response to Flight 370's disappearance, the International Civil Aviation Organization adopted new standards for aircraft position reporting over open ocean, extended recording time for cockpit voice recorders, and, as of 2020, new aircraft designs must have a means to recover the flight recorders (or the information they contain) before they sink below the water.

This tragedy aside, the 2010s brought huge changes in the third category of data—passenger services. Early leader Gogo continued to expand and launched an enhanced network called ATG-4 that tripled its data speeds. By 2014, Gogo was installed on 2000 jetliners and an additional 2000 business aircraft. It was the connectivity market leader even though its service remained narrowband and limited to flights over North America.

Airbus continued its push into connectivity in forming a joint venture with SITA called OnAir. The system used Inmarsat's L-band (1–2 GHz) Swift Broadband satellite network. Singapore Airlines and Air Asia became launch customers in 2011. IFE and avionics supplier Thales also got into the connectivity service business with its TopSeries offering. Qatar Airways signed up its 787 fleet in 2010.

The next higher frequency band—Ku (10.7–14.5 GHz)—offered significantly higher transmission speed, but lay dormant for airlines since the failure of Boeing's Connexion service in 2006, although there were services catering to business jets. Panasonic followed Thales's lead in adding services to IFE equipment when it picked up the remnants of Connexion, updated it, and offered a Ku-band service called eXConnect. Former Connexion customer Lufthansa was an early customer and began operations in 2010. In the same year, Southwest Airlines introduced another Ku system, Row 44, to its fleet a year after it selected the new connectivity supplier to outfit its fleet of 550 737s. Its introductory pricing was just $5 per flight.[8] This was a game-changer for several reasons. First, whereas Southwest's US competitors had opted for ground-based Gogo, it chose Row 44 due to its speed, bandwidth, and flexibility that allowed it to offer a custom portal with a wide range of entertainment choices and new ways to generate revenue in flight [22]. Second, if a highly respected low-cost carrier (LCC) that values equipment simplicity was connecting its cabins with advanced satellite communications, could other airlines be far behind? Shortly after Southwest's decision, European LCC Norwegian Air Shuttle announced plans to equip 60 737s with Row 44. With momentum in hand, Global Eagle, a company founded by two former Hollywood executives, acquired Row 44 in 2012. Other new Ku-band connectivity providers joined the fray. Gogo introduced Ku satellite service in 2013 with speeds of up to 30 Mbps.

The next frontier was Ka-band (20–30 GHz), which promised speeds of up to 50 Mbps and smaller antennae than Ku systems. Two new providers threw their hat into the ring. In 2010, US-based Viasat announced development of the first Ka-band inflight broadband system with JetBlue as its launch customer.

[8]"Southwest Airlines Announces Onboard WiFi Pricing," Southwest Airlines Press Release, 28 Oct. 2010.

Viasat launched its first satellite, which focused on North America, in 2012, and inflight broadband service was introduced several years later. UK-based Inmarsat had an even more ambitious goal to launch four Ka-band satellites providing global coverage. The satellites were launched between 2013 and 2015, and Inmarsat made its service available to IFEC suppliers. Resellers of Inmarsat's new service included Gogo, OnAir, Thales, and SES. Gogo would add a 2Ku service (two Ku antennae) in 2016. By 2017, about 6800 air transport aircraft were outfitted with an inflight connectivity service, out of an active fleet of about 28,000 aircraft [23]. North America, thanks to the legacy of Gogo, had the highest penetration of major markets, and Europe had the lowest thanks to its shorter flight distances and the difficulty of establishing ATG services across so many national jurisdictions.

The explosion of connectivity changed the nature of IFE, and the two began to converge. Thales positioned itself for this trend by moving from IFE equipment into connectivity services, and then purchasing LiveTV from JetBlue Airways in 2014. By combining its Avant IFE system with passenger connectivity, live television, and wireless video services, Thales was now a "full service provider for inflight entertainment and connectivity," said Thales VP-Marketing Duc Tran [24].

The Interiors Segment Comes of Age

In October 2016, avionics supplier Rockwell Collins stunned the aerospace world by announcing the acquisition of B/E Aerospace for $8.3 billion, including $1.9 billion in net debt. Although many were surprised that Collins, primarily known for its avionics products, would make such a large move into interiors with relatively few synergies, others noted the potential upside of the deal. First, the combination would provide a complete nose-to-tail interiors portfolio (Fig. 11.12) combined with Collin's significant electronics and communications technologies—setting the stage for new-generation aircraft interiors concepts. Second, the acquisition provided increased exposure for Rockwell Collins to the buyer-furnished equipment (BFE) market, which allows the company to market and sell products directly to the airline, instead of through aircraft OEMs. BFE customers also don't exert the same pricing pressure as aircraft OEMs do, as exemplified in Boeing's Partnership for Success program [25].

Just three months later, Safran continued the mega-merger spree with the acquisition of Zodiac Aerospace, the chief competitor to B/E Aerospace, for $9.1 billion. At the time, the combination created the second largest aircraft systems supplier, and the product portfolio highly resembled that of UTC Aerospace Systems, thereby creating a mirror supplier across the Atlantic. However, as a result of Zodiac's operational challenges and profit warnings in

Company	Passenger Seats	Galleys	Inserts	Lavatories	Crew Rest	Stowage Units	Lighting	Interior Panels	Luggage Bins	IFE & Connectivity
AIM Altitude		✓								
Boeing					✓	✓		✓	✓	
Bruce Aerospace							✓			
Diehl		✓		✓	✓	✓	✓		✓	
FACC								✓	✓	
The Gill Corporation								✓		
Global Eagle										✓
GOGO										✓
Iacobucci HF			✓							
Jamco		✓	✓	✓						
Luminator							✓			
Nordam								✓		
Panasonic Avionics										✓
Recaro	✓									
Rockwell Collins (B/E Aerospace)*	✓	✓	✓	✓	✓	✓	✓			
STG Aerospace							✓			
Thales										✓
UTC Aerospace Systems							✓			
Zodiac Aerospace (Safran)*	✓	✓	✓	✓	✓	✓	✓	✓	✓	

Source: Author's analysis
Excludes minor and secondary suppliers
* Rockwell Collins acquired B/E Aerospace in 2017; Safran acquired Zodiac Aerospace in 2018

Fig. 11.12 Leading aircraft interiors suppliers: 2018.

2017, the deal was reexamined, and Safran negotiated a 15% lower purchase price in May 2017 [26].

When everyone thought major supplier consolidation was over, United Technologies announced the $30-billion acquisition of Rockwell Collins in September 2017. The announcement came on the heels of the Rockwell–B/E Aerospace integration, and just five years after UTC's own record $18-billion acquisition of Goodrich in 2012. Like the Rockwell–B/E merger, UTC's acquisition of Rockwell broadened the company's product breadth, and the resulting entity would be renamed Collins Aerospace Systems. Although Rockwell's product portfolio and margins were attractive, as were its lean operations, another major reason for the deal was to respond to the three-front aircraft OEM assault on supplier business models: demands for double-digit price concessions (e.g., Partnering for Success and Scope+), OEM vertical integration, and an aggressive push into their aftermarkets [27].

In early 2018, Boeing announced that it was vertically integrating in interiors by starting a joint venture with automotive seating giant Adient (formerly known as Johnson Controls). "Seats have been a persistent challenge for our customers, the industry and Boeing, and we are taking action to help address constraints in the market," said Boeing executive Kevin Schemm. Boeing estimated that the aircraft seating market (production and aftermarket) was worth $4.5 billion and would grow to $6 billion by 2026 [28].

These actions underscored the point that aircraft interiors had come of age. The desire for airlines to achieve service differentiation coupled with new cabin connectivity solutions accelerated the growth of, and interest in, this once sleepy equipment segment. The value of aircraft interiors reached $7.6 billion for production aircraft in 2016, and aftermarket spending for retrofits added billions to this figure. The 2016 Aircraft Interiors Expo, the segment's premier event, attracted a record 16,000 attendees from 180 airlines and 530 exhibitors. This was a long way from Boeing's studio in Manhattan with 707 interior mock-ups 60 years earlier.

References

[1] Christian, G., "Comfort in Airline Cabin Design Makes Dollar Sense," *Aviation Week & Space Technology*, 14 May 1956, pp. 92–97.

[2] "Boeing Shows Luxurious 707 Interiors," *Aviation Week & Space Technology*, 14 May 1956, pp. 92–97.

[3] "Boeing Presses for Standard 747 Interior," *Aviation Week & Space Technology*, 12 Dec. 1966, p. 43.

[4] Carter, J., "Inflight Entertainment Spread Expected," *Aviation Week & Space Technology*, 7 Dec. 1964, p. 28.

[5] "Boeing Presses for Standard 747 Interior."

[6] Watkins, H., "TWA Finds Few TriStar Cabin Problems," *Aviation Week & Space Technology*, 12 Feb. 1973, p. 31.

[7] Pace, E., "How Airline Cabins Are Being Reshaped," *New York Times*, 24 May 1981, https://www.nytimes.com/1981/05/24/travel/how-airline-cabins-are-being-reshaped.html. [retrieved 2 February 2018].

[8] Perry, B., "The 16-g Seat Controversy," *Business & Commercial Aviation*, 1 July 1995, http://aviationweek.com/awin/16-g-seat-controversy [retrieved 13 January 2018].

[9] Cabin Interior Supplier Seeks Synergy Among Acquired Companies," *Aviation Week & Space Technology*, 25 May 1992, p. 67.

[10] Proctor, P., "Cabin Focus Shifts to Convenience Costs," *Aviation Week & Space Technology*, 15 Aug. 1994.

[11] Velocci, A., "Operating Costs Drive Cabin Product Design," *Aviation Week & Space Technology*, 5 Sept. 1994, pp. 91–93.

[12] Shifrin, C., "BA to Revamp First Class Sections," *Aviation Week & Space Technology*, 25 Sept. 1995, p. 39.

[13] "American Air to Put More Room in Coach," *New York Times*, 4 Feb. 2000, http://www.nytimes.com/2000/02/04/business/american-air-to-put-more-room-in-coach.html [retrieved 17 Feb. 2018].

[14] Morrocco, J., "British Airways Adds New Premium Class," *Aviation Week & Space Technology*, 7 Feb. 2000, pp. 45–46.

[15] Croft, J., "Three Airlines to Fly Boeing Connexion," *Aviation Week & Space Technology*, 18 June 2001, p. 100.

[16] Mecham, M., "Boeing Makes Its Internet Connexion," *Aviation Week & Space Technology*, 23 Dec. 2002, pp. 42–43.

[17] "Connexion by Boeing Continues Evolution of Award-Winning High-Speed In-Flight Internet Service," Boeing Press Release, 11 Jan. 2006.

[18] Mecham, M., "Nothing Is Free," *Aviation Week & Space Technology*, 27 July 2009, p. 41.

[19] Kirby, M., "Rockwell Collins to Re-enter In-Seat IFEC Market," *Flight Global*, 13 Sept. 2011, https://www.flightglobal.com/news/articles/rockwell-collins-mulls-in-seat-ife-for-single-aisle-333557. [retrieved 2 February 2018].

[20] Ranson, L., "Air Canada CEO Praises High-Density 777s," *Runway Girl Network*, 15 April 2014, https://runwaygirlnetwork.com/2014/04/15/air-canada-ceo-praises-high-density-777s. [retrieved 6 February 2018].

[21] Bruno, M., "In Adient, Boeing Finds a New Seat Maker," *Aviation Week & Space Technology*, 27 March 2017, http://aviationweek.com/awincommercial/adient-boeing-finds-new-seat-maker. [retrieved 20 January 2018].

[22] Sharkey, J., "Speedy In-Flight Wi-Fi, Even During a Wild Ride," *New York Times*, 18 Oct. 2011, p. B9.

[23] Foster, C., "Connectivity Panel Discussion: Did the Promises Come True?," Valour Consultancy, 14 Sept. 2017.

[24] Reals, K., "Going Short Haul," *Aviation Week & Space Technology*, 31 Aug.–13 Sept. 2015, p. 34.

[25] Michaels, K., "Why the Rockwell Collins-B/E Aerospace Merger Could Work," *Aviation Week & Space Technology*, 1 Nov. 2016, http://aviationweek.com/commercial-aviation/opinion-why-rockwell-collins-be-aerospace-merger-could-work. [retrieved 15 January 2018].

[26] Michaels, K., "Europe's Super Supplier: Reasons to Cheer Safran-Zodiac Deal," *Aviation Week & Space Technology*, 25 Jan. 2017, http://aviationweek.com/commercial-aviation/opinion-europe-s-super-supplier-reasons-cheer-safran-zodiac-deal. [retrieved 10 January 2018].

[27] Michaels, K., "Mega Merger," *Aviation Week & Space Technology*, 18 Sept.–1 Oct. 2017, p. 14.

[28] Nagel, K., "Adient to Make Airplane Seats in New Joint Venture with Boeing," *Cranes Detroit Business*, 17 Jan. 2018, http://www.crainsdetroit.com/article/20180117/news/650281/adient-to-make-airplane-seats-in-new-joint-venture-with-boeing.[retrieved15 January 2018].

Chapter 12

UNSUNG HEROES

If the avionics are the brains of the aircraft, to evoke the human body metaphor, aircraft components and systems are the *organs*, providing vital functions for the aircraft's operation and passenger safety and comfort.

Hydraulic pumps push high-pressure fluid around the aircraft to power the aircraft's flight controls, landing gear, and brakes. Pneumatic systems, fed by air bled off aeroengines and mixed with outside air, provide cabin air conditioning and anti-icing. Engine-mounted electrical generators provide power to aircraft components, avionics, and the cabin. Auxiliary power units provide electrical power and cabin air; they also are used to start the aircraft's engines. Fuel systems carry kerosene from fuel tanks to the aircraft's engines, which in turn feature sophisticated controls. There are other important components, including landing gear, brakes, sensors, safety equipment, and fire suppression.

As of 2017, the value of jetliner component and systems production was \$12.9 billion (Fig. 12.1); adding in the value of the large aftermarket more than doubles this figure.

Historically, the aircraft equipment suppliers were very fragmented. By the early 2010s, however, the sector consolidated as aircraft original equipment manufacturers (OEMs) adopted the Tier 1 supply chain model and daring equipment supplier business leaders pursued economies of scale and integrated systems. How and why did this happen?

AIRCRAFT EQUIPMENT: THE EARLY YEARS

Aircraft in the 1920s and 1930s featured relatively simple systems. Flight controls were operated by mechanical linkages to the cockpit. Fuel and electrical systems reflected the modest performance requirements of the time. Hydraulics were used for aircraft braking and landing gear operation. The

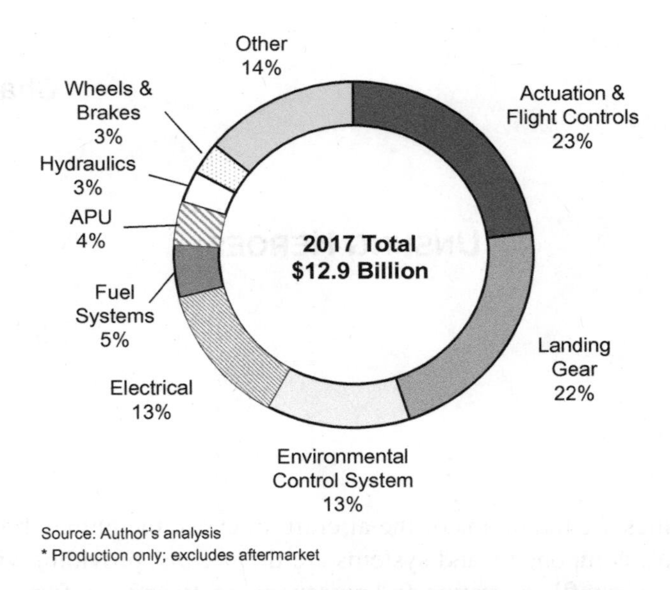

Fig. 12.1 Aircraft systems and component segment size: 2017.

DC-3 ushered in wing flaps operated by hydraulics in the 1930s. Aircraft performance advances in World War II and the early jet age drove aircraft system innovation to handle higher speeds and altitude requirements. The Lockheed Constellation and P-80 fighter had hydraulic power for flight control surfaces; the pressure of these early systems was 1500 psi or less. Another new system mandated by these new aircraft was the auxiliary power unit (APU), a small gas turbine mounted to the tail of the aircraft. Having a second aircraft power source wasn't a new idea. British aircraft and blimps used APUs in World War I, and large US World War II military aircraft, including the B-24 and B-29 bombers, also featured APUs, which were affectionately called *putt putts*. The Germans used APUs for a new purpose: engine starting for its revolutionary jet fighters. These early APUs were two- and four-stroke reciprocating engines.

During the conflict, US entrepreneur Cliff Garrett moved his eponymous company, Garrett Corporation, to a bean field near what would become Los Angeles International Airport to focus on "air research" for pressurizing aircraft. Garrett would go on to produce the pressurization system for the B-29. Recognizing the importance of his work, the US Army asked that he establish a facility inland from the coast. Thus began the Phoenix-based AiResearch Manufacturing Company in 1942. Looking for new sources of revenue following the war, Garrett focused on the design of small gas turbine engines. His work would ultimately lead to the first widely used gas turbine APU, the GTCP85 (see Fig. 12.2). In 1963, the 727 became the first jetliner to feature this innovation; the BAC111 and Mercure would soon follow.

Credit: Honeywell

Fig. 12.2 The first major jetliner APU: Garrett's GTCP85.

The early focus of the APU was aeroengine starting and air conditioning for ground operations as well as takeoffs and landings; over time, airlines became interested in continuous operation of APUs [1]. Soon, they became a standard system on jetliners.

Another innovation coming out of AiResearch, which needed to cope with higher altitude operation, was an air management system that regulated the air pressure and temperature in the aircraft. The heart of these systems was the air cycle machine, a component that uses a reverse Brayton cycle—the thermodynamic cycle of a gas turbine engine—and heat exchangers to cool engine bleed air from 500°F to near-freezing temperatures (if required). Garrett delivered its first air cycle machine on the Lockheed P-80, and it soon became a fixture on jetliners [2].

One of the leaders in fuel and pneumatic systems was Parker, a Cleveland-based supplier founded by Art Parker in 1918 to build pneumatic brake systems for vehicles, machinery, and airplanes. Two years later, he would enter bankruptcy when a trailer carrying his entire inventory fell off a cliff on the way to a trade show. Undeterred, Parker would start a new business and introduce a flared two-piece tube fitting that was well-suited for airplanes. The *Spirit of St. Louis* would use a Parker fuel system, and like many suppliers, its growth was accelerated by World War II. It established a vision of becoming the "General Electric of fluid power" in the early 1950s. In 1978, it expanded its hydraulics capability when it purchased Bertea Corporation, adding cylinder and valve capability. It also bought VanSickle Industries, makers of flight controls and wheel brake equipment for smaller jets [3]. By the late 1970s, corporate revenue approached $1 billion, and Parker established itself as a major aircraft equipment supplier.

Credit: Liebherr

Fig. 12.3 German entrepreneur Hans Liebherr broadened Europe's component capability.

One of Parker's major rivals was nearly a mirror image of itself: Eaton Corporation. Also founded in Cleveland (but seven years earlier), it focused on hydraulic and pneumatic systems. Like Parker, it served automotive, industrial, and aerospace customers. Its aerospace division was a few miles away from Parker's in Orange County, California. It recruited the same types of employees and pursued the same types of acquisitions as Parker. Their rivalry would only grow with time.

Most aircraft equipment capability was based in North America in the first decades of the jet age. The arrival of the Concorde program and Airbus served as catalysts for new European suppliers. Liebherr is an example. In post–World War II Germany, many parts of the country needed to be rebuilt. Hans Liebherr (Fig. 12.3), who managed his parents' building firm, recognized the business opportunity and developed the first mobile tower crane in 1949. The crane business took off, and Liebherr soon expanded into construction machinery and such industries as machine tools, maritime cranes, and refrigerators. At the end of the 1950s, Liebherr began its commitment in aircraft technology. The company's breakthrough as an aircraft equipment supplier came when, with a British partner, it won the contract to develop and produce the slat/flap actuation system on the A300 and A310 programs. It also manufactured the A310 nose landing gear under license.[1] Liebherr would

[1]Alex Vlielander (President—Liebherr USA), interview with author, 5 May 2015.

develop a strong reputation for quality and win even larger positions on future Airbus programs. In 1989, it would broaden its portfolio when it purchased French air management systems supplier ABG-Semca in Toulouse.

The Concorde program, with its demanding technical requirements, also facilitated European supplier development. French supplier Snecma was a beneficiary, with its subsidiaries Messier and Hispano Suiza making the electronically actuated braking system and landing system controls. It also partnered with Rolls-Royce for the Olympus 593 engines. This was arguably the arrival of Snecma as a major equipment and aeroengine supplier.[2]

Airbus played an important role in systems innovation. One of the major innovations—fly by wire—is addressed in Chapter 9. Mechanical linkages to flight control actuators were replaced by wires on the A320 in 1988. A few years earlier, Airbus pioneered carbon brakes on the A310-300. Why the switch? A jetliner needs a stack of discs for braking, and the engineering challenge is immense. Approximately 1 million J of energy is required to stop a car traveling at 200 km/hr. For an Airbus A340, the comparable figure can be over 1 *billion* J for a rejected takeoff.[3] Steel was favored material for most of the jet age, but it is heavy. In the 1960s, some military aircraft, including the C-5A transport, experimented with beryllium as a lighter material. This gave way to a friction material composed of a composite of carbon fiber in a graphite matrix that became known as "carbon-carbon." This new material offered twice the temperature capability—up to 4000°F—compared to steel, with two to three times the heat absorption capability and just one-third of the density. It was used on the leading edge of the Space Shuttles' wings and intercontinental ballistic missile (ICBM) warheads, and came to jetliners with carbon brakes made by Dunlop on the Concorde [4]. Airbus brought the technology into the mainstream in the mid-1980s when it made Messier-Hispano-Bugatti carbon brakes standard on the A310-300, saving more than 1000 lb per aircraft (Fig. 12.4). An added advantage was durability. Carbon brakes could last 2000–3000 cycles (flights) between overhauls, compared to less than 1000 cycles for steel brakes. The tradeoff was their higher cost.

Boeing responded to the A310-300 by making carbon brakes an option on the 767-300. It selected Bendix Aircraft, which cut its teeth on carbon brakes for F-15 and F-18 fighters, as the supplier. On the 747-400, carbon brakes were standard, and Boeing named BFGoodrich and Bendix as the suppliers. Investing in carbon capability was a big move for BFGoodrich. "Until the 747-400, we were a niche supplier of steel brakes to mostly US airlines," recalled former UTC Aerospace Systems executive Ernie D'Amico. "This program enabled us to build our own carbon brake factory. We were taking on AlliedSignal,

---

[2]Safran, "Timeline," https://www.safran-group.com/timeline [retrieved 25 June 2017].

[3]Safran, "Carbon Brakes," https://www.safran-landing-systems.com/wheels-and-brakes/technologies/carbon-brake [retrieved 3 June 2017].

Credit: Aviation Week & Space Technology Archive

Fig. 12.4 Messier-Hispano-Bugatti introduced carbon brakes to jetliners on the A310-300.

which led in steel and was an early adopter of carbon brakes. They were in the pole position."[4] AlliedSignal may have been at the pole position, but its carbon design experienced oxidation problems on its early production units. This would ultimately open the door for the competition. By the mid-1990s, Aircraft Braking System Corporation became the fifth supplier of carbon brakes when it won a position as an option on the A321. The bets by these suppliers on the new technology was significant. A new production facility to make carbon friction material could cost $100 million, and it could take 8–10 years or more to break even on an aircraft program.[5] The intense competition begged the question: Were there too many aircraft brake suppliers?

Aircraft equipment suppliers in the 1980s conducted most of their value creation activities in their home country for reasons highlighted in Chapter 8. But there were a few globalization pioneers. One was William (Bill) Moog (Fig. 12.5), who developed the electrohydraulic servo valve and cofounded the Moog Valve Company in East Aurora, New York, in 1951. Servo valves provided precise control of how hydraulic fluid was ported to an actuator. Moog would ride the wave of more demanding hydraulics requirements, and by 1954 his servo valves were standard equipment on more than half of all US fighter planes and 70% of guided missiles.[6] As a free-thinking visionary, Moog foresaw the growth of Asia and the need to create a global value chain. Legend has it that he initially favored Singapore, but soured on the idea after

[4]Ernie D'Amico (former Vice President – UTC Aerospace Systems), interview with author, 23 May 2016.

[5]D'Amico interview, 23 May 2016.

[6]Moog Inc., "Moog Timeline," http://www.moog.com/about-us/timeline.html [retrieved 11 June 2017].

Credit: Moog

Fig. 12.5 Bill Moog: a pioneer in flight controls and globalization.

he was asked to leave the Raffles Hotel bar because of his long hair.[7] Moog developed a strong affinity for the Philippines, and chose Baguio, a university town in Northern Luzon, as the site of a new manufacturing facility in the mid-1980s. This was new ground—no one had made a significant aerospace investment in the country, let alone in the agrarian north and a six-hour drive from Manila. “We initially met a lot of skepticism,” recalled former Moog executive Sash Eranki, “but we learned how to manage the transfer and put our most sophisticated equipment in the country. We surprised a lot of people.” Moog doubled down on Asia by establishing a manufacturing and software development center in Bangalore, India in 1989.[8]

Arguably, the first “mega-supplier” of aircraft equipment systems emerged in the mid-1980s when Allied Corporation—a conglomerate with chemicals, energy, electronics, and health-care businesses—bought Bendix, a famous aerospace and automotive component supplier, in 1983. Allied then merged with Signal Corporation, which owned Garrett Aviation. In 1987, a new entity—Allied-Signal Aerospace—was created with revenue of about $3.5 billion. Its divisions included the Los Angeles operations of Garrett AirResearch Group and Garrett Auxiliary Power Division; Bendix Avionics Group in Olathe, Kansas; Bendix Fuel Controls Division in South

[7]Sash Eranki (former Vice President—Moog Aircraft), interview with author, 23 May 2016.
[8]Eranki interview, 23 May 2016.

Bend, Indiana; Bendix Wheels and Brakes in South Bend; and a Garrett Engine Division in Tucson, Arizona [5]. Its capabilities spanned the aircraft. Building on Cliff Garrett's legacy, its APUs were a standard feature on most jetliners.

In response to AlliedSignal's APU dominance, Hamilton Standard (United States) and Labinal (France) created Auxiliary Power International Corporation in 1989, a rare European–US joint venture focused on APUs for single-aisle aircraft.

Aircraft Equipment in the 1990s

Aircraft equipment development in the early 1990s was shaped by the Boeing 777 and Airbus A330/340 development programs, which brought new OEM–supplier collaborations. "The 777 was a game changer and was fundamentally different from the approach Boeing took on the 757 and 767," said former Eaton Aerospace President Brad Morton. "Instead of telling us how to design the components they gave us more independence. It was very refreshing and the most enjoyable program I ever worked on. And it was on time."[9] Airbus behaved the same way, giving its equipment suppliers significant design latitude.

These new twin-aisles also ushered in new component technologies. The A330 and A340 brought Airbus's usage of fly-by-wire technology to twin-aisles. Boeing responded by integrating a fly-by-wire flight control system in the 777. "Boeing studied Airbus closely and took a step-by-step approach," recalled former Moog executive Sash Eranki. "It isn't clear that the 777 would have been by fly by wire without the push from Airbus."[10] Japanese supplier Teijin Seiki won the primary flight control contract for the aircraft. This was significant because there were few Asian aircraft component suppliers at the time. The 180-min Extended-Range Twin-Engine Operational Performance Standards (ETOPS) capability of the 777 treated the APU as a critical safety device and drove higher levels of reliability, as well as the mandate to be flight-startable at altitudes up to the aircraft service ceiling.

Equipment suppliers in this timeframe were very fragmented, and the plunge in jetliner production rates in the early 1990s discouraged investments and acquisitions. A few leaders were outliers. One was then-BFGoodrich president Dave Burner (Fig. 12.6). In 1993, he announced a $500-million expansion initiative while other suppliers were retrenching. Aerospace revenues were $750 million, and Burner's stretch goal was several billion dollars by the end of the decade. With aerospace revenues

[9]Brad Morton (former President—Eaton Aerospace), interview with author, 24 Aug. 2016.
[10]Eranki interview, 23 May 2016.

Credit: Dave Burner

Fig. 12.6 Consolidation protagonist: BFGoodrich President Dave Burner.

collapsing, Wall Street wasn't impressed and punished the company's stock price [6]. Burner didn't take long to act. In May 1993, BFGoodrich paid $200 million to acquire landing gear manufacturer Cleveland Pneumatic. Burner struck six months later when the company bought Rosemount Aerospace, a leading sensor and air data computer supplier, for $300 million—its largest acquisition to date [7].

The mid-1990s brought some supplier consolidation (Fig. 12.7). Europe's two major landing gear suppliers, Dowty (owned by the United Kingdom's TI Group) and Messier-Bugatti (owned by Snecma) merged in 1994. "Messier-Bugatti and Dowty have known each other for many years, and we both wanted to end the French–British [marketing] battle," said Snecma chairman Gerard Renon. The move combined Messier's strength in Airbus programs with Dowty's presence on major US military platforms. Combined revenues were $367 million [8]. This left BFGoodrich (Cleveland Pneumatic) and Canadian supplier Menasco Aircraft Division as the other two jetliner landing gear suppliers.

In the United States, The Triumph Group, an industrial conglomerate serving multiple industries, began to focus on aerospace in 1993 when executive Richard Ill led a management buyout. He would go on to buy numerous aerospace component and aerostructures companies in the 1990s, headlined by the acquisition of actuation supplier Frisby Aerospace. Another new conglomerate, Transdigm, was formed in 1993, and would go on to acquire more than 20 companies over the next several decades. Its focus was on small companies with significant aftermarket content.

In the mid-1990s some companies also divested noncore offerings as they adopted the mantra of GE's CEO Jack Welch to be a top-two supplier—or divest. This led AlliedSignal (the new brand for Allied-Signal) to sell its actuation business to Moog and its landing systems business to Coltec.

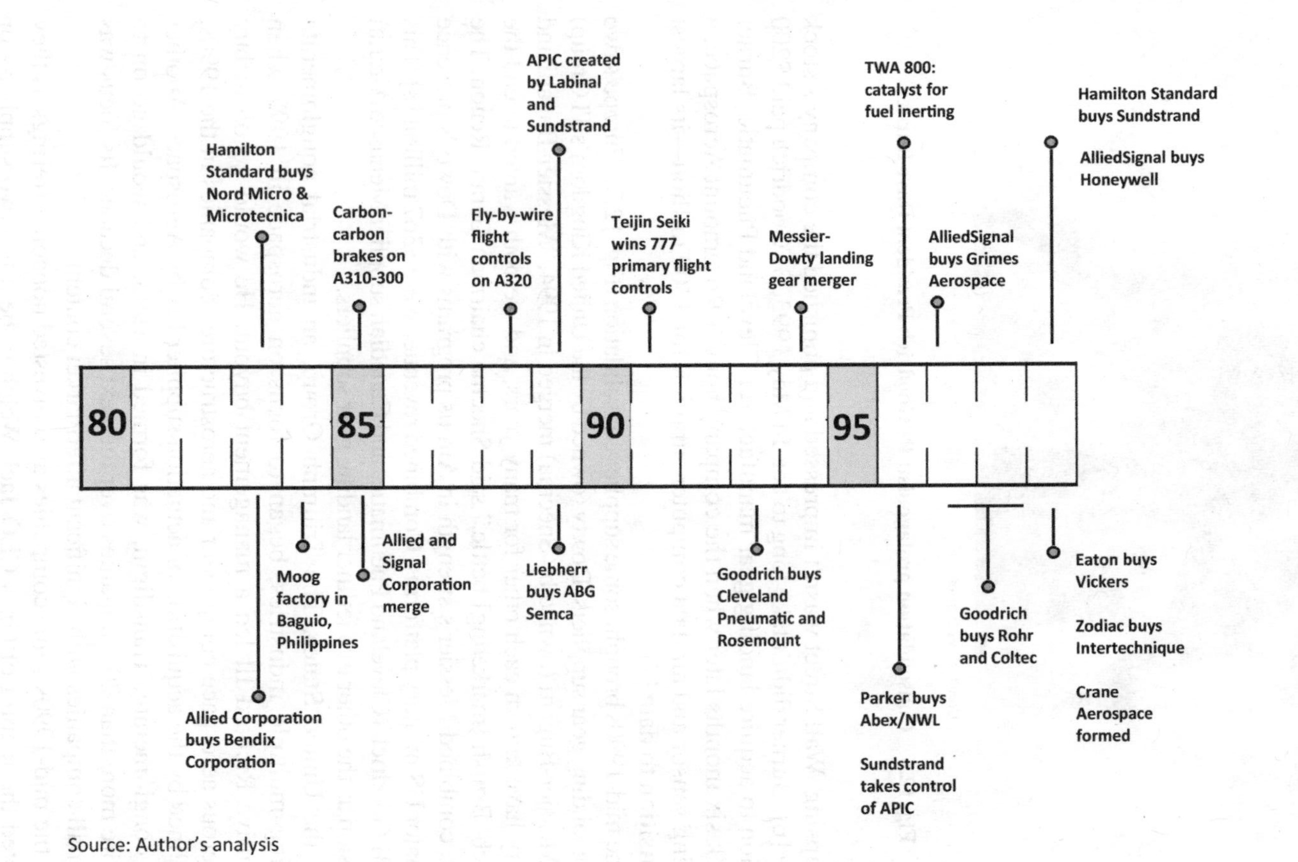

Source: Author's analysis

Fig. 12.7 Aircraft systems milestones: 1980–1999.

Aircraft systems upgrades sometimes come from tragedy. TWA Flight 800 took off from New York's JFK airport on a warm evening 17 July 1996. The aircraft exploded after just 12 min of flight over the Atlantic Ocean, killing all 230 on board. Although investigators never pinpointed the exact cause of the accident, they suspected that the most likely cause was an explosion of flammable vapors in the aircraft's center fuel tank ignited by wiring in the fuel quantity indication system. The explosion of a Philippine Airlines 737 six years earlier and a Thai Airways jet in 2001 occurred under similar circumstances—a nearly empty center fuel tank on a warm day. After 10 years and more than 100 airworthiness directives, the Federal Aviation Administration (FAA) mandated in 2008 that all new aircraft should be equipped with an inerting system that injected nitrogen into the fuel tank to reduce oxygen content and the potential for explosion. Retrofits would be required for an additional 2700 aircraft. The cost would be $92,000 to $311,000 per aircraft [9]. Inerting fuel tanks wasn't a new concept; combat aircraft had long used inerting systems, but the technology was rejected for transport aircraft due to cost and weight considerations. Parker Aerospace, Intertechnique (owned by Zodiac), and Liebherr-Aerospace would ultimately become major suppliers of fuel inerting systems' key components.

The emergence of Tier 1 supply chain strategies (explained in Chapter 8) began to shape the growth strategies of equipment OEMs in the second half of the 1990s. As Bombardier and Embraer asked for larger work packages—including entire systems—many companies concluded that they needed to acquire end-to-end system capability rather than depending on outside suppliers for vital components. Parker Aerospace won early Tier 1 contracts with Bombardier and took on a new challenge in integrating complete fuel and hydraulics systems. "We acquired Abex/NWL in 1996, and they had the Bombardier Global Express hydraulic systems contract," recalled former Parker Aerospace president Bob Barker. "We also won the fuel system on that aircraft—we had to do fuel pumps, tank size, center of gravity computational stuff—things we had never done before. We knew we would lose money, but it would be an investment in creating system capability."[11] Parker would benefit from this experience by winning systems on the Embraer E-Jet—a program with just 38 major suppliers. Other equipment companies would win contracts for entire systems on the E-Jet, although many would lose money as they followed the systems supplier learning curve.[12]

After pausing for a few years, BFGoodrich resumed its acquisition campaign by acquiring nacelle and thrust reverser specialist Rohr in 1997 for $790 million. "By combining our two outstanding companies, we will expand

[11]Bob Barker (former President—Parker Aerospace), interview with author, 7 April 2015.
[12]Morton interview, 24 Aug. 2016.

our capabilities to provide customers with integrated aircraft systems and services," said CEO Dave Burner. "As a larger company in an industry that continues to consolidate and become more highly focused, we will become an even stronger competitor" [10]. In the same year, AlliedSignal would add heft by acquiring Grimes Aerospace, a leading lighting supplier. The following year, BFGoodrich raised eyebrows with a $2.2-billion acquisition of landing gear supplier Coltec. Combined with its Cleveland Pneumatic business, BFGoodrich was the clear leader in landing gear. It was rivaled in capability only by Messier Bugatti Dowty. Landing gear became the first major equipment segment to consolidate.

The 1990s ended with a burst of acquisition activity as equipment suppliers changed their perspectives of critical mass. In early 1999, Eaton purchased Aeroquip-Vickers for $1.7 billion and significantly expanded its aerospace hydraulics capability. "At $2.2 billion, [the company] is too small to realize its strengths," said Aeroquip-Vickers CEO Darryl Allen. The move pushed Eaton closer to its goal of $10 billion in revenue by 2000 [11].

Weeks later, Hamilton Standard made a $4.3 billion acquisition of Sundstrand Corporation, a leading supplier of electrical and mechanical systems and APUs. Hamilton had been eying Sundstrand since the early 1990s. "I drew a picture of all of the Hamilton components, the Sundstrand components and engine accessories," recalled former President Bob Kuhn. "It all fit together, and I argued that it was key to the aftermarket." Parent company UTC, however, wasn't ready for such a major move in the early 1990s.[13] Seeing the synergy, Sundstrand approached UTC to acquire Hamilton Standard in 1999; then-UTC CEO George David responded, "We're not interested in selling Hamilton, but we'd like to buy you."[14] UTC had its way, and the merged company was branded Hamilton Sundstrand. Some 47% of Sundstrand's $1.2 billion in aerospace revenue was from the aftermarket; combined, the companies averaged $700,000 worth of systems per jetliner. "As other suppliers consolidate, we need to pay attention to economies of scale issues," said David. "We've got to keep pace and not get left behind. This means we've got to play in this consolidation game" [12].

This set the stage for an even larger combination in mid-1999. The AlliedSignal–Honeywell merger (as highlighted in Chapter 9) would create more than an avionics leader; it also created an aircraft equipment superpower. It boasted $10.5 billion in aerospace revenue and leading positions in avionics, APUs, wheels and brakes, air management systems, electrical systems, engine controls, and lighting. Moreover, Honeywell's latest integrated modular avionics products included software control functionality for many of these systems. The merger was widely praised by aerospace analysts. "The merger

[13]Bob Kuhn (former President—Hamilton Standard), interview with author, 21 May 2015.
[14]Dave Hess (former President—Hamilton Sundstrand), interview with author, 13 Nov. 2015.

is a good move by both companies," said JSA analyst Paul Nisbet. "Both have superb complimentary technology." *Aviation Week* proclaimed, "...the new Honeywell as a package meant to please primes" [13].

Another equipment conglomerate, Crane Aerospace, formed in 1999 when Crane Corporation combined its four aerospace companies—ELDEC Corporation, Lear Romec, Hydro-Aire, and Interpoint. The $400-million company boasted sensing, power, fuel, and braking systems. It was particularly known for its anti-skid brake control systems.

2000S: "MOONSHOTS" DRIVE INNOVATION

The "moonshots" for the first decade of the new millennium—the A380, 787, and A350 XWB—would drive significant aircraft system technology development to meet demanding performance requirements.

The A380's large size and volume required new approaches to reduce the weight of hydraulic distribution and electrical systems. It became the first civil aircraft to use electrohydrostatic actuators (EHA), an approach that used a local hydraulic fluid supply rather than depending on a central hydraulic fluid reservoir and numerous hydraulic lines. The aircraft also became the first jetliner to use a 5000-psi hydraulic system rather than the 3000-psi standard in use for more than 50 years (Fig. 12.8). Why the upgrade? Higher pressure

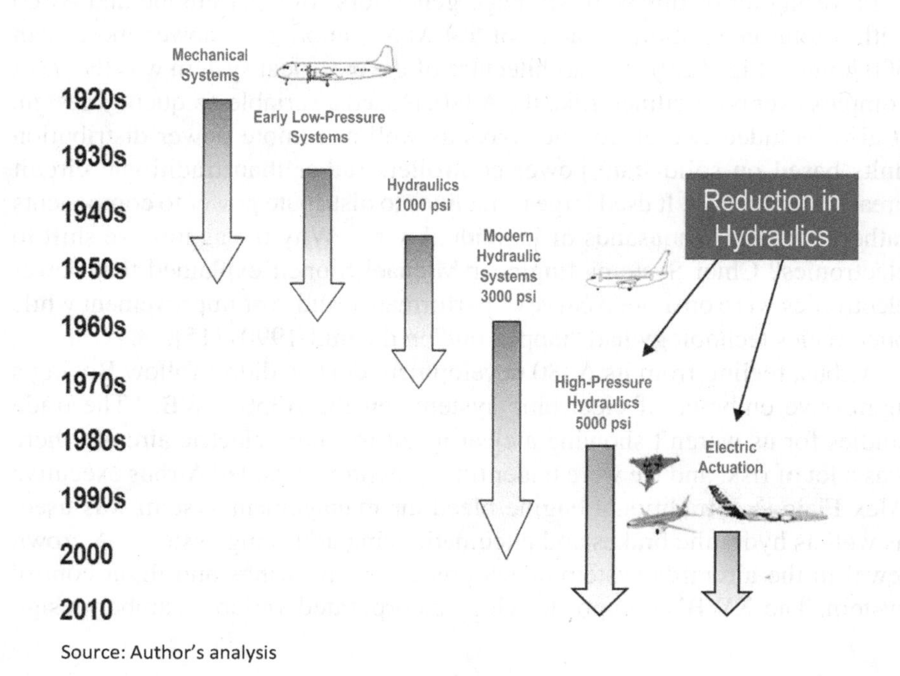

Fig. 12.8 The evolution of hydraulic systems.

enabled smaller diameter hydraulic lines, yielding important weight savings. Another source of weight savings came from the aircraft's massive landing gear, which were made mostly from titanium rather than high strength steel. This was a first for jetliners.

The A380 also used a variable-frequency electrical power system with four 150-kVA generators. This approach eliminated the need for constant-speed drives—heavy and maintenance-intensive devices that ensured a steady frequency output from generators. The A380's wiring debacle aside, it reduced weight by using aluminum power cables (instead of copper) and solid-state contactors and breakers. Still, it had more than 100,000 wires with a length of 292 miles and a weight of 12,500 lb [14].

The 787 would be even more revolutionary in overhauling its electrical, hydraulic, and pneumatic systems. Boeing's aggressive design and performance goals, coupled with advances in electronics technology, empowered Boeing to make groundbreaking decisions. These included:

- Cabin air provided by electronic motors rather than bleed air from the engines
- Electronic anti-icing for the composite wings instead of pneumatics
- Electronic start capability rather than relying on pneumatic starters
- Electronic braking systems—a first for jetliners

Powering all of this were six huge generators (two per engine and APU) with a total generation capacity of 1.4 MW—enough to power more than 500 homes (Fig. 12.9). The architecture of the electrical system was the most complex ever on a jetliner. Like the A380, it used a variable-frequency design. It also included two electronics bays as well as remote power distribution units based on solid-state power controllers rather than traditional circuit breakers and relays. It used large trunk lines to distribute power to components rather than using thousands of individual wires. Why the aggressive shift to electronics? Chief Systems Engineer Michael Sinnett explained that power electronics were on a steep curve of performance and cost improvement while pneumatics technology had "tapped out" in the mid-1990s [15].

Airbus, reeling from its A380 development delays, didn't follow Boeing's aggressive embrace of electronic systems on the A350 XWB. "The trade studies for us weren't showing a clear result for more electric aircraft, there was a lot of risk, and we were under time pressure," recalled Airbus executive Alex Flaig.[15] A traditional engine bleed air management system was used, as well as hydraulic brakes and pneumatic wing anti-icing systems. A crown jewel in the aircraft's system development was its wings and flight control system. The XWB's composite wings incorporated variable camber design

[15]Axel Flaig (Sr. Vice President—Research & Technology, Airbus), interview with author, 10 June 2016.

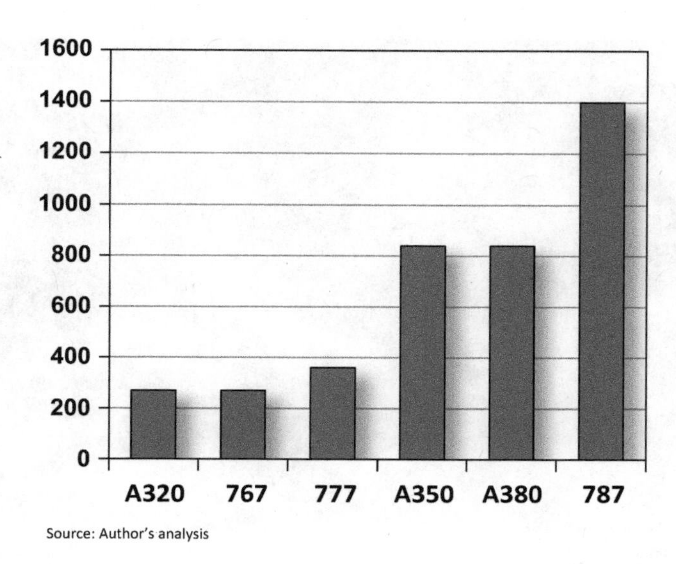

Fig. 12.9 Electrical power-generating capability by aircraft (kVA).

and used differential flaps setting (Fig. 12.10), enabling pilots to use the flaps not only for takeoff and landing, but also while cruising to reduce wing drag. It incorporated an active control system to adjust the wing's moving surfaces using on-board computers. Probes on the front of its wings could perform calculations and send commands to flight controls on the rear of the wing in just 40 ms to ensure smooth flight. "We call it the Airbus spoiler piano," said Airbus's Flaig. "The control surfaces are constantly moving. We have a stiffer wing than Boeing, and believe that a stiff wing plus active control means better aerodynamics. Our design minimized wave drag on the front of the wing, which is a remarkable accomplishment."[16]

The new millennium started off with the gargantuan $44-billion GE bid to buy Honeywell (discussed in Chapter 9), which was ultimately rejected by the European Commission. The shock of this failure and the exodus of leadership would haunt Honeywell for at least a decade.

While European regulators worried about excessive concentration, aerospace suppliers had different thoughts. More consolidation was needed—not just for critical mass, but also to produce complete systems. Goodrich struck again with a major acquisition in 2002 when it bought TRW's aircraft systems business for $1.5 billion. This was Goodrich's second largest acquisition and its 40th since the early 1990s when it began its acquisition campaign [16]. Through this transaction, it gained actuation, engine controls, power generation, and cargo systems businesses.

---

[16]Flaig interview, 10 June 2016.

Credit: Airbus

Fig. 12.10 A350 XWB wing.

Former Eaton Aerospace President Brad Morton recalled the pressure for consolidation. "In the early 2000s, Boeing put leading component suppliers in a room and encouraged us to collaborate to optimize systems. It was hard when each firm was looking out over its interests." Eaton would try several joint ventures with suppliers to create complete hydraulic and fuel systems, but without success. "By 2004, I realized that we needed acquisitions to be successful with systems."[17] Eaton followed with three acquisitions in 2005–2007—the aerospace fluid and air division of Cobham plc, PerkinElmer's aerospace division, and Argo-Tech. The significance of this string of acquisitions was twofold: It moved Eaton from hydraulics and pneumatics into fuel systems, meaning it could now offer systems solutions in each (Fig. 12.11). Additionally, it increased its aerospace revenue by more than 50% to $1.5 billion.

Not surprisingly, Eaton's moves intensified its rivalry with Parker Hannifin, which was also a fuel systems and pneumatics supplier. Parker also bulked up in aerospace when it purchased ABX/NWL (1996) and Shaw Aero Devices (2007). Competitions for jetliner hydraulic systems were typically a battle between these two rivals. Parker Hannifin built a beautiful new corporate headquarters in suburban Cleveland in 1997; years later, Eaton built a much larger headquarters just a few miles away. The rivalry continued.

[17]Morton interview, 24 Aug. 2016.

Company (HQ)	Actuation	Air Management	Electrical	Airframe Fuel	Auxiliary Power Unit	Hydraulics	Landing Gear	Wheels & Brakes	Fire Protection	Engine Accessories*
Major Supplier = ✓ Secondary Supplier = ✓✓										
UTC Aerospace Systems (US)	✓✓	✓✓	✓✓				✓✓	✓✓	✓✓	✓✓
Safran - incl. Zodiac (France)		✓✓	✓✓	✓✓		✓	✓✓	✓✓		✓✓
Honeywell Aerospace (US)		✓✓	✓✓		✓✓			✓✓		✓✓
Liebherr (Switzerland)	✓✓	✓✓		✓		✓	✓✓			
Moog (US)	✓✓									
Parker Aerospace (US)	✓✓	✓		✓✓		✓✓				✓
Eaton Aerospace (US)	✓			✓✓		✓✓	✓			✓
Meggitt (UK)		✓						✓✓	✓✓	✓
Crane Aerospace & Elect. (US)			✓	✓			✓	✓		
Pratt & Whitney Canada (Canada)					✓✓					
Heroux Devtek (Canada)							✓			
Woodward (US)	✓					✓				✓✓
GE Aviation (US)			✓							✓✓
Triumph Integrated Systems (US)	✓					✓				✓
Nabtesco (Japan)	✓									
Esterline (US)			✓							
Curtiss Wright (US)	✓					✓				
Transdigm (US)	✓	✓	✓	✓		✓				
BAE Systems (UK)	✓									✓✓
Ametek (US)	✓	✓								
Circor Aerospace (US)	✓					✓				

Source: Author's analysis

Fig. 12.11 Major jetliner component and system suppliers: 2017.

Engine accessories manufacturer Woodward Governor Corporation, the company that pioneered engine fuel controls and was a leading fuel nozzle supplier, also participated in consolidation when it expanded into actuation and cockpit controls by buying MPC Products Corporation and HR Textron in 2008/2009 (Fig. 12.11). In the same timeframe, Goodrich and Rolls-Royce formed Aeroengine Controls, a joint venture to produce full authority digital engine controls, the brains of modern aeroengines. The joint venture (JV) focused on Rolls-Royce engines and competed with Hamilton Sundstrand and FADEC International, a JV between BAE Systems and Sagem.

THE MEGA-SYSTEMS AWARD

The power of systems capability would be on full display with the 787, and the newly enlarged Hamilton Sundstrand would take full advantage of the situation. Its chief engineer, Joe Ornelas, believed that Boeing was primed for a mega-systems approach. Ornelas was a highly credible messenger given his 33-year career at Boeing and his previous role as the

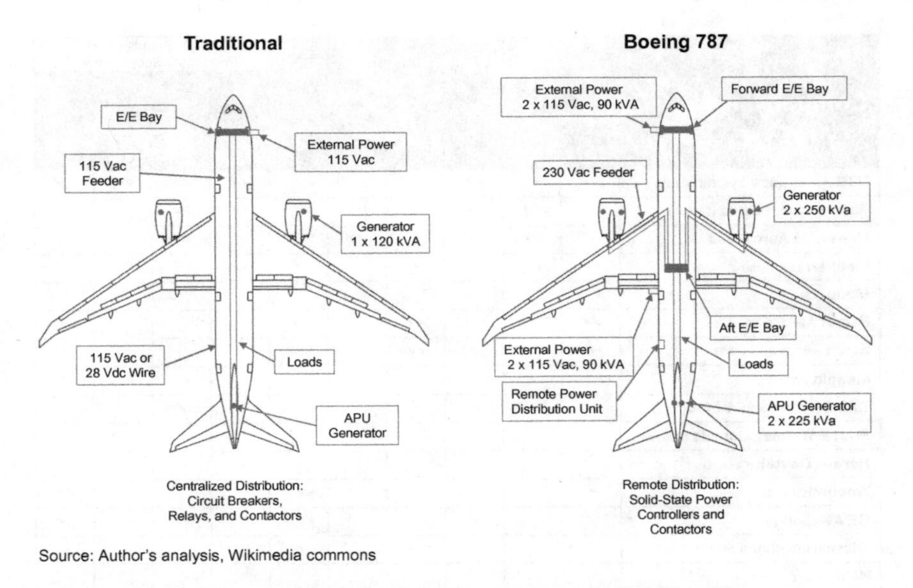

Fig. 12.12 The 787's electrical design used remote distribution; Hamilton Sundstrand was the system integrator.

777's chief engineer.[18] After the Sundstrand acquisition, the company had air management, APU, and electrical systems capability in-house and could conduct trade studies optimizing the interactions of each. Moreover, like many suppliers, it had cut its teeth on systems integration on the E170 program.

Serious trade studies on systems optimization began in 2001 when the Sonic Cruiser program was active, and they had advanced considerably by the 787 program launch in early 2003. Hamilton Sundstrand submitted bids for four major work packages for the aircraft's APU, ECS, power generating/starting, and secondary power distribution systems (Fig. 12.12). Betting on its hunch, it also submitted an unsolicited proposal to tie these four systems together and integrate related systems. "We did something unusual," recalled former executive Bob Guirl. "Instead of delivering just a proposal on a compact disk, we delivered in a large truck of full-scale mock-ups of APUs, electrical distribution racks, generators, as well as simulators and videos." Joe Ornelas and then-President Ron McKenna received the good news in 2005 that its "system of systems" unsolicited proposal was selected by Boeing.[19] Hamilton Sundstrand would add on other systems, including the ram air turbine, electrical motor pumps, and fuel inerting system. It also teamed

[18]Bob Guirl (former Hamilton Sundstrand Vice President—Technology), interview with author, 16 Oct. 2015.

[19]Ibid.

with Zodiac to win the primary power distribution. It was a stunning win, quite possibly the most decisive of any equipment supplier in the jetliner era. Hamilton Sundstrand had hundreds of components on each 787 with a shipset value of more than $2 million. The lifetime value was estimated to be more than $15 billion.[20]

The shock of Hamilton Sundstrand's 787 sweep reverberated around the industry but was particularly strong in Honeywell Aerospace's Phoenix headquarters. "The 787 was a wake-up call for us. We began to think about extended perimeters—not just looking at the APU, but whole installation—ducting, doors, and other components," recalled former Honeywell executive Jeff Johnston. "And we made the A350 our mission." Honeywell Aerospace dispatched a team to Toulouse to work with Airbus on defining the new systems. It came up with a "light engine bleed" architecture that it felt was more efficient than the 787.[21] In 2007, Honeywell won major content on the A350 XWB including the HGT1700 APU; environmental and cabin pressure control systems, supplemental cooling, ventilation, and fans; bleed air systems; and wing anti-ice systems. It also had major avionics content as well. Honeywell estimated the value of the contract to be worth more than $16 billion over the life of the program.[22] This was comparable in value to Hamilton Sundstrand's win on the 787.

While Hamilton Sundstrand and Honeywell won major content on the moonshots, European equipment suppliers continued to grow and gain content on new platforms. Liebherr took major positions on the E-Jet, C Series, A380, and A350 XWB; on the latter, it won its first landing gear position with Airbus. British supplier Meggitt made two acquisitions—Dunlop Standard Aerospace (2004) and Aircraft Systems Braking Corporation (2007)—to position itself as a major braking supplier. In 2009, it won an electronic braking system on the C Series—the first time this technology was deployed on a single-aisle aircraft. In 2011, it added $378 million in revenue when it acquired Danaher's Pacific Scientific Aerospace business, a supplier of electric power, actuation, safety, and security products.

Meanwhile, Snecma made several moves to position itself as the leading European equipment supplier. In 2000, it purchased French supplier Labinal, adding the company's cabling and braking businesses as well as its Turbomeca and Microturbo subsidiaries—important manufacturers of small gas turbines and APUs. The following year, it created Hurel-Hispano, a dedicated engine nacelle and thrust reverser business to challenge US supplier Goodrich. And in 2005, Snecma merged with French defense electronics supplier Sagem to form a new European super-supplier with more than €10 billion ($12.5 billion) in

[20]Hamilton Sundstrand, "This Is Hamilton Sundstrand," July 2007, 8.

[21]Jeff Johnston (former Vice President—Honeywell Aerospace), interview with author, 1 March 2016.

[22]Honeywell Inc., "Honeywell Signs $16 Billion Systems Contract with Airbus on A350 Extra Wide-Body Plane," 19 Sept. 2007.

revenue and 55,000 employees. The new company was named SAFRAN—the French word for the rudder blade on a boat.

RISING CHINA

The year 2008 was pivotal in China for several reasons. First, it marked the creation of the Commercial Aircraft Corporation of China (COMAC) and the launch of the C919 program, as discussed in Chapter 6. Airbus opened its final assembly facility in Tianjin that year. It was also when China consolidated most of its aerospace value chain into a single company, the Aviation Industry Corporation of China (AVIC). China's aerospace suppliers were historically fragmented, decentralized, and balkanized—a legacy of Chairman Mao's desire to reduce China's vulnerability to military attack. Since the Mao era, its aerospace industry had endured numerous reorganizations and was falling behind Western suppliers. In November 2008, China's central government created AVIC, an integrated entity with 10 business units, nearly 200 subsidiaries, and more than 400,000 employees. Its capabilities included avionics, equipment, aeroengines, helicopters, and business aviation—just about everything outside of COMAC's jetliner mandate. Much of its revenue was derived from the military sector, but the reorganization combined with indigenous programs aimed to change this.

The C919 program, launched in 2008, created a unique opportunity for China and AVIC to create jetliner equipment capability in-country and for Western equipment suppliers to expand their presence in-country. Like its approach with avionics on the aircraft, COMAC required major equipment suppliers to establish in-country JVs with Chinese partners to be on the program. These included:

- Eaton Aerospace and Shanghai Aircraft Manufacturing Co. for fuel and hydraulic conveyance systems
- Honeywell and AVIC Engine for the APU, starter, and generator; AVIC Electronics for flight control computer; and Hunan Boyun New Materials Company for braking systems
- Moog and AVIC Systems for high lift systems
- Parker Aerospace and AVIC Systems for primary flight control actuation systems, fuel inerting systems, and fuel systems
- Liebherr and Chinese AVIC member Landing-Gear Advanced Manufacturing Co., Ltd. (LAMC) for the development and production of the landing gear system

The C919 presented suppliers with a dilemma: Should they risk transference of intellectual property and possibly the creation of new Chinese competitors by entering JVs, or avoid the program? The answer for some, according to industry analyst Richard Aboulafia, was to hold back on providing the latest aircraft systems technology [17].

2010S: CONSOLIDATION CONTINUES

The second decade of the new millennium would be characterized by another round of consolidation and the emergence of four mega-suppliers. In 2010, Safran attempted to acquire Zodiac Aerospace—a move that would have positioned Safran to provide 75% of jetliner systems. "There is a worldwide trend toward globalization of the Tier 1 supply base," noted Safran CEO Jean-Paul Herteman. He also saw an opportunity to leverage the shift from hydraulic to electronic systems. "We see an annual market of $3–4 billion a year in electric systems and room for two or three players. We want to be one of them" [18]. Safran's attempted takeover of Zodiac ultimately failed, but it would not be its last attempt.

An even larger merger would follow, this one successful. On the heels of its huge 787 win, Hamilton Sundstrand started to look at complimentary acquisitions. After examining several large suppliers, it settled on Goodrich as its first choice in 2010. Dave Gitlin recalled the rationale:

> When we looked at a landscape chart of Hamilton Sundstrand's systems and Goodrich's capabilities there was very little overlap. We had electric, environmental control systems, engine accessories, and Goodrich had the nacelle, landing gear, wheels and brakes, sensors, and actuation. It was a perfect marriage. Both companies also had common customers and complimentary cultures.[23]

There was also the issue of scale. "If you run a $60-billion company and you do deals that are $200–500 million, it's a yawner," said then-UTC CEO Louis Chênevert (Fig. 12.13). "So, I had to go after a big property…I was really focused on Goodrich as a prize." In September 2010, Chênevert phoned Goodrich Chairman and CEO Marshall Larsen and offered him $100 per share, a 40% premium over Goodrich's market value, which valued the company at over $14 billion. Larsen rejected the offer, a position that he had steadfastly maintained for years. UTC made the same offer three months later, and was again rebuffed [19]. Chênevert sensed that he was in a race against time. "Marshall Larsen was 63 at the time, and nearing retirement. I believed that he would name his successor in his late 2011 board meeting, and if this happened, the deal would be over. The new leader would want to define his own direction."[24] In June 2011, Chênevert phoned Larsen and offered $110/share, and the talks were renewed. After haggling, both parties agreed to a

---

[23]Dave Gitlin (President—UTC Aerospace Systems), interview with author, 29 March 2017.

[24]Louis Chenevert (former Chief Executive Officer—UTC), interview with author, 1 Sept. 2015.

Fig. 12.13 UTC CEO Louis Chênevert championed the acquisition of Goodrich.

price of $127.50, and UTC's acquisition of Goodrich was announced on 21 Sept. 2011. The value of the deal was $18.4 billion—the largest in aerospace history. After government approvals and minor divestitures in 2012, a new $13.3 billion enterprise, named UTC Aerospace Systems, became the world's largest aerospace equipment supplier.

UTC Aerospace Systems immediately undertook a massive post-merger integration operation to combine two major companies with global footprints, 170 facilities, and strong cultures. It initially separated the various businesses into two groups—one led by a former Hamilton Sundstrand executive, the other by Goodrich. It also transferred its APU business to Pratt & Whitney Canada (PWC). PWC was the world's leading manufacturer of small aeroengines and produced the APU for the 747-400. PWC could now amortize its investments over its entire gas turbine product line and achieve greater economies of scale and improve its competitiveness versus arch-rival Honeywell.

Other suppliers continued to pursue systems capability through acquisition (Fig. 12.14). Meggitt's 2011 acquisition of Pacific Scientific combined their fire detection and fire suppression capabilities to create an integrated system. Meggitt CEO Terry Twigger, who engineered deals for two brake companies in the prior decade, called it a "bread and butter" Meggitt business with its numerous sole source positions and high aftermarket content [20]. The $685-million deal increased its content on the 787 and A380 programs, expanded its electronics capabilities, and pushed it into the ranks of major equipment suppliers.

Following an eventful 2011, the equipment merger trend subsided as suppliers settled in for a wave of reengined aircraft programs, including the A320neo, 737 MAX, and Embraer E2. To minimize nonrecurring engineering

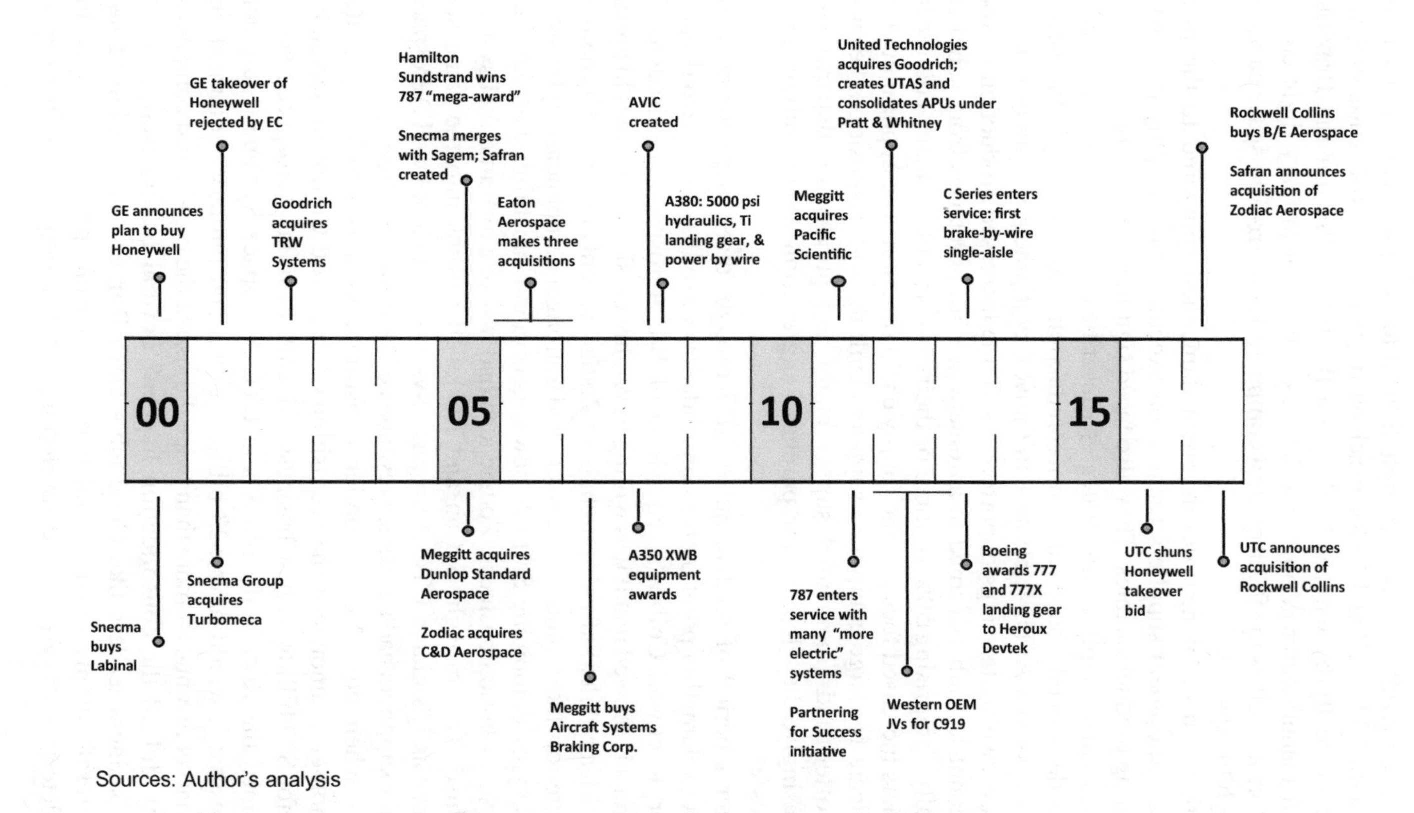

Sources: Author's analysis

Fig. 12.14 Aircraft equipment milestones: 2000–2017.

for these programs, aircraft OEMs often reused components and systems from the legacy aircraft designs. Although difficult to estimate precisely, it's a safe bet that the 737 MAX and A320neo achieved greater than 70% reuse. Embraer did embrace fly-by-wire technology on the E2—a first for the Brazilian aircraft manufacturer. Moog won the E2 contract for primary flight control actuators as well as the flight control computer; Liebherr-Aerospace captured the high lift system.

With the pace of new development dampened compared to the prior decade, equipment suppliers focused on execution and keeping pace with escalating production rates. They also faced significant pricing pressure from new customer supply chain strategies (described in Chapter 8). As aircraft OEMs demanded double-digit price reductions from Boeing's Partnering for Success and Airbus's Scope+ initiatives, equipment suppliers searched for their own cost savings opportunities. Some increased production in low-cost countries; others turned to automation. With purchases from suppliers typically comprising 50% or more of their cost structures, many equipment suppliers increased focus on the supply chain function. The opportunity was significant for large equipment suppliers built through acquisition, because they often had fragmented supply chain and purchasing departments. Increasingly, they looked for purchase aggregation opportunities across businesses.

After a period of relative quiet, 2017 brought three mega-mergers and put an exclamation point on the consolidation wave that began two decades earlier. Rockwell Collins's acquisition of B/E Aerospace, as discussed in Chapter 11, showed that it was willing to venture outside of avionics to pursue scale. Meanwhile, Safran's purchase of Zodiac not only expanded its domain into interiors like Collins, but also created a stronger components and systems company by combining Safran's power-generation capability with Zodiac's power-distribution business. Zodiac also had a strong airframe fuel systems franchise. This brought Safran and Rockwell Collins into the same size category as US giants UTC Aerospace Systems and Honeywell Aerospace. All four suppliers had aerospace revenues of $8 billion or more.

Just when equipment supplier consolidation appeared over, United Technologies announced the acquisition of Rockwell Collins in September 2017 for $30 billion. A new business unit called Collins Aerospace Systems combined the Rockwell Collins and UTC Aerospace Systems equipment businesses; it would have $23 billion in civil and military equipment and aftermarket revenues—more than twice the size of the next largest competitor.

Why did all the consolidation happen? Several factors were at work. Bold business leaders like Goodrich's Dave Burner had a vision of how the aircraft supply chain would evolve and were proactive in creating complete aircraft systems and developing a global footprint. Jetliner OEMs, anxious to reduce suppliers and shed responsibility with Tier 1 supply chain strategies, supported the movement—at least until the early 2010s when they adopted new supply chain strategies to reduce the power of mega-suppliers. Changes in aircraft equipment technologies favored systems architectures, as evidenced by the 787 mega-systems award. Finally, equipment suppliers recognized that scale could enhance their bargaining leverage versus jetliner OEMs and their aggressive supply chain practices such as *Partnering for Success* and *Scope+*.

The aircraft equipment sector, once composed of dozens of small equipment suppliers—unsung heroes—transformed into a sophisticated, consolidated, and indispensable element of the jetliner supply chain.

REFERENCES

[1] "New BAC111 Version Proposed," *Aviation Week & Space Technology*, 7 Sept. 1964, p. 41.

[2] McClellan, J. M., "Fire and Ice," *Flying*, October 1986, p. 17.

[3] Salpukas, A., "Parker-Hannifin Buys Growth," *The New York Times*, 28 Dec. 1979, http://www.nytimes.com/1979/12/28/archives/parkerhannifin-buys-growth-40-acquisitions-swell-volume.html [retrieved 13 June 2017].

[4] Currey, N., *Aircraft Landing Gear Design: Principles and Practices*, AIAA, Reston, VA, 1988, p. 139.

[5] Vartabedian, R., "Allied-Signal Picks L.A. as Site of New Aerospace Group," *Los Angeles Times*, 18 Nov. 1987, http://articles.latimes.com/1987-11-18/business/fi-14853_1_aerospace-group [retrieved 11 June 2017].

[6] Velocci, A., "Goodrich Best $500 Million on Long-Term Industry Outlook," *Aviation Week & Space Technology*, 15 March 1993, p. 33.

[7] "Goodrich to Acquire Rosemount Aerospace," *The New York Times*, 11 Nov. 1993, https://www.nytimes.com/1993/11/11/business/company-news-goodrich-to-acquire-rosemount-aerospace.html [retrieved 11 June 2017].

[8] Schifrin, C., and Sparaco, P., "TI Group, Snecma Combine Landing Gear Businesses," *Aviation Week & Space Technology*, 14 March 1994, p. 28.

[9] Crawley, J., "U.S. to Require Fuel Tank Safety System on Jetliners," Reuters, 16 July 2008, https://www.reuters.com/article/us-airlines-fueltank-idUSN1646155220080716 [retrieved 3 June 2017].

[10] Fisher, L., "Goodrich Plans to Buy Rohr in $789 Million Stock Deal," *The New York Times*, 23 Sept. 1997, https://www.nytimes.com/1997/09/23/business/goodrich-plans-to-buy-rohr-in-789-million-stock-deal.html [retrieved 24 June 2017].

[11] "World News Roundup," *Aviation Week & Space Technology*, 8 Feb. 1999, p. 18.

[12] Velocci Jr., A., "Sundstrand Deal May Herald New Consolidation Wave," *Aviation Week & Space Technology*, 1 March 1999, pp. 24–25.

[13] Mecham, M., and Velocci Jr., A., "New Honeywell Is Package Meant to Please Primes," *Aviation Week & Space Technology*, 14 June 1999, pp. 60–61.

[14] Pichevant, C., "WRC-15 Agenda Item 1.17—Industry's Motivation," EC–CEPT Workshop on World Radiocommunications Conference 15, 10 Dec. 2013.

[15] Dornheim, M., "Electric Cabin," *Aviation Week & Space Technology*, 28 March 2005, p. 48.

[16] Mecham, M., "Goodrich's TRW Buy Sets Stage for Act 2," *Aviation Week & Space Technology*, 24 June 2002, p. 46.

[17] Cendrowski, S., "China's Answer to Boeing and Airbus Takes Its First Flight," *Fortune International*, 4 May 2017, http://fortune.com/2017/05/05/china-c919-comac-flight [retrieved 4 July 2017].

[18] Taverna, M., "Tied to Tier 1," *Aviation Week & Space Technology*, 2 Aug. 2010, p. 39.

[19] Anselmo, J., Norris, G., and Velocci Jr., A., "High Gear," *Aviation Week & Space Technology*, 2 Jan. 2012, p. 44.

[20] O'Doherty, J., "Deal Lifts Meggitt's Aircraft Role," *Financial Times*, 18 Jan. 2011, https://www.ft.com/content/2c9aa4b6-232e-11e0-b6a3-00144feab49a?mhq5j=e2 [retrieved 15 July 2017].

Chapter 13

LIFEBLOOD

MAINTENANCE, REPAIR, AND OVERHAUL: A LARGE AND CRUCIAL SECTOR

Jetliners are complex and expensive machines that operate between 8 and 15 hours per day at the highest levels of safety and reliability. They aren't just dream machines; they are flying factories. Each hour of downtime is incredibly expensive. It isn't surprising that spending on maintenance, repair, and overhaul (MRO) is significant. In 2017, MRO spending on the global fleet of more than 25,000 jetliners was a whopping $70 billion, composed of five major elements [1]:

- $21.7 billion spent on aeroengine overhauls to restore the engine to designed operational condition, according to guidelines established by the engine manufacturer.
- $19.0 billion for the maintenance of aircraft systems and components, including avionics, electronics, fuel systems, pneumatics, landing systems, auxiliary power units, and other key aircraft systems.
- $16.7 billion for line maintenance—light and regular maintenance checks that are carried out to ensure that the aircraft is fit for flight, including troubleshooting, defect rectification, overnight maintenance, and component replacement.
- $5.9 billion for airframe heavy maintenance—detailed inspections of the airframe and components, including any applicable corrosion prevention programs and comprehensive structural inspection and overhaul of the aircraft.
- $6.6 billion for modifications, including avionics upgrades, painting, interior upgrades, passenger-to-freighter conversions, and airworthiness directive and service bulletin compliance.

Maintenance spending is typically 8–10% of an airline's cost structure and one of its largest controllable costs. For component and system original equipment manufacturers (OEMs), MRO parts and services—often referred to as the *aftermarket*—can comprise 50% or more of revenues (see Fig. 13.1).

The MRO sector is not only big and a necessary evil, but also lucrative. For many suppliers, it is the key source—the lifeblood—of profitability. Consider aeroengines: Manufacturers spend billions of dollars to develop new models, deliver the engine at a substantial discount (sometimes below cost) to the customer, and hope to achieve their return on investment from four or five overhauls over the engine's 25- to 30-year life. Aftermarket services are 40–60% of the revenue and most or all of the profits for today's aeroengine OEMs. Some refer to aeroengines as a "razor/razor blade" business model—sell the razor at a loss to gain access to the razor blade revenue stream. The same is true for component and system OEMs, which depend on aftermarket revenue to underpin their business models, although there are variations by system type. Digital avionics, for example, generate less aftermarket activity than wheels and brakes. But overall, most component and system OEMs today sell their wares at modest and sometimes negative profit margins to gain access to the aircraft platform and thus the aftermarket activity. Ironically, aircraft manufacturers are the least dependent on this revenue stream, which typically accounts for 10% of their revenues and perhaps one-quarter of their profits. The aftermarket for aerostructures is relatively modest due to the infrequent need to replace structural parts.

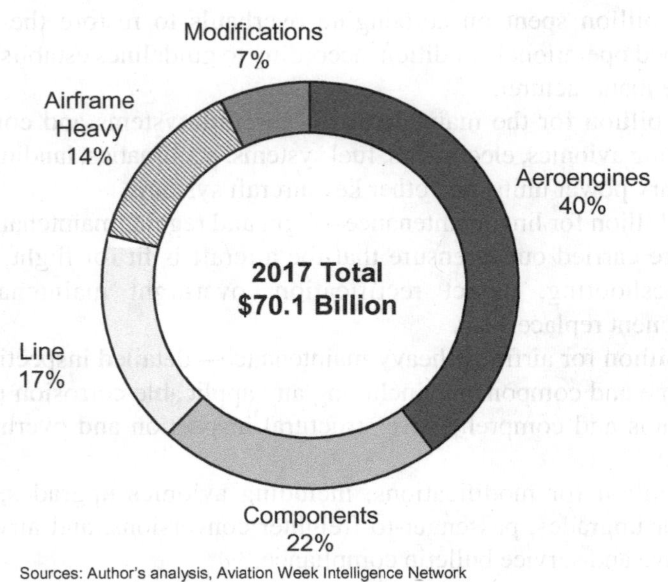

Fig. 13.1 2017 jetliner MRO spending.

If the aftermarket is the commercial lifeblood to manufacturers, it is the operational lifeblood to operators. Dispatch reliability—the percentage of departures that leave within a specified period of a scheduled departure time—can make or break an airline. The cost of an inoperable twin-aisle, when including lost revenue, can exceed $100,000 per day. Thus, in-service and customer support performance can have a major influence on an airline's purchasing decision.

Despite its clear importance, MRO received scant attention from aerospace manufacturers for much of the jet age. For many, the aftermarket was an *afterthought*. It was viewed as a burden rather than a business opportunity. And there were few meetings to bring the maintenance community together.

This changed in the 1990s as MRO growth exploded. Independent service providers prospered, and OEMs woke up and seized opportunities. This period saw the birth of MRO as a new aerospace industry sector with a unique identity. What drove the metamorphosis, and how does MRO influence the jetliner business?

MRO ORIGINS

In the first few decades of aviation, maintenance programs for basic piston-powered aircraft were developed by mechanics; these programs were simple and generally without analytical basis. The creation of airlines in the 1920s and 1930s led to new regulations and broader regulatory involvement in maintenance requirements. With the entry of large jet airplanes into the commercial market in the 1950s and their spotty safety record, the airplane manufacturer became the source of maintenance program development. The prevailing maintenance philosophy was to overhaul everything at a specified time. One underlying assumption in this philosophy was an intuitive belief that because mechanical parts wear out, the reliability of any equipment is inversely related to operating age. It therefore followed that the more frequently equipment was overhauled, the better protected it was against the likelihood of failure. At the time, most air travel was regulated, so maintenance activities were generally completed in-house with costs passed on to passengers. With little incentive to economize, maintenance expenses grew rapidly, and by 1960, airline representatives and the Federal Aviation Administration (FAA) had formed a task force to identify approaches to reduce expenses. This led to the concept of "on condition" maintenance.

In 1968, the US Air Transport Association published a 747 Maintenance Steering Group (MSG) document, MSG-1, which applied reliability-centered maintenance concepts to the Boeing 747. A few years later, a new revision (MSG-2) guided maintenance programs for the Lockheed L-1011 and Douglas DC-10. MSG-2 was process-oriented and analyzed failure modes from the part level up. It included the theory that all airplanes and their components reach a period when they should be zero timed or overhauled and restored to new condition. MSG-2 led to a huge reduction in scheduled maintenance activities versus legacy aircraft like the DC-8. The studies proved that the fundamental assumption of design engineers and maintenance planners—that every airplane and every major component in the airplane (such as its engines) had a specific lifetime of reliable service, after which it had to be replaced (or overhauled) in order to prevent failures—was wrong in nearly every specific example in a complex modern jet airliner. The impact of MSG-2 was significant: "Hard time" (fixed-interval) removal of engine components by leading European airlines was just 10% in 1973 compared to 95% a decade earlier [2].

In 1978, United Airlines developed a new methodology, MSG-3, for designing maintenance programs based on tested and proven airline practices. The MSG-3 methodology included a task-oriented approach to maintenance that analyzed failure modes from a top-down, systems-level framework. Failures were not necessarily linked to age, and energy should be redirected from predicting life expectancies to managing the process of failure and condition monitoring. MSG-3 remains the current standard for aircraft maintenance. Revisions to its philosophy have provided added methodology for improving coverage of all modes of failure, such as inclusion of the corrosion prevention and control programs.

Airline Maintenance Alliances

Throughout the 1960s, most jetliner maintenance was conducted in-house by airline maintenance and engineering organizations. North American airlines invested in significant maintenance capability including large maintenance hangers, engine overhaul facilities, and component maintenance shops. The primary maintenance base for United Airlines was in San Francisco; Northwest did most of its maintenance in Minneapolis–St. Paul, American Airlines in Tulsa, and Air Canada in Montreal. The scale of the North American market, which in 1970 operated 59% of the global fleet, underpinned the bias toward self-sufficiency.

The same was not true in Europe, where the fleet was relatively small and fragmented across numerous flag carriers. This made significant investment in maintenance equipment hard to justify; the learning curve also was a challenge. Scandinavian Airlines System (SAS) and Swissair began pooling some of their maintenance operations in 1959 and 1960; Swissair maintained both carriers' Sud Caravelles and JT8D engines, while SAS had responsibility for DC-8 maintenance. However, it was the advent of expensive and complex twin-aisle aircraft, including the DC-10, L-1011, A300, and 747, that really increased the imperative to collaborate. In 1968, KLM, SAS, and Swiss

ATLAS (1969)	KSSU (1970)
Air France **L**ufthansa **A**litalia **S**abena	**K**LM Royal Dutch Airlines **S**candanavian Airlines System **S**wissair **U**nion de Transports Aeriens

*UTA joined KSSU after originally joining ATLAS; it was replaced by Alitalia.
Iberia joined ATLAS in 1972

Fig. 13.2 Maintenance alliances.

created the KSS consortium for aircraft purchasing, training, and maintenance (Fig. 13.2). The three carriers ordered a total of seven 747s with the same configuration and a common cockpit. They purchased a single 747 flight simulator and based it in Amsterdam. KLM would be responsible for 747 airframe maintenance, SAS would maintain the aircraft's JT9D engines, and Swissair would overhaul the JT8D engines on DC-9s. Each airline would be responsible for its line maintenance. This approach not only yielded major savings to the three carriers, but also allowed them to pool their experience and improve dispatch reliability [3]. French operator Union de Transports Aeriens (UTA) joined in 1970, and the consortium was renamed "KSSU." In 1972, KSSU placed a huge order for 36 DC-10-30s (for all members except Sabena), and in a surprise move, it chose GE CF6-50s over Pratt & Whitney JT9Ds to power the fleet. KLM assumed engine maintenance responsibility, Swiss focused on airframe maintenance, and UTA maintained landing gear and other components.

Atlas, a second European consortium, formed in 1969 for the same reasons as KSSU. Originally composed of Air France, UTA, Lufthansa, Alitalia, and Sabena, the group focused on cost reductions for twin-aisle aircraft. UTA left the group to join KSSU but was soon replaced by Iberia. Again, an order for the 747 was a catalyst for dividing responsibilities, and by 1972, the consortium operated 26 of the new model. It also purchased DC-10s and Airbus A300s. Air France maintained CF6 engines and the 747 airframe, Lufthansa handled JT9D and A300 repair, Alitalia repaired the DC-10 airframe, and Iberia and Sabena were slated for Concorde maintenance. There was also a pooling of flight simulators. A study by Lufthansa concluded that it saved 15–20% on 747 maintenance thanks to the Atlas consortium [4].

While Europe reaped the benefits of collaboration, 747 operators in other regions, including Pan Am, Braniff, TWA, Japan Airlines, El Al, Qantas, and United Airlines, handled their own maintenance with significant OEM support. Boeing established a pool of 100 technicians and assigned a team of four to each new 747 operator. It also stationed maintenance vans at major US airports to meet every incoming 747 flight. Aeroengine OEM Pratt & Whitney spent $200 million on JT9D-3A engine support to overcome early teething problems. Boeing achieved mechanical schedule reliability of 95.4% based on

616,000 hours of operation by early 1972—below its 97% goal but impressive considering the new technology and sheer scale of the 747 rollout [5].

THE 1980S: AFTERMARKET = AFTERTHOUGHT

If the 1970s were about supporting new, high-technology twin-aisles, the 1980s were about supporting larger and more diverse fleets and phasing out the first generation of jet transports. The global jetliner fleet during this decade would grow from 5751 to 8912, with much of the growth coming from the US market, which deregulated in 1978. The US airline sector now had price competition, which stimulated travel demand and began to increase airline focus on costs in the late 1980s. Major airline maintenance bases continued to expand to cope with 1700 aircraft added during the decade. Several independent maintenance suppliers also rode the wave of growth. Ryder Systems was one of the first major independents. In 1987, it purchased Caledonian Airmotive, the engine maintenance division of British Caledonian Airways. At the time, it had revenue of $50 million and was Europe's largest independent engine maintenance supplier [6]. By the late 1980s, Ryder's MRO revenue reached $1 billion.

Another major independent maintenance supplier was Allen Aircraft Radio, a firm founded in 1951 by Ira Eichner (Fig. 13.3), an entrepreneur who started by selling parts from the back seat of his Studebaker. He expanded into Europe in the late 1960s and would grow the Chicago-based company into a major

Credit: AAR Corporation

Fig. 13.3 Ira Eichner: AAR Corporation.

independent MRO supplier with significant parts distribution and maintenance capabilities. It was rebranded as AAR Corporation in 1969.

In Europe, the KSSU and Atlas consortia continued to grow and mature. Within Atlas, Sabena was given responsibility for the A310, and JT9D capability was updated with split responsibilities: Lufthansa for the JT9D-7As, Iberia the JT9D-7Qs, and Sabena the JT9D-7Rs. In KSSU, KLM became the largest engine shop maintaining GE aeroengines outside of the United States, SAS grew its JT9D footprint, and UTA focused on aircraft components. KSSU also brought in third-party maintenance work from the likes of Garuda, Cargolux, and Avianca. Several new independent maintenance organizations also were created. One example was aeroengine manufacturer Motoren-und Turbinen-Union (MTU), which formed a separate engine maintenance division in Hannover, Germany.

Asia-Pacific began the 1980s with a fleet of just 690 jetliners, which expanded to 1160 by 1990. This enabled some airlines to expand maintenance capabilities, including Singapore Airlines, Thai Airways, Qantas, Japan Airlines, and All Nippon Airways. A major independent in the region was Hong Kong Aircraft Engineering Company (HAECO), which specialized in twin-aisle and RB211 engine maintenance in the 1980s.

For most OEMs, the aftermarket during this period was an afterthought from a commercial perspective. Most were willing to sell parts and provide maintenance services *if asked*; their cultures were reactive despite the aftermarket's financial lure. Maintenance facilities often shared space with factories, and only a few OEMs had a global maintenance service footprint. The primary focus was on technical and customer support. By 1990, the clear majority of engine overhauls, some 65%, were performed internally by airline maintenance organizations. Twenty-five percent were completed by independent suppliers and airlines overhauling other airlines' engines (airline third party). Engine OEMs had just a 10% share of engine MRO; Pratt & Whitney had the strongest maintenance position because of its huge installed base, but Rolls-Royce and General Electric were beginning to catch up.

Component maintenance followed a very similar pattern to engines: airlines performed 60% in-house, followed by independents and airline third party (25%) and OEMs (15%). Some component OEMs, including Parker Aerospace and Collins, had dedicated aftermarket functions, but the vast majority left maintenance to the operators and focused on spare part sales and technical support.

Airframe heavy maintenance was heavily insourced with 80% captive and 20% completed by independent and airline third-party suppliers. Aircraft OEMs avoided this activity altogether due to its cost structure, which was 70% labor. OEM labor costs were significantly higher than those of airlines or independent MROs, and they couldn't compete even if they were interested. Consequently, aircraft OEMs focused on product and technical support.

Boeing was the undisputed leader in providing such support and enjoyed a competitive advantage over Douglas and Airbus. Spare parts sales were also available, and Douglas was the most aggressive in pursuing this revenue stream. The remaining two maintenance activities, line maintenance and modifications, were also mostly performed in-house by airlines. The heavy dependence on internal maintenance capabilities reflected the regulated nature of the global airline industry at the time. Airfares in most countries were regulated, and airlines could pass on the cost of maintenance equipment, hangars, and engine test cells to their customers, and OEMs did little to convince them that this wasn't the appropriate course.

THE INFLUENCE OF ETOPS

The next phase of MRO's development was sparked by something no one could have foreseen. It arose from concerns about the reliability of piston engines, which led the US FAA in 1953 to prohibit aircraft with two or three engines from flying on routes that took them more than 60 min from an adequate alternate airport. The "60-Minute Rule," as it was called, effectively limited transoceanic flights to four-engine propeller airplanes like the Lockheed Constellation or Douglas DC-4. The International Civil Aviation Organization (ICAO) enforced a similar rule for non-FAA countries, which featured a 90-min limit. When more reliable jet engines began to proliferate, the rule became outdated. In 1964, the FAA 60-min rule was waived for aircraft with three engines, opening the way for development of twin-aisle intercontinental aircraft such as the DC-10 and L-1011, as well as the four-engine 747. By 1976, Airbus A300 twinjet aircraft were operated across the North Atlantic, the Bay of Bengal, and the Indian Ocean under a 90-min ICAO rule.

In 1985, the FAA created a 120-min Extended-Range Twin-Engine Operational Performance Standards (ETOPS) rule that allowed aircraft meeting its criteria to operate routings that are up to two hours (single-engine) flying from an alternate airport. The criteria included [7]:

- An engine inflight shutdown rate of 0.02 shutdowns per 1000 hours of operation
- Three sources of electrical power (e.g., three generators)
- An auxiliary power unit (APU) that can start at the maximum operating altitude of the airplane or 45,000 ft (whichever is lower) and run for the remainder of the flight
- Engine condition monitoring capability

In 1988, the ETOPS limit increased to 180 min.

ETOPS approval was a two-step process. First, the airframe–engine combination had to satisfy the basic requirements during its type certification, including special tests where engines were shut down in the middle of the ocean. Second, operators needed to satisfy their own country's aviation regulators about their ability to conduct ETOPS flights. ETOPS was generally granted long after the introduction of a new aircraft. As outlined in Chapter 3, the 777 would break this pattern by achieving 180-min ETOPS certification at entry into service in 1995. This contributed to its commercial success and its large sales advantage over the four-engine A340.

At the 2002 Farnborough Air Show, Airbus unveiled a controversial advertising campaign with a 300-ft-long sign at the edge of the runway that said, "A340: 4 engines 4 long-haul." The campaign was a toned-down resurrection of a similar effort that Airbus backed away from in 1999 that said, "If you're over the middle of the Pacific, you want to be in the middle of four engines." To many, the ad implied that four-engine aircraft are inherently safer than twins like the 777 and 767, and the imagery reminded many of the alternative, slang definition of ETOPS: *Engines Turn Or People Swim*. "We are vehemently opposed to what Airbus is doing here," said GE President and CEO David Calhoun. "An ad like that is the last thing this industry needs right now" [8].

Airbus had good reason to be concerned, because the impact of ETOPS on aircraft reliability was undeniable. Between 1985 and 2002 there were more than 3 million ETOPS flights, of which nearly 90% were completed by Boeing twinjets. Twinjets proved to be as reliable as three- and four-engine aircraft. One key measure of progress was the striking improvement in inflight shutdown rates of aeroengines. In the early 2000s, the average inflight shutdown rate of the 180-min ETOPS fleet was 0.01 per thousand engine hours, twice the required reliability [9]. Based in part on these results, 207-min ETOPS flights became available in 2000 on a case-by-case basis for 777 operators flying the North Pacific.

In 2009, the Airbus A330 became the first airliner to obtain beyond-180-min ETOPS certification, when it was granted an ETOPS 240-min certification by the European Aviation Safety Agency (EASA). A few years later, Boeing received ETOPS 330-min certification for its latest 777s and the 787. This created new routes, particularly over the Pacific Ocean. Air New Zealand became the first to leverage the new certification by operating an Auckland–Buenos Aires route with 777-200ERs.[1] In 2014, the A350 XWB became the first new airliner ever to be approved for ETOPS beyond 180 min *before* entry into service, with diversions of up to 370 min allowable under certain circumstances.

It is difficult to overstate the impact of ETOPS on aircraft maintenance. Airlines with mixed ETOPS and non-ETOPS fleets saw the dramatic

[1]"Boeing, Air New Zealand Celebrate First Flight Approved for 330-Minute ETOPS," Boeing press release, 1 Dec. 2015.

difference in reliability, and many began applying ETOPS practices to the entire fleet. No longer were unreliable aeroengines viewed as the major risk for long-haul transoceanic flights. ETOPS forced airlines to upgrade their engineering and maintenance skills, and was a catalyst for engine condition monitoring and aircraft health management programs. It would prove to be one of the most significant factors shaping MRO in the 1990s and 2000s.

THE 1990S: THE BIRTH OF THE MRO SECTOR

The 1990s were a turning point for the aircraft maintenance sector. Liberalization drove not only airline growth with concomitant increased maintenance activity, but also greater competition and more focus on cost reduction, which led to more outsourcing. New maintenance suppliers emerged. OEMs became forward-leaning participants. And aircraft maintenance acquired an identity as a distinct, important, and lucrative industry called maintenance, repair, and overhaul (MRO).

The global jetliner fleet development in the 1990s reflected the economic activity and maturity of aviation liberalization in each region (Fig. 13.4). The

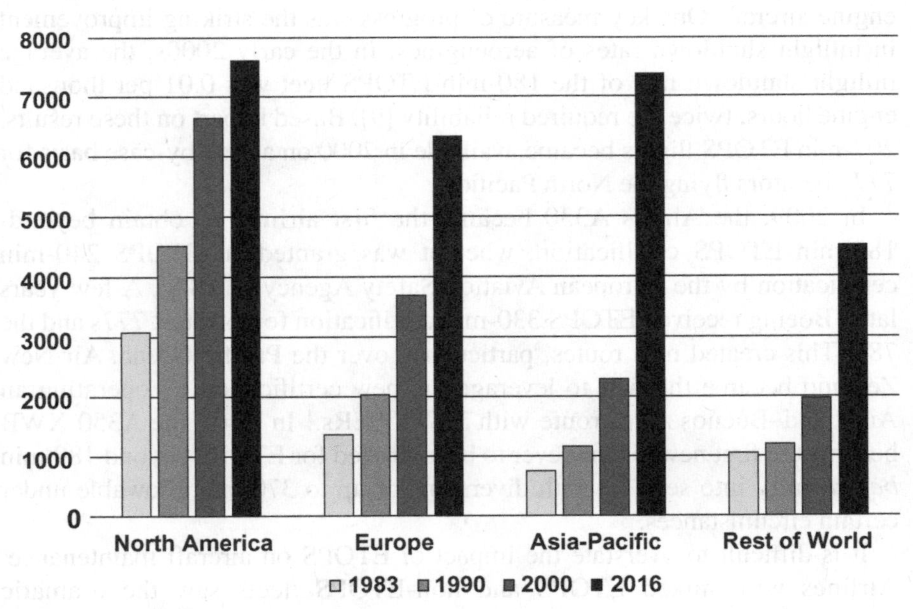

Fig. 13.4 Jetliner fleet growth: 1983–2016.

US fleet, deregulated since 1978, experienced a tremendous period of growth as competition heated up, airfares fell, and regional jets entered the market by the hundreds. It was also a period of significant economic growth under President Bill Clinton. The North American jetliner fleet increased by over 2000 units, from 4515 in 1990 to 6650 in 2000, and comprised 46% of the global fleet. Europe liberalized during the 1990s, and by the late 1990s new low-cost carriers including Ryanair and easyJet had begun to challenge the region's flag carriers. By 2000, the European fleet was 3694 units—about 25% of the global figure. Finally, Asia-Pacific (including China) remained highly regulated, and the bias of governments was to keep airfares high and protect flag carriers. The Asia-Pacific fleet was 2295 in 2000—just 16% of the global total.

The split of maintenance activity broadly reflected the fleet distribution. The United States was the largest market by far, followed by Europe with Asia-Pacific a distant third in overall size. Consultancy AeroStrategy estimated that in 2000 global spending on jetliner maintenance was $35 billion [10].

Despite the size of the maintenance market, few airlines viewed it as a global business opportunity. This would change as a result of German flag carrier Lufthansa, one of the most capable and respected airlines in the world. It was a founding member of the Atlas maintenance alliance, a launch customer for several new aircraft, and enjoyed an outstanding reputation for engineering and maintenance capability. Its maintenance organization not only supported its Atlas partners, but also supported other airlines on an arms-length basis. Third-party revenue as a percentage of total grew from single digits in the 1960s to 50% by 1991.[3,4] In the early 1990s, Lufthansa perceived a major opportunity in the global aircraft maintenance business. During a period when OEMs were passive participants and independents were small and fragmented, it believed that aircraft maintenance would one day become a large and lucrative global business opportunity. In the early 1990s it created a separate maintenance business, Lufthansa Technik, as part of a broad restructuring of the airline that broke it into six operating units. Its leader was Wolfgang Mayrhuber (Fig. 13.5), a visionary and energetic executive. Lufthansa Technik undertook an in-depth study of the aircraft maintenance market and developed growth plans in the early 1990s but needed to work through a major recession and complete its privatization in March 1994 before it could pursue growth. "We defined a growth strategy even as we went through a global crisis," recalled former Lufthansa Technik Head of Engine Maintenance Wolfgang Moerig.[5] A major twin-aisle overhaul center

[3]Klaus Mueller (former Market Research Manager—Lufthansa Technik), interview with author, 5 June 2016.

[4]Lufthansa Technik, "Lufthansa Technik: The History," 2005, http://www.lufthansa-technik.com [retrieved 5 Sept. 2016].

[5]Wolfgang Moerig (former Head-Sales—Lufthansa Technik), interview with author, 6 June 2016.

Credit: Sueddeutsche Zeitung Photo / Alamy Stock Photo

Fig. 13.5 Lufthansa Technik's Wolfgang Mayrhuber foresaw the globalization of MRO.

expansion in Hamburg, Germany was completed just as the airline was forced to park several dozen aircraft and downsize its workforce in Hamburg by 1200 employees.

Under the leadership of Mr. Mayrhuber, and with the full support of Jürgen Weber, the parent company's CEO, Lufthansa Technik got to work. It started operations as a separate stock corporation within Lufthansa AG and created its own balance sheet and P&L; no longer would it be merely an airline cost center. It also stood up well-resourced sales and marketing organizations and began investing in its brand. Lufthansa Technik was formally launched with high hopes on 2 Jan. 1995. It landed a major contract later that year with United Airlines to overhaul 100 of its V2500 engines. It also developed several new integrated maintenance services covering complete technical support of aircraft, engines, and components. Additionally, it created Shannon Aerospace, an aircraft maintenance joint venture in Ireland with Swissair and Guinness Peat Aviation. By 1997, third-party maintenance revenue exceeded \$2.2 billion. As it grew, the logic of remaining in the Atlas maintenance alliance eroded.

Lufthansa would not be the only supplier to achieve unprecedented scale in the 1990s. On the other side of the Atlantic, General Electric CEO Jack Welch identified a new growth opportunity for his industrial conglomerate in the mid-1990s: services. "With the exception of our medical business, most of the people making the heavy hardware in the company thought of services as the aftermarket—supplying spare and replacement parts for the aircraft engines, locomotives, and power generation equipment we sold," he said in his autobiography *Jack: Straight from the Gut*. "We might have scores of executives debating whether we'd sell 50 or 58 gas turbines or several hundred aircraft engines a year while we routinely handled the service opportunities for an installed base of 10,000 existing turbines and 9,000 jet engines. This had to change" [11].

GE's aeroengine services revenues were $2.2 billion in fiscal year 1994. This included revenues from a former British Airways engine overhaul facility in Wales (Fig. 13.6), which it had purchased to close a deal for the airline's purchase of GE90 engines in 1991. Under Welch's leadership, GE rolled out an objective for its business to double its service revenue in four years. Internally, this was known as "4 × 4," denoting the goal of $4 billion in services revenue within four years. In 1996, GE made the business a separate profit center and named Bill Vareschi as vice president of engine services. It also acquired Celma, an overhaul facility in Brazil. In late 1996, it watched independent engine maintenance supplier Greenwich Air Services buy Aviall's engine overhaul business, and in early 1997, Greenwich bought UNC, another independent supplier. Concerned that a competitor would scoop it up, Welch and Vareschi met with Greenwich CEO Gene Conese in March 1997 and quickly agreed to a $1.5-billion purchase price. GE now had $5 billion in service revenues [12]. It had surpassed its 4 × 4 goal with one year to spare and along the way acquired significant overhaul capability on its competitors' engines.

GE's acquisitions were viewed as a blow to competitor Pratt & Whitney, which was also emphasizing services growth. Pratt had staked out a goal of expanding its services revenue from $800–$850 million in 1997 to $1.5 billion by 1999. "Pratt needed Greenwich business a lot more than GE did," opined JSA research analyst Paul Nisbet. "Pratt is belatedly growing this portion of their business and soon they'll fall further behind GE because they couldn't move fast enough when a market opportunity presented itself" [13]. In 1997,

Credit: GE

Fig. 13.6 GE acquired British Airways' engine maintenance facility in 1991.

Pratt made up some ground when it purchased Dallas Aerospace, a major distributor of used and serviceable material (USM), to improve its positioning with operators of mature engines. Used aircraft parts could be acquired for a substantial discount compared to list prices for new parts, and were popular with operators of older aircraft. In 1998, Pratt entered a joint venture called Eagle Services Asia (ESA) with Singapore Airlines. It included Singapore's former engine shop and focused on PW4000 and JT9D engines. The overhaul center complemented two joint ventures that Pratt had previously set up in Singapore for engine component repair. Following ESA, Pratt would go on to start other parts repair joint ventures in Asia including Turbine Coating Service Pte. Ltd. (2000) and International Aerospace Tubes—Asia Pte. Ltd. (2002).

Rolls-Royce pursued a very different path to service revenue expansion. Like its competitors, it did not view repair and overhaul as a mainstream activity until the 1990s; however, its commercial success in winning RB211 and Trent positions on a variety of different aircraft platforms resulted in a broad and global customer base, while it had just a few civil repair centers. In 1993, it created a separate engine services division and began building a global service network through joint ventures with strategic customers. In 1997, it started Hong Kong Aero Engine Services Limited (HAESL) in Hong Kong, a 50:50 joint venture with Hong Kong Aircraft Engineering Company (HAECO). HAECO had 20 years of experience in maintaining RB211 engines, primarily for Cathay Pacific and Dragonair. HAESL created Asia's first Trent maintenance facility capable of running engines up to 130,000 lb of thrust. The next year it created Texas Aero Engine Services LLC (TAESL) with American Airlines to service its RB211-535 and Tay engines with an eye towards Trent 800s, which American ordered that year. The following year, it stood up Singapore Aero Engine Services Pte. Ltd (SAESL), another Asian joint venture (JV) focused on Trent overhaul. To avoid conflicts of interest, HAESL and TAESL purchased equity in each other's JV. By 2001, this expansion transformed the Rolls-Royce network from six facilities and $400 million in service revenue in 1993 to 16 facilities and $1.6 billion. In just eight years it had quadrupled its service revenue and doubled its repair and overhaul share of its engines to 50% [14].

After creating a global network, Rolls-Royce then restructured its service value propositions. Historically, most engine overhaul was "time and material," where operators paid for overhauls at the time of service. The cost of overhauls could easily reach several millions of dollars and was "lumpy"—customers could be surprised by massive bills when the unexpected occurred—and airline profitability suffered accordingly. There was a potential conflict of interest with this arrangement: The more work performed by the OEM, the more expensive the overhaul; yet there was no guarantee that more expense resulted in greater time on wing. Operators absorbed this financial risk. In the late 1990s, Rolls-Royce introduced a new contractual approach called *TotalCare*, where the payment

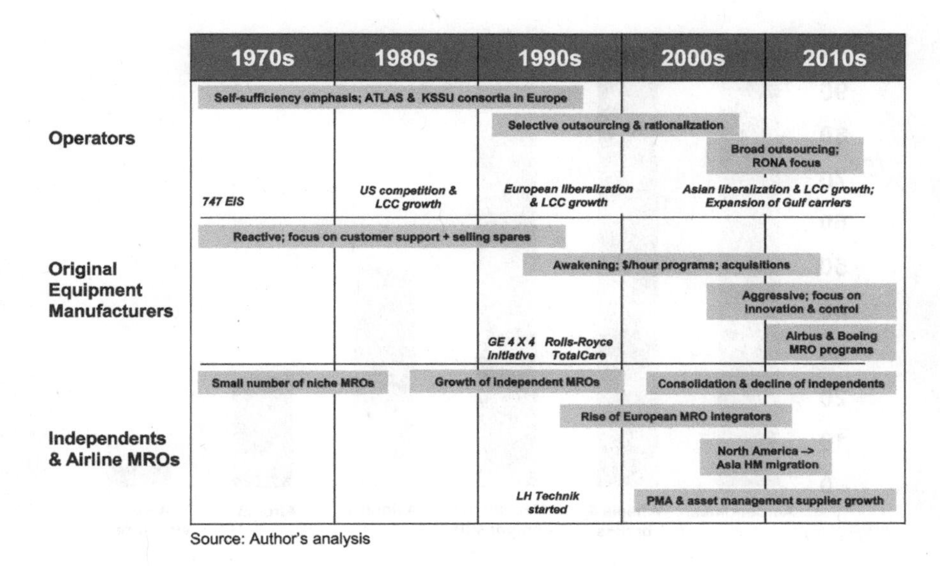

Fig. 13.7 MRO sector development.

mechanism was on a cost per flying hour basis (Fig. 13.7). This approach rewarded reliability and transferred both time on wing and shop visit cost risks back to Rolls-Royce. It also included predictive maintenance services. The new value proposition gained immediate acceptance, and the company signed major TotalCare contracts with British Airways, American Airlines, Continental Airlines, and Cathay Pacific, worth approximately $3 billion in 2001/02 alone.

While engine OEMs awakened to the aftermarket's potential, revenue opportunities for aircraft OEMs were more limited because most aircraft OEM service parts are structural components that rarely require replacement. In contrast, aeroengines and components are loaded with parts that wear out thanks to rotation, friction, and harsh conditions. As a result, the aftermarket comprises less than 10% of a typical aircraft OEM's revenue and 20–30% of its profits. In contrast, the aftermarket can drive 50% or more of an aeroengine or component OEM's revenue and all its profits (see Fig. 13.8). Success in the aftermarket can be life or death from a financial perspective.

Beyond dollars and cents, there are critical nonrevenue customer support activities that are expected of OEMs including technical support, reliability, warranty administration, contracts, quality, and aircraft on ground (AOG). *AOG* means that a problem is serious enough to prevent an aircraft from flying and requires rapid response to put the aircraft back into service to prevent further delays or cancellations. Boeing was particularly adept with customer support and built a world-class organization honed through supporting thousands of jetliners since the late 1950s. Customer support was a critical competitive advantage versus Douglas and Airbus, and often tipped the balance in its

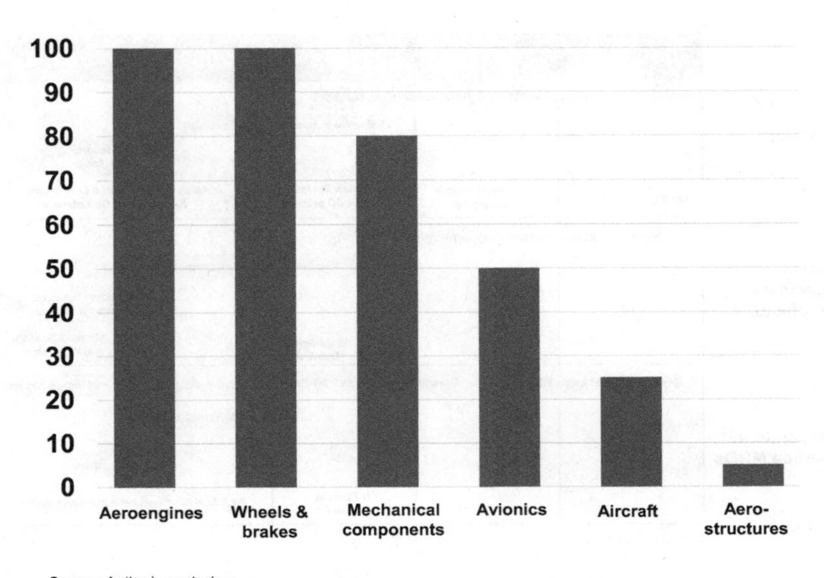

Fig. 13.8 Percentage of profits from aftermarket.

favor during sales campaigns. This advantage ebbed in the 1980s and 1990s as Airbus focused on closing the gap and stood up regional customer support organizations with talented leadership. Eliminating Boeing's competitive advantage in customer support helped to position Airbus to surpass its rival in aircraft orders in the late 1990s.

The regional jet revolution of the 1990s had interesting implications for maintenance services (Fig. 13.9). In many instances, turboprop operators took delivery of more technologically advanced regional jets with glass cockpits and advanced aeroengines. The cost of regional jet maintenance was also significantly greater than for legacy aircraft, and there was significant uncertainty about the reliability and true maintenance costs of these new jets. Regional operators were interested in value propositions to cope with these uncertainties, and OEMs delivered. Rockwell Collins, the avionics supplier for Bombardier's CRJ, created *Dispatch 100*, a cost-per-hour maintenance program. Rolls-Royce and GE did the same with engines. Lufthansa Technik offered *Total Technical Support*. And Bombardier and Embraer were actively involved in airframe maintenance of their own aircraft.

Despite new value propositions and the sheer growth in aircraft maintenance spending, maintenance lacked an industry identity. "Maintenance was

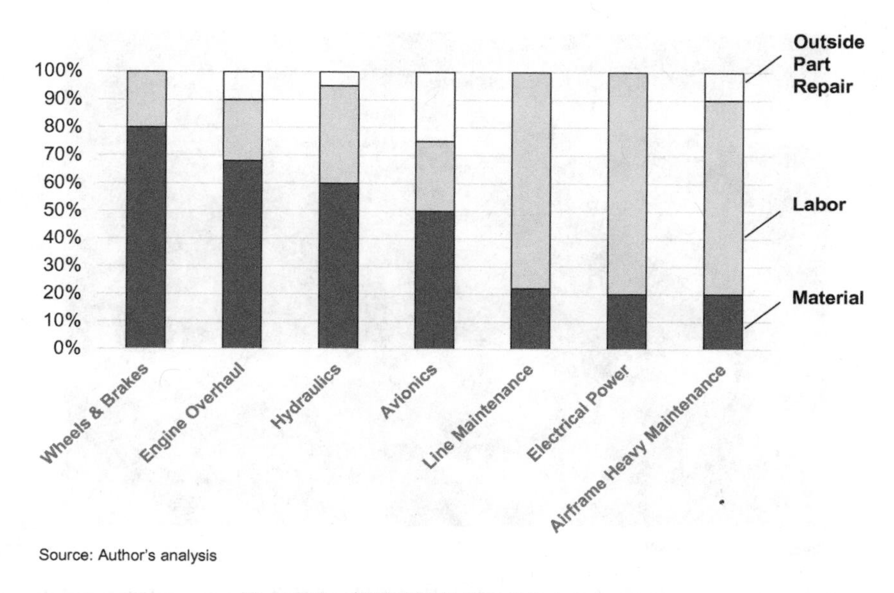

Fig. 13.9 MRO activity cost structures.

viewed as a necessary evil but not a business," said former MTU executive Klaus Mueller. Although engineering and technical directors communicated regularly and participated in forums to exchange ideas, there was no venue where operators, OEMs, and maintenance suppliers could exchange ideas and focus on the commercial aspects of maintenance. One person who wanted to close this gap was George Ebbs, a longtime industry consultant. In 1987, he started the Canaan Group, a consultancy focused on the aviation aftermarket. Ebbs and his small team in Park City, Utah—a ski resort—worked on many significant engagements including Lufthansa Technik's globalization strategy, GE's 4 × 4 initiative, and Rolls-Royce's aftermarket strategy. He was convinced that the aftermarket was a major business opportunity that required an event to bring together the ecosystem. At the same time, *Aviation Week & Space Technology* (*AWST*) writers Tony Velocci and Don Fink had discussed the possibility of increasing their magazine's exposure to aircraft maintenance. George Ebbs traveled to Tony's New York office in 1993 to share his ideas, and the two quickly found common ground. *AWST* publisher Ken Gazzola got behind the idea of a maintenance conference, and in 1996 *AWST* and the Canaan Group launched the MRO Conference in Dallas–Ft. Worth. Conference organizers conceived of *MRO* as an acronym for maintenance, repair, and overhaul, and soon it took hold as the name of the entire segment. The first event was a success, and MRO conferences were held in Dallas–Ft. Worth and Ft. Lauderdale in subsequent years. In 1998, MRO Europe was launched, and three years later MRO Asia joined the mix. The conferences combined thought-provoking speeches with a large exhibition where

Credit: Aviation Week Network

Fig. 13.10 MRO conferences attracted thousands of visitors and underscored its identity as a distinct segment in aviation.

operators, OEMs, and maintenance suppliers could interact (Fig. 13.10). All three conferences grew rapidly, and suddenly aircraft maintenance had its own identity within the aerospace and aviation industries: the MRO segment.[7,8]

MRO in the New Millennium

The new millennium continued the momentum of the 1990s. Significant orders by easyJet and Ryanair made it clear that increased airline competition in Europe was going to drive faster air travel growth, larger fleets, new business models, and increased maintenance spending. The Atlas and KSSU alliances phased out as European operators gained critical mass and key members pursued their own third-party maintenance outside the alliance. Swissair, a former member of KSSU, gained significant third-party maintenance work under its SR Technics brand, as did former consortia members SAS, KLM, Air France, Sabena, and Alitalia. In the United Kingdom, FLS Aerospace built a successful independent airframe and component maintenance business based out of London's Stansted Airport. Lufthansa pushed into Eastern Europe in 2000 when it established a new maintenance center in Budapest. Among European OEMs, Snecma, GE's partner in CFM International, made its own push to capture engine overhaul business by establishing several joint ventures including ones in Belgium and Morocco.

[7]George Ebbs (former President—The Canaan Group), interview with author, 20 June 2016.
[8]Tony Velocci (former Editor—*Aviation Week & Space Technology*), interview with author, 25 Sept. 2015.

There was also a recognition that maintenance demand in Asia was poised to explode and that the region lacked adequate maintenance capacity. In 2000, MTU opened an engine maintenance joint venture in Zhuhai, China with China Southern Airlines to specialize in CFM engines; Lufthansa also announced it would open a composite repair JV in Shenzen, China. In a surprising move, Lufthansa took over the Philippines Airlines (PAL) maintenance facility in Manila and outlined plans to make it its global A330/340 center of excellence. It would fly these aircraft around the world to take advantage of dramatically lower labor rates. Other factors in the decision included the English-speaking workforce, excellent buildings, demand from PAL's own fleet, and the huge potential in Asia. "Globally, the MRO sector in Asia will experience the strongest growth in MRO over the long term," said CEO Thomas Gockel [15]. The two new Lufthansa Technik maintenance bases complimented a JV (Ameco) that Lufthansa had established with Air China many years earlier in 1989. Snecma also established an engine maintenance partnership with Air China in Chengdu.

The new millennium also brought high hopes that the Internet could reduce transaction costs and improve capacity utilization. The Internet bubble, which underpinned dizzying appreciation in global equities and the formation of thousands of "dot com" firms, also touched MRO. One of the major initiatives was *Aeroxchange*, a neutral purchasing portal founded by 13 leading airlines. Its service would cover repair, replenishment, sourcing, inventory pooling, and other critical MRO operations; it would also automate the exchange of documents and other information for commercial transactions. In 2000, United Technologies, Honeywell, and I2 Technologies formed *My Aircraft*, another e-marketplace aimed at the aftermarket. BFGoodrich would later join. Finally, the major aircraft OEMs and defense primes entered the arena when Boeing, BAE Systems, Raytheon, Lockheed Martin, and IT specialist Commerce One formed their own e-marketplace. The idea of these exchanges was to accelerate and automate transactions between customers and suppliers. At the time, the facsimile machine and telephone were the most popular MRO transaction channels. *My Aircraft* would ultimately fail, whereas *Aeroxchange* would grow and achieve commercial success.

The giddiness of 2000/2001—underpinned by growth, globalization, and e-commerce—quickly gave way to a global recession that was exacerbated by the 9/11 terrorist attacks. Europe and the United States were particularly affected. As outlined in Chapter 3, Boeing's production rates were halved, hundreds of aircraft were parked, and MRO demand contracted by 10% between 2001 and 2002, from $42.2 billion to $37.8 billion [16]. Independent MROs, dependent on mature and sunset aircraft, were the first to feel the effects, and many slid into significant financial losses. AT Kearney Vice President Andy Schmidt predicted that the number of independents could be reduced to 6–8 from the current 20–25 because of failures and acquisitions by larger firms [17]. Aviation Sales Company, one of the largest used serviceable material (USM) dealers, declared bankruptcy in 2002. Swissair and Sabena—two former members of the KSSU alliance—also failed. This meant that SR Technics, Swissair's engineering arm, was on its own. It was acquired by private equity and became fully independent in 2002. Elements of Sabena's maintenance department became Sabena Technics, an independent MRO supplier.

In North America, five major carriers (Air Canada, Delta, Northwest, United, and US Airways) went through bankruptcy restructuring. As a result, United Airlines closed its Indianapolis and Oakland aircraft maintenance bases and negotiated a clause with its labor unions allowing it to outsource aircraft maintenance. Northwest Airlines announced plans to curtail its maintenance activity in Minnesota and cut 2000 aviation technicians, and Air Canada Technical Service—the maintenance arm of Air Canada—became a separate company. Maintenance outsourcing became the mantra of most North American carriers.

Not all the news during the recession was negative. Rolls-Royce established a joint venture with Singapore Airlines to broaden its capabilities in Asia. Lufthansa Technik established an airframe maintenance base in Malta in 2003 and created an engine maintenance JV with Rolls-Royce the following year. In North America, Embraer expanded its airframe maintenance network when it bought Celsius Aerotech in 2003.

This period also saw a major milestone with the creation of the European Aviation Safety Agency (EASA) in 2002. EASA replaced the Joint Airworthiness Authorities and upgraded the objective of creating common safety and environmental rules at the European level. EASA took over responsibility for certifying maintenance technicians and organizations from national authorities, and importantly, it gave Europe greater influence in setting global aviation maintenance and safety standards. It also made it easier for European MROs to grow as maintenance standards enabled more mobility for maintenance technicians across the continent.

PMA PARTS: THE OEM ALTERNATIVE

The lean times strengthened interest in a cost savings alternative: parts manufacturer authority (PMA) parts. PMA-holding manufacturers are permitted by the US Federal Aviation Administration to make replacement parts for aircraft, even though they are not the original equipment manufacturer. PMA suppliers reverse-engineer an OEM's original part and prove identicality and safety to the FAA. The financial benefits were significant, and PMA prices were 30% or more cheaper than list prices for new parts.

Although the FAA's PMA regulation had been around since 1947, it did not really gain significant popularity until the 2000s. The leading PMA supplier

Credit: HEICO

Fig. 13.11 HEICO popularized PMA parts, a cost-saving alternative to OEM parts.

was HEICO (Fig. 13.11). Based in Florida, HEICO built a niche business in the 1960s and 1970s on the back of JT3D burner cans. A British Airtours disaster in 1985 caused by a faulty JT8D burner can design led to a surge in orders for HEICO's PMA alternative after the FAA issued an Airworthiness Directive. Still, the business remained small.

In 1990, a new leadership team led by Laurans Mendelson took over. Laurans put his son Eric in charge of growing the PMA parts business. In 1992, aggressive marketing paid off, and HEICO convinced United Airlines to accept PMA parts as an alternative to OEM parts. The business expanded but still faced the burden of convincing airlines—particularly those outside of the United States—to accept the PMA alternative. Many operators and lessors were concerned that they were not the same quality as OEM parts or would negatively impact the residual value of their aircraft because a subset of operators refused to accept PMA. At the first MRO conference in 1996, Eric Mendelson participated in a session with Lufthansa Technik (LHT) CEO Wolfgang Mayrhuber where both expressed their frustration at "excessive" annual price increases by OEMs on service (spare) parts. "We met that night after our MRO conference session, and we saw eye to eye," recalled Eric.[9] Mayrhuber's interest in PMA reflected the changing nature of LHT, which had evolved from an internally focused technical organization to a $4 billion global MRO giant with an eye on the bottom line and the cost-savings opportunity represented by PMA OEMs, which had once been partners and were increasingly viewed as competitors—particularly as they expanded their service and aftermarket operations. General Electric, for example, once counted on LHT to be its engine MRO partner in Europe. Now GE owned its own facilities in Wales and Scotland, LHT's home turf.

In 1997, LHT acquired 20% of HEICO. This was a groundbreaker, because it was considered by many to be the most technically competent MRO in aviation. Now it was a user, promoter, and stakeholder of PMA parts. HEICO had a critical ally and pursued an aggressive acquisition campaign to consolidate the fragmented PMA parts industry. Another major domino fell in 2001 when

[9]Eric Mendelson (President – HEICO), interview with author, 9 Sept. 2016.

HEICO established a JV with American Airlines to develop PMA parts, and the following year it would conclude a partnership with United Airlines.

RETURN TO GROWTH

MRO spending returned to growth in 2004, and the single European market would play a prominent role. Air France and KLM combined their maintenance operations when the two airlines merged in May of that year. SR Technics increased its heft when it bought UK-based FLS Aerospace. European airline liberalization, completed in 1997, began to have a major impact as low-cost carriers (LCCs) placed large orders and gained market share from incumbents, and passengers demanded lower fares. Ryanair placed an order for 155 737-800s, and easyJet made a major A319 purchase in the early 2000s. This was good news for maintenance suppliers because low-cost carriers outsourced most of their maintenance activities. In 2005, easyJet awarded a £552 million ($1 billion), 10-year maintenance contract to SR Technics to provide full technical management (excluding engines) for its fleet of A319s. This was one of the largest deals ever awarded by an airline for maintenance of a single aircraft type [18]. "As European LCCs achieved scale and outsourced maintenance in large packages, it changed the maintenance market," opined industry consultant David Stewart. "The chief beneficiaries were large European maintenance suppliers including Lufthansa Technik, SR Technics, and Air France/KLM. The LCC outsourcing enabled them to achieve significant scale without internal investment."[10] As they grew, these large maintenance suppliers increased their market share while airline maintenance outsourcing accelerated.

The Middle East was another postrecession maintenance bright spot. In 2006, Emirates Airlines built a massive engineering center on a 136-acre site to support its burgeoning fleet and to prepare for the A380. Etihad and Qatar also grew their maintenance operations. As these airlines developed extensive hub-and-spoke networks, the Gulf region—particularly the United Arab Emirates (UAE)—became an attractive location for maintenance operations and parts distribution. In 2007, Goodrich Corporation opened a major maintenance campus in the UAE. Governments also participated: a consortium of Mubadala Development Corp (a strategic investment company owned by the Abu Dhabi government), Dubai Aerospace Enterprise (backed by the Dubai government), and Istithmar (a Dubai government investment fund) bought SR Technics in 2006. The following year, DAE bought highly regarded Canadian MRO Standard Aero and US MRO Landmark Aviation for $1.8 billion. "The UAE is quickly establishing itself as a significant player in the global aerospace industry," said Bob Johnson, DAE's CEO. "The acquisition of Standard Aero and Landmark Aviation provides a critical

[10]David Stewart (Partner—Oliver Wyman), interview with author, 6 Sept. 2016.

platform for DAE to take advantage of growth opportunities in the MRO business around the world."[11] Notice had been served: the Gulf region aspired to be a globally competitive MRO cluster.

Pratt & Whitney took the aviation world by surprise in 2006 when it announced the launch of a new division, Global Material Solutions, to manufacture PMA replacement parts for CFM56-3 engines. For many years, aeroengine OEMs had publicly criticized PMA parts as not being manufactured to the same testing and quality standards as OEM parts. Pratt & Whitney's move changed the nature of aeroengine competition, because no major OEM had replicated its competitors' parts. "A lot of customers have asked us for this," said Pratt executive Matthew Bromberg. "There is a strong demand for lower-cost maintenance solutions, and we've done a significant job of [expanding] our capability on the CFM56" [19]. United Airlines would be the launch customer for Global Material Solutions. Another reason for the move was to develop relationships with operators in the hopes that one day Pratt would sell them new engines. "There are many operators that operate the CFM56 exclusively, and we want to be able to sell engines to them. That's the hope" [20]. This move would ultimately unleash a counterreaction from CFM International when it developed a new MRO network strategy. Instead of trying to service engines exclusively in its own facilities, CFM licensed partners to overhaul its engines on the condition that they would limit usage of PMA parts. Soon, a growing number of CFM overhaul shops were part of the network, which limited demand for Pratt's competing parts. More defensive actions would follow in the years ahead.

RETURN ON ASSETS

After a five-year period of growth following the early 2000s recession, MRO was again thrown into reverse by the Great Recession, which began in 2008. Global airline revenues contracted significantly and airline expenses followed suit, declining by 17% from $571 billion to $476 billion. The US airline industry consolidated into four major airlines controlling more than 80% of capacity and was run by more financially oriented leadership. Their drive for cost savings led to maintenance outsourcing to low-cost regions, particularly to China, which enjoyed a 40% or more labor cost advantage versus North American independent MROs and an even larger advantage versus more expensive airline internal maintenance operations. By the early 2010s, some 20% of North American heavy maintenance for twin-aisle aircraft was performed in Asia [21]. A growing portion of single-aisle maintenance also migrated to Central America, taking advantage of competitive suppliers in Costa Rica and El Salvador (Fig. 13.12).

---

[11]"Dubai Aerospace Enterprise Completes Acquisition of Standard Aero and Landmark Aviation," Standard Aero press release, 1 Aug. 2007.

Credit: Banku

Fig. 13.12 US airlines begin to outsource airframe heavy maintenance to Asia and Latin America in the early 2010s.

Airlines in Asia went through their own adjustments due to the Great Recession. Japan Airlines emerged from bankruptcy in 2010 after slashing its workforce by one-third, abandoning unprofitable routes and aircraft, and embracing maintenance outsourcing. Although legacy airlines were under pressure, LCCs in the region boomed, headlined by Air Asia's order for 200 A320s in 2011. Asian LCCs, like elsewhere, outsourced most of their maintenance. China's major carriers—Air China, China Southern, and China Eastern—also powered through the recession and increased their capacity by more than 50% between 2009 and 2012. Thus, Asian maintenance facilities filled up, and investment continued to flow into the region.

While airlines restructured, OEMs of all varieties finally got serious about the MRO business. Their aftermarket businesses were battered by the recession as airlines cut aircraft operations, deferred maintenance, and conserved cash. Aircraft and engines due for expensive maintenance events were often simply parked. Aeroengine OEMs Pratt and General Electric experienced dramatic decreases in revenue as engine shop visits were deferred. Component OEMs also had double-digit decreases in aftermarket activity. One exception to this phenomenon was Rolls-Royce, which derived most aftermarket revenue from its *TotalCare* service. This cost-per-hour service linked its aftermarket revenue with aircraft operations. Its peers took notice.

Component and system OEMs developed sophisticated aftermarket strategies and began to align their organizations with customer needs. For them, aftermarket revenue took on added importance as they sought payback on investments in new programs including the A380, 787, and A350 XWB. UTC Aerospace Systems, for example, offered an integrated maintenance program called *CARE*, covering its large number of components on the 787. Rockwell Collins did the same with *Dispatch 100* for its avionics on the same aircraft.

Engine OEMs, which built out their service center networks a decade earlier, focused on new value propositions and partnerships to broaden their influence. Rolls-Royce continued to grow its share of customers signed up under *TotalCare* programs and authorized new independents to overhaul its engines. It also created a new variant of *TotalCare*, *TotalCare Flex*, tailored toward the needs of operators with aircraft approaching retirement, and further broadened its service portfolio in 2015 with the launch of *SelectCare*, which provided shop visit cost risk transfer, but allowed operators to retain the time-on-wing risks.

Snecma and GE benefited from surging CFM-56 aftermarket revenues thanks to growing A320 and 737 fleets, and GE promulgated a build standard called *TRUEngine* to halt the growth of PMA parts and unauthorized repairs. Its argument was that aeroengines maintained in the OEM configuration had higher residual value versus those engines maintained with PMA content. Pratt & Whitney also focused on its own cost per hour programs, particularly when it rolled out its new geared turbofan engines in the mid-2010s.

A large change was the aggressiveness of aircraft OEMs, who targeted the MRO service market for growth. In 2014, Boeing Commercial Airplanes' president called growth in services "strategic," citing his own company's estimate that $2.5 trillion would be spent on aircraft services through 2033 [22]. Boeing promoted *GoldCare*, a cost-per-hour maintenance program that included aircraft health management, asset management, and maintenance for its 787s. It also collaborated with Air France/KLM to maintain hundreds of 737, 777, and 747-8 aircraft under long-term maintenance programs. Airbus created *Flight Hour Services* (FHS), a program that provided component maintenance and asset management services. It also expanded its maintenance footprint when it established Sepang Aircraft Engineering, a JV in Kuala Lumpur, Malaysia in 2011. It would take full control six year later. Airbus also expanded its distribution capability when it purchased Satair, a leading distributor, in 2011, mirroring Boeing's purchase of Aviall in 2006. Embraer and Bombardier were also successful in expanding their regional jet MRO participation.

Aircraft OEM MRO ambitions hit a crescendo in 2017. Airbus established a goal to triple its air transport services revenue to €6 billion ($7.3 billion) by 2020. Embraer created an integrated services business unit that combined its operations across business units. Boeing announced a monumental goal of achieving $50 billion in civil and military services revenue. At the time, its services revenue was $15 billion [23]. It created a new Global Services business unit in Plano, Texas, to facilitate focus and a more agile, service-oriented culture. All OEMs focused on enhanced aircraft health monitoring services, leveraging enhanced onboard health monitoring systems, cheaper communications links, and advanced machine learning algorithms. The goal was to improve aircraft reliability, enhance prognostics (the ability to predict failures in advance), and hopefully be paid for the service. This was a long way from the "aftermarket as afterthought" sentiment of the early 1990s.

The growth of the OEMs changed the maintenance supplier landscape considerably. Airlines performed 80% of their own airframe heavy maintenance in 1990; by 2010 it was 45%. Aeroengine maintenance experienced an even bigger shift, with engine maintenance collapsing from 65% in-house in 1990 to 20% by 2010. Aeroengine OEMs picked up the lion's share of this work with a 45% market share by 2010. The same trend influenced component maintenance with OEMs, independents, and airline third-party shops picking up 70% of global maintenance spending by 2010 (Fig. 13.13).

After the tumult caused by the Great Recession, the objectives of airline management shifted as they realized that pursuit of market share was a failed strategy. Instead they began to focus on return on invested capital (ROIC) as a key metric. Delta Air Lines CEO Richard Anderson was a vocal proponent of this shift in mindset, declaring a goal of 15% ROIC for his airline, which he achieved in 2013. Many major North American and European airlines followed suit. The upshot for maintenance was that these airlines—much like low-cost carriers—were loath to invest in new maintenance facilities or acquire supporting inventory of spare components and parts with new aircraft acquisitions. Instead, they were interested in

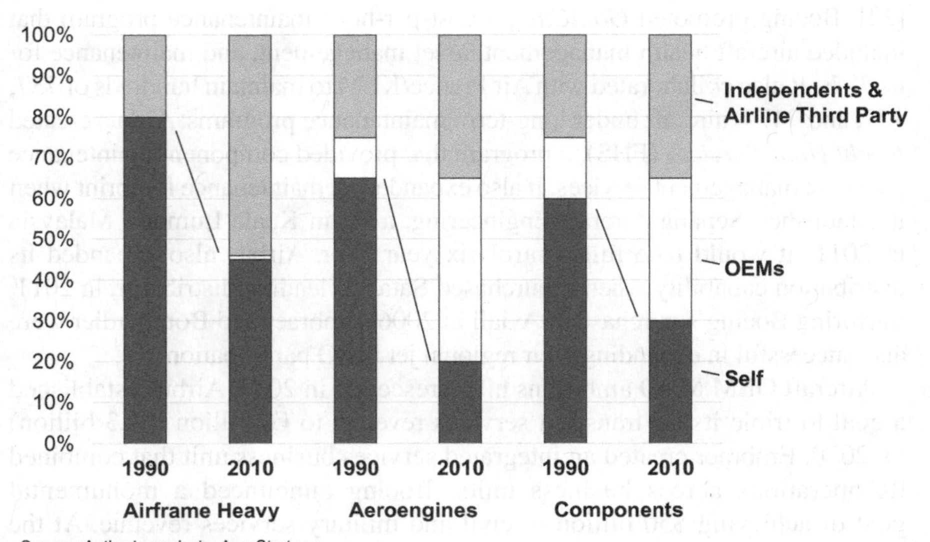

Fig. 13.13 Airline maintenance outsourcing.

being part of asset pools where a third-party supplier held spare components on behalf of multiple airlines, increasing asset efficiency and boosting the airline's ROIC. This opened the door for a group of major European MROs, including Lufthansa Technik, Air France/KLM, and SR Technics (all with Atlas and KSSU lineage), to offer integrated programs that combined component maintenance services and access to asset pools. They captured a significant share of the new-generation twin-aisles delivered in the 2010–2015 timeframe and did well with the many LCCs taking delivery of A320s and 737s during this time. Airbus and Boeing also gained traction with their new programs, and by 2015 had hundreds of aircraft each under integrated maintenance programs; they also enjoyed distribution revenue streams from their respective subsidiaries, Aviall and Satair.

Another outcome from the new airline financial orientation was greater focus on alternatives to OEM parts. One alternative was used and serviceable parts: reusing parts from a parted-out aircraft or buying excess parts from an airline's or MRO's inventory. A record number of aircraft retirements due to high fuel prices created ample numbers of aircraft part-outs, and by the mid-2010s, spending on used and serviceable material (USM) exceeded $3 billion and achieved a double-digit share of maintenance parts spending. Moreover, the USM segment—once considered a backwater—attracted interest of OEMs, lessors, and financial institutions. OEMS, including GE, CFM, Pratt & Whitney, Honeywell and Rockwell Collins, offered USM as an alternative alongside new OEM parts. Major independent dealers included AAR, AJ Walter, GA Telesis, Unical Aviation, and VAS Aero Services. The other alternative, PMA parts, captured an additional 2–3% of parts spending. This left OEMs with an 85% share for their own aftermarket parts [24].

As the tumult of the Great Recession faded, maintenance spending returned to growth in 2013. By 2017, it would reach $70 billion globally (Fig. 13.14)—more than three times larger than 1990. It yielded billions of dollars of profits, which served as the lifeblood of aerospace OEMs. The MRO conference, launched with sparse attendance in 1996, attracted a staggering 15,000 attendees by 2015. Aircraft maintenance was a big global business and now had its own identity as a major sector in aviation.

Another important trend was the dramatic growth of the Asian MRO market underpinned by a tripling of the in-region jetliner fleet from 2,295 in 2000 to 7,383 in 2016. Chinese airlines were behind much of this growth. While Asia added 5,000 aircraft over this timeframe, airlines in the mature North American market added less than 1,000 (Fig. 13.4). As a result, Asia-Pacific MRO spending reached $19 billion by 2017—on par with North America. Europe was the largest market with just over $20 billion in MRO spending.

Although the size of the MRO sector is notable, even more impressive is its role in improving air travel safety. By 2014, engine inflight shutdown rates fell to 0.01 per 1000 hours—a 20-fold improvement over the 1970 figure of 0.20

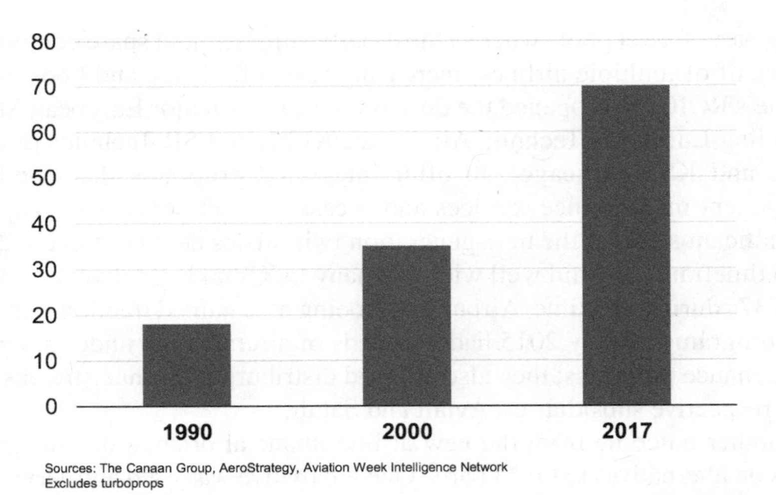

Fig. 13.14 Global airline maintenance spending growth ($ billion).

per 1000 hours [25]. Improved maintenance practices and ETOPS played an important role in this transformation. In 2017 there was not a single fatality in global jetliner operations with more than 37 million flights. This was a first, and airline maintenance technicians and MROs played a critical role. "The credit for 2017's safety record does not belong to one person, but was the work of thousands of people over decades," said safety consultant John Cox. "Airline regulators, manufacturers, maintenance crews, dispatch crews and others all played a role." When asked what the most dangerous part of flying can be, Cox's reply was telling: "Driving to the airport" [26].

Not bad for an activity that was once an afterthought.

References

[1] Berger, J., "MRO Forecast and Market Outlook," *Aviation Week & Space Technology* MRO Latin America Conference, Lima, Peru, 21 Jan. 2016. Author excluded turboprop estimates.

[2] Winston, D., "European Carriers Expand Cooperation," *Aviation Week & Space Technology*, 22 Oct. 1973, p. 69.

[3] Kolcum, E., "SAS, KLM, Swissair in Maintenance Pact," *Aviation Week & Space Technology*, 22 July 1968, p. 34.

[4] "European Carriers Expand Cooperation."

[5] "Mechanical Reliability Improving," *Aviation Week & Space Technology*, 22 Oct. 1973, pp. 34–35.

[6] Abruzzese, L., "Ryder System Acquires Caledonian Airmotive," *Journal of Commerce*, 5 May 1987, https://www.joc.com/ryder-system-acquires-caledonian-airmotive_19870305.html. [retrieved 1 Sept. 2016].

[7] Federal Aviation Administration, "Extended Operations (ETOPS) of Multi-Engine Airplanes, Rules and Regulations," Vol. 72, No. 9, *Federal Register*, 16 Jan. 2007, pp. 1872–1874.

[8] Lunsford, L., "Critics Say Ad Implies Rival Jet Is Unsuitable for Long Flights," *Wall Street Journal*, 26 July 2002, https://www.wsj.com/articles/SB1027622698744765040. [retrieved 5 Sept. 2016].
[9] Ekstrand, C., and Pandey, M., "New ETOPS Regulations," *Boeing Aero*, 2nd quarter 2003, pp. 5–6.
[10] Velocci, A., "Near-Term Market Offers Little Growth," *Aviation Week & Space Technology*, 6 Jan. 2003, p. 41.
[11] Welch, J., and Byrne, J., *Jack: Straight from the Gut*, Warner Business Books, New York, 2001, pp. 317–319.
[12] Ibid., pp. 319–321.
[13] Velocci, A., "GE Purchase of Greenwich Sets Stage for MRO Overhaul," *Aviation Week & Space Technology*, 17 March 1997, p. 29.
[14] Pugh, P., *The Magic of a Name: The Rolls-Royce Story—Part Three*, Icon Books, Cambridge, UK, 2002, pp. 194–196.
[15] Dennis, W., "Lufthansa Technik Uses Manila as Second Asian Overhaul Base," *Aviation Week & Space Technology*, 10 Sept. 2001, p. 46.
[16] Rosenburg, B., "MRO Providers in Europe Prepare for Consolidation, Prepare for Recovery," *Aviation Week & Space Technology*, 16 Sept. 2002, p. S1.
[17] Wall, R., "Airline Downturn Exacerbating Independent MROs' Woes," *Aviation Week & Space Technology*, 10 Sept. 2001, p. 47.
[18] "easyJet Signs SR Technics Maintenance Deal," *Irish Times*, 17 Aug. 2005, https://www.irishtimes.com/news/easy-jet-signs-sr-technics-maintenance-deal-1.1181438. [retrieved 14 Sept. 2016].
[19] Thurber, M., "Pratt & Whitney Enters Market for PMA Parts," Aviation Industry News online, 19 Sept. 2006, https://www.ainonline.com/aviation-news/aviation-international-news/2006-09-19/pratt-whitney-enters-market-pma-parts. [retrieved 14 Sept. 2016].
[20] Ibid.
[21] Baldwin, H., "Airframe MRO Opportunities," *Aviation Week & Space Technology*, 14 April 2014, p. 124.
[22] Broderick, S., "Aftermarket Aspirations," *Aviation Week & Space Technology*, 9 June 2014, p. 20.
[23] Michaels, K., "OEMs Focus on Mature Aircraft for Aftermarket Growth," *Aviation Week & Space Technology*, 16 Jan. 2018, http://aviationweek.com/commercial-aviation/opinion-oems-focus-mature-aircraft-aftermarket-growth [retrieved 27 Jan. 2018].
[24] Michaels, K., "Alternative Reality," *Aviation Week & Space Technology*, 8 Sept. 2014, p. 14.
[25] Allianz Global Corporate and Specialty and Embry Riddle Aeronautical University, "Global Aviation Safety Report," Dec. 2014, p. 27.
[26] Horton, A., "Air Travel Was Miserable in 2017, but At Least Nobody Died in a Commercial Jet Crash," *Washington Post*, 2 Jan. 2018, https://www.washingtonpost.com/news/dr-gridlock/wp/2018/01/02/air-travel-was-miserable-in-2017-but-at-least-nobody-died-in-a-commercial-jet-crash [retrieved 27 Jan. 2018].

Chapter 14

FOUR CLUSTERS AND A FUNERAL

Why does aerospace activity flourish in some areas and not in others—particularly in an era of globalization and mobile capital? It is a question that has challenged government officials, corporate leaders, financiers, and development agencies for decades. Comparative advantage for any industry is ephemeral, and the jetliner business is no exception.

As outlined in Chapter 8, the unbundling of the value chain coupled with sourcing from low-cost countries and the rise of emerging economies as major customers transformed how and where aircraft were developed, produced, and supported. New clusters emerged, which often involved close collaboration among industry, governments, universities, and research institutions. There were also losers, as regions lost competitive advantage in aerospace. Studying the rise and fall of clusters is therefore instructive and helps to explain the new landscape of the jetliner supply chain. This chapter highlights the fall of what was once the industry's biggest cluster—Southern California—and the rise of four new clusters in unlikely places: Singapore, Mexico, Morocco, and the Southeastern United States.

THE FUNERAL: SOUTHERN CALIFORNIA

Southern California is synonymous with aerospace and jetliners. During World War II, some 2 million people built aircraft in the area. During the Cold War, it remained the undisputed global center of aerospace, boasting scores of leading aircraft original equipment manufacturers (OEMs) within 15 miles of Los Angeles City Hall, which were linked to suppliers by an emerging system of highways. There has never been, and probably never will be, such a concentration. As recently as 1990, Southern California had 271,000 aerospace employees—a figure that collapsed to 88,000 by 2011, with only a fraction dedicated to jetliners [1]. How did this happen?

Aircraft production was very much a cottage industry in the United States in its early years, with new producers sprouting up weekly. In 1925, there were 44 US aircraft manufacturers. The following year there were 67, which produced an aggregate of 1125 aircraft. New York boasted the most of any state with 12 producers. Thirty-six new aircraft OEMs entered the field in 1927, and 15 of these were in California [2]. Opportunity was in the air, and Southern California was determined to seize it. "There is going to be a Detroit of the aircraft industry. Why not here in Los Angeles?" declared local businessman E. J. Klapp. Political and business leaders got behind the dream. The Los Angeles (LA) chamber of commerce created an aviation department. The boosters' watchword was "airmindedness" [3]. The first Academy Award from the region's other growth industry—film—went to *Wings*, a World War I drama. In 1929, there were 53 airfields within 30 miles of downtown LA. The following year, two major airports opened—United Airport in Burbank and Mines Field, which would later become Los Angeles International Airport (LAX). Aircraft and parts production, which stood at $5 million in 1929, would explode to $100 million in Southern California by 1939—44% of US aircraft production [4]. Aircraft suppliers were attracted by the area's fair weather, abundant land and labor, and backing by government and business leaders. One example was Cleveland-based Parker Appliance Company, which set up a new aircraft division in Los Angeles in 1939. A cluster was born.

World War II drove an explosion of activity. Aircraft manufacturing became the largest industry in the world, and Southern California plants employed 2 million people who built 300,000 aircraft during the war [5]. The aircraft manufacturers even defined the urban contours of Los Angeles: Lockheed and Vega in North Hollywood/Burbank; Douglas and Hughes Aircraft in Santa Monica/Culver City; North American, Douglas, and Northrop in Inglewood/El Segundo; Vultee in Downey; and Douglas in Long Beach (see Fig. 14.1). Like Detroit's automotive cluster, Los Angeles attracted waves of economic migrants, including "Aviation Okies" from the nation's heartland. By the end of World War II, Southern California accounted for 60–70% of the US aircraft industry [6].

There was a major employment crash following the war, but soon the Cold War created another wave of demand—including for rockets and missiles. The aircraft industry morphed into the aerospace industry.

By 1965, 15 of the 25 largest US aerospace companies were concentrated in California, most of them in Southern California. Orange County became a bedroom community for aerospace workers. Further south, San Diego also boomed. One exception to Southern California's magnetic pull was the mid-1950s decision by Lockheed to move production of the C-130 military

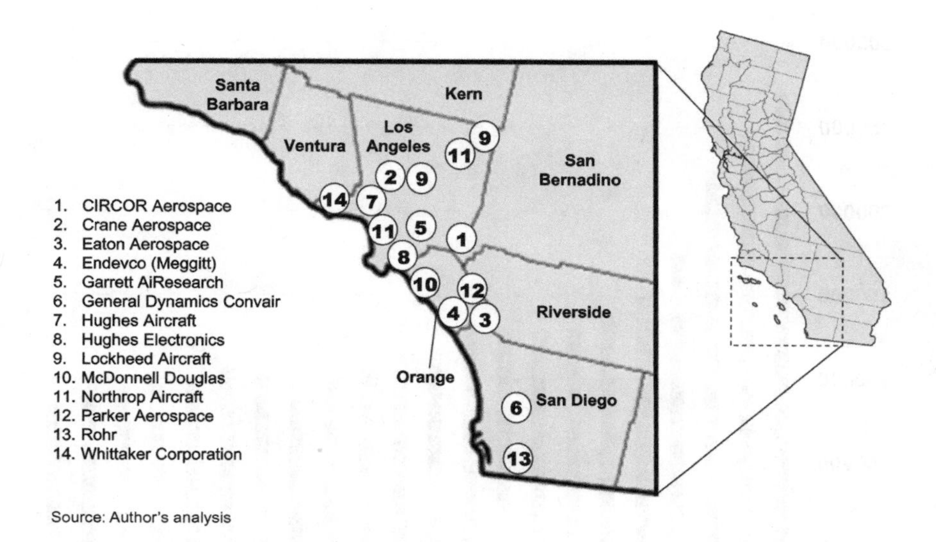

Fig. 14.1 Major aerospace companies in Southern California circa 1975.

transport from its Burbank facility to Marietta, Georgia. At the time, the move seemed innocuous but it would prove to be a harbinger.

The region was still well positioned for the jetliner era that emerged in the 1960s. Two of the largest global producers were anchored there with Douglas Aircraft in Long Beach and Lockheed in Burbank. The Vietnam War kept production lines humming for defense suppliers. Douglas Aircraft, however, depended on the jetliner market and was strained by the development costs of the DC-8 and DC-9. In 1967, it merged with military-oriented McDonnell to create McDonnell Douglas Corporation. The 1970s would bring more trouble, with a simultaneous bust in civil and military demand and the end of the Apollo program. This impacted the area's other jetliner anchor, Lockheed. In 1981, Lockheed announced the termination of the L-1011 program after suffering billions in losses. It shuttered most of its buildings at its massive Burbank facility where it had operated since 1928. Nearly 10,000 jobs were lost. Its aircraft division, the Lockheed Aeronautical Systems Company, was moved to Georgia. The exception was its Skunk Works operation, which remained in the area. There was one more important event in the 1970s that barely attracted notice but would shape the cluster's future: the founding of Airbus.

The boom–bust pattern continued, and the 1980s were a boon to California. Under President Reagan, defense expenditures skyrocketed, and the Golden State was a major beneficiary. The previously cancelled Rockwell B-1 bomber program was brought back to life, and scores of new programs—classified and unclassified—ramped up. Northrop produced the secretive B-2 Stealth Bomber in Palmdale, and to the south General Dynamics produced the Tomahawk cruise

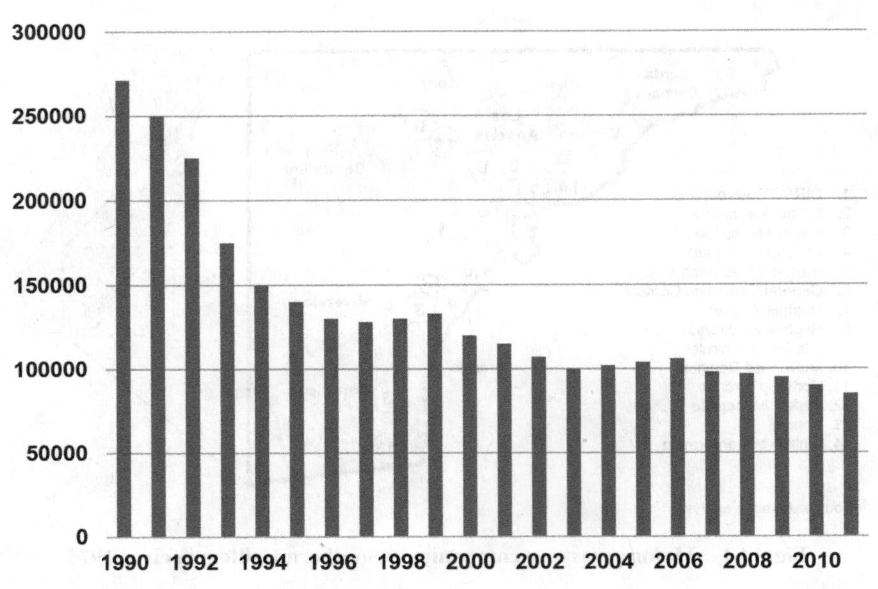

Fig. 14.2 Aerospace manufacturing in Southern California, 1990–2011.

missile in San Diego. McDonnell Douglas spooled up MD-80 production in Long Beach and was producing more than 100 per year by the end of the decade, with C-17 military transport also in production and the MD-11 nearing entry into service. Space also boomed. California aerospace was a $30-billion industry with 28% of US aerospace establishments and 29% of all US aerospace personnel [7]. And the clear majority of this was in Southern California (Fig. 14.2). Then the world changed.

By 1992, California aerospace was under assault on three fronts. Defense spending crashed with the end of the Cold War. The United States was mired in a deep recession, and the state's business climate was increasingly uncompetitive in a world of rebooted value chains. Aerospace executives had complained for years about state regulations that were burdensome to manufacturers including the cost of litigation, workers' compensation, high taxes, and stringent environmental regulations. These warnings were largely unheeded during the 1980s defense boom as military programs were generally "cost-plus." Contractors could leverage defense contracts to keep capacity utilization high and paper over the growing structural issues. When military demand crashed, the realities of competing in the jetliner business were brutally exposed. A November 1992 article in *Aviation Week & Space Technology* captured the new competitive dynamic: "Aerospace companies are leaving California in growing numbers because of the increasing importance of reducing operating costs in today's highly competitive defense and commercial markets" [8]. A Council on California Competitiveness was established to stem the losses, but to no avail. Even as the defense crash settled, it became clear that California had a competitiveness problem and that other US states, such as Oklahoma, Arizona, Utah, Nevada, and Georgia, were seizing the opportunity to recruit aerospace firms. Like many defense-oriented firms, GM Hughes Electronics CEO Michael Armstrong focused on diversifying into commercial markets. He also decided to move his missile production to neighboring Arizona. He stated:

> While our corporate headquarters, R&D and some software development can remain in California our manufacturing increasingly is not competitive here...markets have no patience. They demand your output to be competitive at your next bid. So, Hughes needs to redeploy to be competitive. But it's not just a California problem. The global village is here. Things like the North American Free Trade Agreement will continue to unfold [9].

He would prove to be prescient.

The news grew worse. Boeing acquired the region's remaining jetliner OEM, McDonnell Douglas, in 1997. The product overlap with Boeing's own portfolio was gradually purged, and with the last 717 delivered in 2006, jetliner production ended in Long Beach, which had produced more than 15,000 civil transport aircraft in 65 years. Competition with other *states* morphed into competition with other *countries* in the 2000s—particularly Mexico and China. By 2011, aerospace employment in Southern California had fallen to 88,000—a 65% decrease from more than 250,000 jobs in 1990. As the *Orange County Register* noted, "During the Cold War, 15 of the 25 biggest aircraft companies in the U.S. were based in Southern California. Today, they have mostly closed their doors, merged with rivals or moved their headquarters away" [10]. The loss of more than 150,000 high-paying jobs came with great human cost as careers were disrupted and families dispersed. The 1993 movie *Falling Down*, staring Michael Douglas as an angry unemployed engineer, captured the zeitgeist of the era.

By the mid-2010s, California had become primarily an engineering and administrative center with a vastly smaller jetliner manufacturing footprint. There were a few "green shoots" of activity in other aerospace sectors, such as commercial space. Elon Musk, for example, decided to base his SpaceX launcher business in Los Angeles. But this couldn't offset the broader narrative that the world's largest and deepest aerospace cluster contracted violently in just 15 years.

Hecho En México

Mexico and the aerospace industry existed in alternate universes for nine decades following the Wright brothers' first powered flight in 1903. Outside of a few investments by Honeywell, Westinghouse, and Collins in the 1960s and

1970s, capability was sparse despite bordering the world's largest aerospace market. Until the 1980s, Mexico pursued an import substitution trade policy that limited foreign direct investment. The results were predictably abysmal. The 1980s were known as the "La Década Perdida" (The Lost Decade), as the country fought through the worst recession since the 1930s. One of the few economic bright spots was the maquiladora system, where factories import material and equipment on a duty-free basis for assembly and export—often back to the raw materials' country of origin. Another bright spot was the 1983 opening of a component production facility in Mexicali by business jet OEM Gulfstream.

President Carlos Salinas de Gortari changed course after assuming office in 1988 and pursued deregulation, denationalization, and foreign direct investment. His crowning achievement was the North American Free Trade Agreement (NAFTA) with Canada and the United States, which came into force in January 1994. Mexico had executed an economic U-turn; it was open for business. But would aerospace activity follow?

In the immediate aftermath of NAFTA, aerospace activity remained relatively quiet through the period to 1999. In 1997, Mexico dropped tariffs on imported aerospace parts, a measure that could improve its competitiveness for maintenance, repair, and overhaul (MRO) and manufacturing outside of maquiladoras. "NAFTA dropped the tariffs from as high as 20% to duty free," explained Luis Lizcano, the head of Mexico's aerospace industry association (FEMIA). "It also assured intellectual property protection."[1] In 1998, French OEM Labinal added to its footprint when it purchased Aerotec, a wiring supplier in Chihuahua. Mexmil, a small aircraft insulation supplier, announced a new production facility in Baja California Norte.

The investment dam broke in subsequent years as the trickle of investments turned into a steady flow in 1999 when Unison, Senior Aerospace, and Beechcraft all built facilities. GE then set up an engineering center in Querétaro, the first significant aerospace investment in this state. Other aeroengine and gas turbine investments by Precision Castparts, Esco, ITP, and GKN followed. The steady stream of investments then turned into a torrent with several dozen firms establishing operations in Mexico over the 2003–2007 timeframe, including many escaping the punishing environment in Southern California. One of these firms was Eaton Aerospace, which set up a 700-person manufacturing facility in Tijuana in 2007, a two-hour drive from its Orange County aerospace headquarters. Intense pressure from its customers to reduce costs was a key factor in the decision. "We can't look at ourselves as constrained within the borders of the US," said Eaton Aerospace president Bradley Morton, "If we do, we die. It's a matter of survival" [11]. Not surprisingly, Parker Aerospace—Eaton's aerospace rival—arrived in the same year. By the end of 2007, there were 150 aerospace firms in Mexico (see Fig. 14.3).

[1]Luis Lizcano (Director General—FEMIA), interview with author, 12 July 2016.

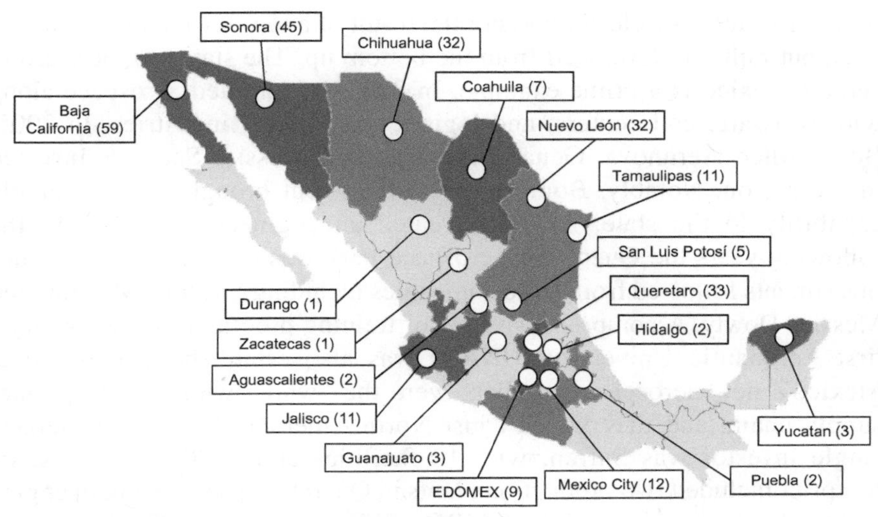

Fig. 14.3 Aerospace facilities in Mexico: 2013.

Another even larger wave followed, with business aircraft OEMs Beechcraft and Cessna and numerous aircraft systems and aerostructures suppliers joining the cluster. Bombardier made a signature investment in 2008 when it opened a major facility in Querétaro to manufacture electrical harnesses and structural components. Another 110 firms would come to Mexico by 2011. Regulatory barriers were eased when in 2008 a US–Mexico Bilateral Aviation Safety Agreement (BASA) was signed that allowed products made in Mexico to receive Federal Aviation Administration (FAA) certification in Mexico rather than being shipped back to the United States for approval. And the economic case remained compelling. Suppliers in Mexico benefited from total labor costs (including overhead) of 32% of US levels; when factoring in higher transportation costs and the need to import raw materials and parts the total cost advantage was 15% or more. Currency exposure was also minimized with the Mexican peso linked to the US dollar [12].

Regional specialization emerged as the cluster grew. Baja California, located just south of California, specialized in interiors, inflight entertainment, electrical and hydraulic systems, and precision machining. Sonora was strong in aeroengine and auxiliary power unit (APU) parts and specialty processes; Chihuahua in aerostructures, aircraft systems and components, wiring harnesses, and emergency systems; and Nuevo Leon (home of Monterrey) in MRO and engineering. To the south, Querétaro focused on aircraft final assembly, aerostructures, aircraft systems, machining, and aeroengine MRO services. The patterns of development mirrored the decentralized nature of Mexico's political system and the relative autonomy of its 31 states. The cluster was not the result of a top-down national master plan, but rather it developed from the bottom up. The state of Querétaro in central Mexico is a prime example. In 2003, it identified aerospace along with software, call centers, and logistics as priority industries. In 2005, Bombardier, Aernnova, General Electric, and Messier Services invested in the region. Notably, Bombardier's investment brought final assembly capability to the state, and GE created an engineering center. In the following years the Querétaro Aerospace Park was established, and major investments followed from large companies including Snecma, Meggitt, and Messier-Dowty. A comprehensive set of training programs at the country's first Aeronautic University (UNAQ) was also established in the state. Mexico's new aerospace suppliers were the "who's who" of the jetliner supply chain, and they weren't just North American OEMs. The largest single investor was Safran, with 10 facilities and 5700 employees. Its footprint included two production plants in Querétaro specializing in engine and landing gear parts and three MRO service centers. In 2000, it acquired Labinal, one of Mexico's aerospace pioneers, and continued to invest. By the 2010s, Safran operated the world's largest aircraft wiring complex in Chihuahua, which designed and manufactured 95% of the wiring used on the Boeing 787.

There was one notable difference between Mexico's industrial strategy and those of other emerging economies: Mexico had no aspiration to build its own aircraft to compete with its investors, and it focused on strong intellectual property protection. Mexico was in stark contrast to China, another investment hotspot, which pursued the transition of intellectual property to participate on the C919 program. Investors that favored China in the early 2000s gave priority to Mexico in the 2010s. Some $4.7 billion in investment flowed into Mexico's aerospace sector in the 2008–2011 timeframe [13].

By 2015, the cluster would comprise more than 350 aerospace facilities with some $7 billion of activity (Fig. 14.4). More than 70% was manufacturing oriented; engineering (13%), MRO (11%), and support services (4%) made up the balance [14]. Twenty years following NAFTA, the Mexican economy changed quickly. The automotive industry was a major catalyst, but aerospace also played an important role. By 2013, more than 115,000 engineering and technology students graduated annually from Mexican institutions—more than the total in Germany [15].

Mexico created a globally renowned aerospace cluster in just 20 years. FEMIA president Benito Gritizewski captured the zeitgeist when in 2016 he said, "Mexico is under the spotlight in the global aerospace industry. We have had significant growth and have not seen a single aerospace project fail."[2]

[2]Benito Gritizewski (President—FEMIA), interview with author, 12 July 2016.

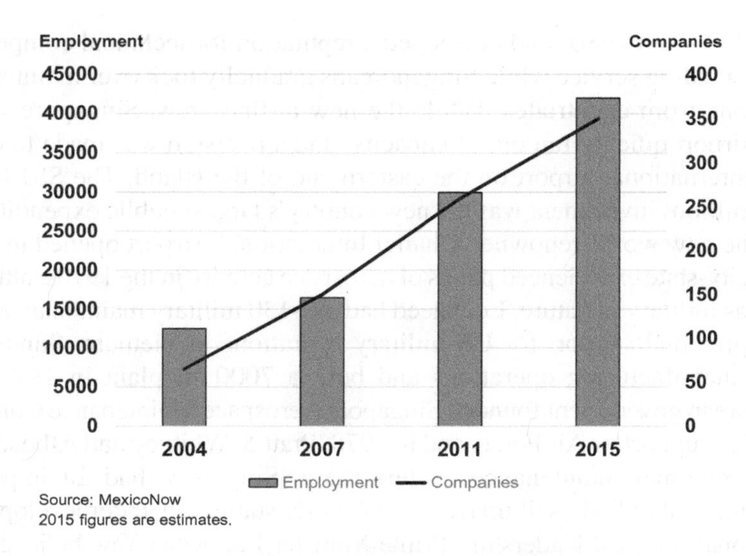

Fig. 14.4 Mexican aerospace cluster growth.

SINGAPORE: ASIA'S ÜBERCLUSTER

There is no apparent reason for Singapore to be a cluster for anything related to aerospace; a city-state with 5.3 million people and 278 sq. miles of territory, roughly the size of Dallas, its only apparent advantage is its strategic location overlooking the Straits of Malacca. If anything, it should be a maritime services location. Fifty years ago, following Singapore's separation from Malaysia and founding in 1965, it was beset by race riots and had a per-capita gross domestic product (GDP) of just $300. Yet in the span of several decades it became arguably the most complete aerospace cluster in Asia, boasting MRO, advanced manufacturing, aviation services, and research and development. How did this happen?

Not much related to aerospace occurred in Singapore in the years after its founding. British forces withdrew in 1968, leaving a major economic void. Its principle airline, Malaysia Singapore Airlines (MSA), ceased operations in 1972 after the two countries couldn't agree on route priorities; Singapore was interested in international expansion whereas Malaysia wanted to focus on domestic routes. Singapore Airlines was created in the aftermath, retaining MSA's 707s and 737s. It grew rapidly, adding cities in Asia and India and expanding its fleet with 727s, 747s, and DC-10s. The 747 addition was a particularly audacious move for the new airline, which depended primarily on British and Australian expatriates for its maintenance operations. "Adding the 747s showed the world that we were a serious airline," recalled a senior Singapore Airlines executive.[3] The airline would continue to expand globally

[3]Anonymous (Senior Executive—Singapore airlines), Interview with author, 6 August 2016.

in the 1970s and 1980s and developed a reputation for technical competence and outstanding service while Singaporeans gradually took over maintenance operations from expatriates. While the new airline grew, Singapore's Paya Lebar airport quickly ran out of capacity, and a decision was made to create a new international airport on the eastern end of the island. The S$1 billion ($500 million) investment was the new country's largest public expenditure to date. The now world-renowned Changi International Airport opened in 1981.

The city-state experienced pangs of aerospace activity in the 1970s, although most was military in nature. Lockheed had a C-130 military maintenance base, which provided support for US military operations in Vietnam. Sundstrand started manufacturing operations and built a 7000 m^2 plant in 1974. The Singaporean government founded Singapore Aerospace Maintenance Company in 1975 to support its Air Force, and in 1978, Pratt & Whitney and AlliedSignal opened regional maintenance centers there. Singapore had an important advantage as it pulled itself up from third-world status and underdevelopment: exceptional political leadership. Prime Minister Lee Kuan Yew believed that the key to Singapore's development would be good governance and strong government institutions, which would attract investment to fuel exports. Singapore would be a corruption-free oasis in a region known for the opposite. It would slash tariffs and business taxes, invest in infrastructure, champion the rule of law, and become the best place to do business in Asia. And at a time when developing economies were eschewing foreign investment, Singapore would do the opposite and pursue Western multinational companies at the same time that it was separating from its former colonial masters, the British. It was a risky strategy. The key institution charged with executing this strategy was the Economic Development Board (EDB). The EDB was a "super-planning" agency with few parallels in western governments. Staffed with talented civil servants (many with PhDs), it would create Singapore's economic master plan, identify potential investors, and then convince them to invest.

The EDB began to focus on aerospace in the early 1980s with the vision of creating a one-stop shop for aerospace, including manufacturing and MRO. An early victory occurred in 1981 when Singapore hosted the Changi International Airshow, the forerunner to today's Singapore Air Show. This event provided an opportunity to attract potential investors and to excite its youth about the possibility of working in aerospace. A second show was held in 1984, and by 1986 the rechristened Asian Aerospace Show attracted 500 exhibitors from 21 countries and established itself as an important event on the aerospace calendar [16]. The EDB also played a role in getting the country's two major players, Singapore Airlines and ST Aerospace, to form a joint venture for JT8D aeroengine maintenance. There were a few other investments in the 1980s, but overall investment was modest. But it was not for lack of effort. "We called on everyone," recalled Jeremy Chan, who was based in the EDB's Los Angeles office in the late 1980s. "We asked them 'Do you have a plan for Asia?' If so, think of Singapore."[4] Meanwhile, Singapore Airlines continued its rapid growth and burnished its reputation as a leading airline when it became one of the first 747-400 operators.

The pace of investment in Singapore accelerated dramatically in the 1990s as it became a magnet for MRO investment. Several factors underpinned this change. Singapore Airlines spun off its maintenance operations into a separate business, Singapore Airlines Engineering Company (SIAEC), which changed its maintenance strategy from self-sufficiency to partnership with OEMs. This led to four separate joint ventures (JVs) with Pratt & Whitney, including Eagle Engine Services, a JV that merged their local engine maintenance operations. Goodrich, Hamilton Standard, and Messier-Dowty would also enter maintenance JVs with SIAEC. Meanwhile, EDB's recruitment efforts also bore fruit as OEMs recognized the need for a regional service center to meet Asia's surging air travel growth. For its part, ST Aerospace began to diversify into civil MRO, but without the backing of a parent airline like SIAEC it needed a hook to create a presence and establish relations with this new customer base. The opportunity came when an airworthiness directive was issued to modify the nose section, known as Section 41, of all in-service 747s. The EDB helped ST Aerospace with calling on potential customers, including British Airways, Japan Airlines, and even Singapore Airlines to garner Section 41 modifications.[5] The fact that the EDB had to intervene to bring the country's two key aerospace firms together highlighted the fact that each was on a parallel and independent development path. ST Aerospace would also start MRO operations in the United Kingdom and United States in the 1990s as it globalized.

The momentum continued in the 2000s as MRO investment flooded into Singapore. After pausing for the Asian financial crisis of 1998 and a global recession in 2001/2002, it returned to export-driven growth in aerospace and other high-technology industries. Per capita GDP increased a whopping 50% between 2002 and 2008, while labor and land costs increased correspondingly. At the same time, it faced growing competition from China, which had dramatically lower labor costs, the world's fastest growing fleet, and the attention of every major OEM. Singapore needed a new game plan—it needed to transition from labor-intensive activities to capital- and innovation-intensive activities, including high technology manufacturing and research and development (R&D).

"At first, we didn't get a strong reception from investors on aerospace manufacturing," recalled former EDB official Kheng Yok Sia. "We'd ask them about it and they would scratch their head."[6] One key obstacle was the fact that Singapore and southeast Asia lacked a genuine manufacturing supply chain.

---

[4]Jeremy Chan (former EDB employee), interview with author, 5 Aug. 2016.

[5]Jeremy Chan and Robin Thevathasin (former executives—ST Aerospace), interview with author, 5 Aug. 2016.

[6]Kheng Yok Sia (former official—EDB), interview with author, 5 Aug. 2016.

Still, the EDB persisted. Its target would be Rolls-Royce, who established a major engine MRO facility with SIAEC in 2001. In 2005, EDB executive Manohar Khiatani visited Rolls-Royce COO John Cheffins and asked him a provocative question: "What's the difference between maintaining engines and assembling engines?" Following the meeting, Rolls-Royce asked the EDB if a very large parcel of land could be made available for a new facility.[7] In truth, Singapore was effectively sold out because the land around Changi Airport was almost fully utilized.

As Rolls-Royce was contemplating a new Asian facility (some 20 locations were under consideration), a debate was raging regarding the future of Seletar Airport. A former RAF station, Seletar was a general aviation airport on the northeast side of the island. Alternatives included housing developments and even a new Disney theme park. The EDB, however, felt that the missing ingredient for continued aerospace growth was land, specifically Seletar land. Where would it accommodate Rolls-Royce, should it choose Singapore? In 2006, it partnered with JTC Corporation to repurpose the area as Seletar Aerospace Park. It would be a brownfield, mixed-use development committed to aerospace manufacturing, business and general aviation MRO, pilot training, and R&D (see Fig. 14.5).

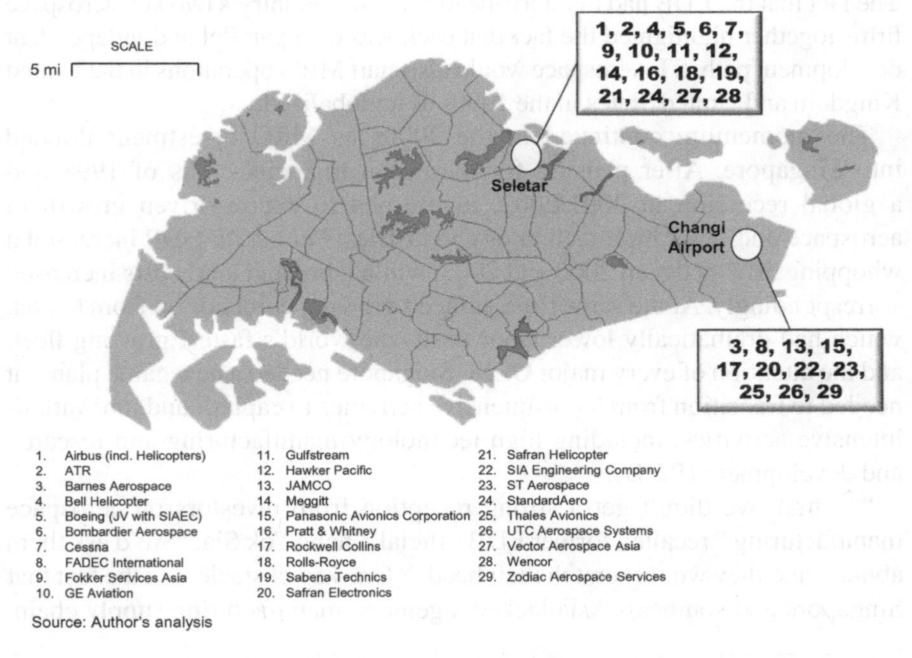

Fig. 14.5 Map of aerospace investments in Singapore.

[7]Kheng Yok Sia interview, 5 Aug. 2016.

Photograph courtesy of ©Rolls-Royce Plc

Fig. 14.6 Rolls-Royce's Seletar campus.

The EDB hit pay dirt in 2007 when Rolls-Royce announced a S$700-million ($450 million) investment in Seletar to build a Trent final assembly facility, a fan blade fabrication facility, a regional training center, and an advanced technology center. This was Rolls-Royce's first Trent final assembly facility outside of the United Kingdom, and it established Singapore as a manufacturing destination in Asia. Other manufacturing investments would follow, including a Pratt & Whitney aeroengine component fabrication facility. Airbus's parent company would also set up a new research facility, EADS Innovation Works, at Seletar. By 2016, 60 companies and more than 5000 employees would work at Seletar[8] (Fig. 14.6).

While Singapore gained momentum in manufacturing, there were several positive developments in the expansion of its aerospace R&D activities. The Agency for Science, Technology and Research (A*STAR), a government organization focused on developing scientific research and talent, began to focus on aerospace. In 2007, it created a consortium of EADS, Boeing, Pratt & Whitney, and Rolls-Royce to focus on precompetitive research. By 2016, it would expand to include 18 participants, including Bombardier and SIAEC. Outside of A*STAR's efforts, several other R&D centers were established by Pratt & Whitney, Thales, and GE. Government support, outstanding human capital, English-speaking skills, and strong intellectual property protection all contributed to Singapore's success.

As manufacturing and R&D flourished, SIAEC continued growth as a global MRO player. It created more than 20 JVs with leading OEMs and expanded

[8]Association of Aerospace Industries (Singapore), "Celebrating 10 Years at Seletar Aerospace Park," Aerospace Singapore, June 2016, pp. 18–20.

operations geographically throughout Asia. Publicly listed as a separate company in 2000, it would reach $1 billion in revenue by 2015. The other major Singaporean company, ST Aerospace, emerged as a global MRO powerhouse with operations in the United States, United Kingdom, and China. In 2013, it accounted for more than 11.5 million labor hours in airframe heavy maintenance, which made it the largest independent MRO in this segment [17].

From the mid-1990s to the mid-2010s, Singapore's aerospace cluster grew by 10% per year to reach S$7 billion ($5.6 billion) in activity by 2014. It had been a remarkable journey, from labor-intensive MRO to skilled labor MRO, and then adding advanced manufacturing and R&D to the mix. Singapore became Asia's aerospace *Über*cluster.

MOROCCO: THE UNLIKELY CLUSTER

Hamid Benbrahim El-Andaloussi (Fig. 14.7) had a problem. His company, Royal Air Maroc, had been a faithful Boeing customer for many decades, yet the aircraft manufacturer had not invested in Morocco. Nor had anyone else up to the early 2000s. Granted, Snecma started a joint venture for engine MRO with his airline in 1999, but Mr. Benbrahim was interested in something more audacious: creating an aerospace manufacturing cluster in Morocco.

Initially he faced skepticism from Boeing executives. But after persistent lobbying, he broke through when he convinced Boeing and Snecma subsidiary Labinal to enter a joint venture with Royal Air Maroc to produce wire harnesses for the 737. Boeing managers initially expected labor productivity of just 30% of industry norms from the new firm, called Matis Aerospace (Morocco Aero-Technical Interconnect Systems). Matis opened a small shop in Casablanca in 2001; within two years it achieved 70%. As Matis grew, job openings attracted floods of highly educated applicants—more than 80%

Credit: GIMAS

Fig. 14.7 Hamid Benbrahim El-Andaloussi spearheaded Morocco's aerospace cluster.

of whom were women. This not only grabbed Boeing's attention, but also impressed parent company Snecma [18]. The first leader of the company was the irrepressible Mr. Benbrahim.

Still, confidence in Morocco's ability to compete in aerospace wasn't widespread. In 2005, global consulting company McKinsey & Co. drafted for the Moroccan government a comprehensive industrial program known as "Emergence," which listed several promising industrial sectors for economic growth. It did not identify aerospace as a target industry [19].

In 2004–2006, Hutchinson Aerospace, Circor Aerospace, and Daher made their own investments, and in subsequent years Safran followed with two large investments including a nacelle production facility for its Aircelle subsidiary. Snecma added to its in-country wire harness capability in acquiring a small firm outside of Rabat. A dozen subtier suppliers also came to Morocco, bringing the total number of aerospace firms to 22 by 2006. In parallel, Benbrahim brought government officials to the Paris and Farnborough Air Shows to meet potential investors, learn the industry, and educate them on the large opportunity in front of them.

As Morocco gained traction, it drew confidence from the fast-expanding Mexican aerospace cluster, which it studied closely. Like Mexico, Morocco had geographic proximity to a major market, a large and energetic talent pool, and preferential market access courtesy of a Morocco–European Union free trade agreement. Garnering just 1–2% of European aerospace production—worth more than $100 billion—would create a substantial success story. And Morocco had an advantage that Mexico lacked: centralization. Politically, Morocco was a constitutional monarchy led by the popular King Mohammed VI, and economically its activity was concentrated in a few cities, led by Casablanca. This meant that it could take a page from the Singapore playbook and offer speed and one-stop shopping that brought together public and private stakeholders. The next step in its evolution was the creation of a strong aerospace industry association to accelerate its development and be the focal point for potential investors. GIMAS (Groupement des Industries Marocaines d'Aéronautique et Spatiales), modeled after the French aerospace industry association, was founded in 2007. Not surprisingly, it was led by Benbrahim.

As Morocco gained momentum, it ran into several roadblocks. The first was a lack of in-country subtier suppliers. With the cost structure of most fabrications dominated by the bill of material—parts, raw materials, and components—Morocco needed more domestic capability to enhance its competitiveness. It needed to create a true ecosystem rather than depending on labor arbitrage. This was specifically called out in the government's Emergent Industry Report in 2009.

The second obstacle was training. With a dearth of trained personnel, the aerospace pioneers faced the prospect of what was happening in China: needing to poach employees from each other, escalating wages, or both.

Some companies even forbade their local staff from riding together on shuttle buses out of fear they might try to recruit each other [20]. It addressed this in 2011 by starting a new aerospace vocational school (Institut des Métiers de l'Aéronautique, or IMA) in a public–private partnership where the government contributed land and buildings and GIMAS members organized and sponsored training for new hires, modeled after French standards. Training programs lasted between 22 and 44 weeks, and the volume of trainees grew rapidly.

As IMA spooled up, geopolitics raged across North Africa in what became known as the Arab Spring. As uprisings toppled the governments in Egypt, Libya, and Tunisia, Morocco remained a relative haven of calm as King Mohammed VI offered a more democratic constitution and new elections, but the tumult reemphasized the importance of the need to create jobs, including its aerospace gambit.

The year 2011 was also when the Shorts Division of Bombardier decided to establish a new manufacturing facility in Morocco. With thousands of aerostructures employees in relatively high-cost Belfast, Northern Ireland, it needed a low-cost manufacturing base to remain competitive. Bombardier already had significant experience operating in Querétaro, Mexico. "We looked at Tunisia, Poland, Romania, Turkey and several other countries," recalled Stephen Orr, Bombardier's Morocco site leader. Ultimately, Bombardier's decision was based on the same factors that lured other investors—political stability, government support, proximity to Europe, the talent pool, costs, and specialized training programs like IMA. There was one other factor: "We didn't want to put all of our eggs in one basket—Mexico."[9] Meanwhile, in France, Christophe Delque was finalizing Hamilton Sundstrand subsidiary Ratier-Figeac's low-cost strategy and settled on Morocco. "The next step was to convince leadership, which was very hard. But the fact that there were so many companies and employees there helped make the case. There was a cluster there."[10] GIMAS's strong position was also helpful. Combined, investments by Bombardier and UTAS totaled $200 million, and the number of aerospace companies in Morocco reached 52 that year (Fig. 14.8).

Two years later, in 2013, the 128-ha MIDPARC industrial park dedicated to aerospace development opened right across the street from the IMA training center near Casablanca's international airport. It was designed with a one-stop-shop approach where investors could set up a business in as little as a week. Again, this was a GIMAS-led initiative with the help of local pension funds. Again, Mr. Benbrahim was in charge, making him the leader of the industrial park, the training institute, and the industry association. In October of that year, King Mohammed VI appointed Moulay Hafid Elalamy as Minister of Trade

[9]Stephen Orr (Vice President—Morocco Manufacturing Centre), interview with author, 12 Aug. 2016.
[10]Christophe Delque (Site Director—Morocco), interview with author, 11 Aug. 2016.

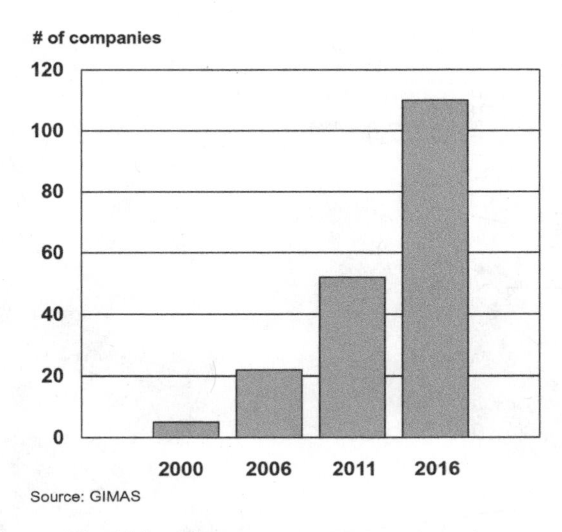

Fig. 14.8 Moroccan aerospace companies.

and Industry. Elalamy was a powerhouse; he was not only one of Africa's wealthiest men and founder of a very successful insurance company, but also a former senior advisor to Quebec's Ministry of Finance. He soon introduced the 2014–2020 Industrial Acceleration Plan, which created a blueprint for 500,000 new jobs. Aerospace was a key industry in this plan, with goals of creating 23,000 new jobs, $1.6 billion in new revenue, 100 new investors, and $1.3 billion of investment. To strengthen this ecosystem, the target of 35% local sourcing was also included [21]. With the leadership combination of Messrs. Elalamy and Benbrahim and the experience of investors, investments continued to pour in during subsequent years, including from Alcoa Fastening Systems, Hexcel, Stelia, and Aerolia (Fig. 14.9). The new manufacturing facilities focused on more sophisticated products than in prior years; Thales even announced plans to build an additive manufacturing facility in Midparc. At the same time, existing companies continued to expand. Employment at Matis Aerospace, the original aerospace manufacturing investor, hit 850 as it won many new electrical harness contracts. Missing from the roster of investors were Moroccan firms.

By 2016, Mr. Benbrahim's outlandish dream of creating an aerospace manufacturing center in North Africa took shape. Morocco's aerospace cluster reached 110 companies and 11,500 employees, with $1 billion in exports, and IMA broke ground on a new expansion to train 1200 people per year. This time, global media took notice. In a May 2016 feature on the Arab world, *The Economist* magazine cited Morocco as "a rising star in the west," and featured the aerospace cluster as one reason for optimism [22].

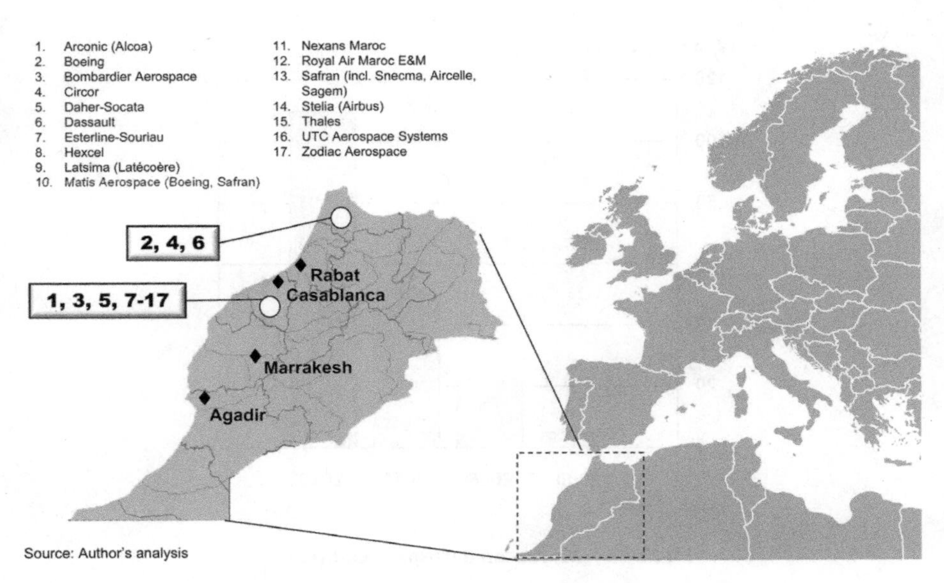

Fig. 14.9 Morocco's major aerospace companies: 2016.

The Southeastern United States: Aerospace Manufacturing Hotspot

With the demise of Southern California and emergence of new clusters in low-cost countries in the 1990s and 2000s, one could be excused for thinking that jetliner manufacturing was destined to shift away from high-cost locations in North America and Europe. A funny thing happened on the way to the future: In the 2010s, a new manufacturing cluster emerged in the Southeastern United States. How did this happen?

The Southeastern United States, roughly bounded by North Carolina and Tennessee to the north and the Mississippi River to the west (a broader definition includes Virginia, Kentucky, Louisiana, and Arkansas), had a strong aviation and military heritage but limited aerospace manufacturing presence through the 1990s. Gulfstream did build business jets in Savannah, Georgia, and Lockheed Martin produced F-22 fighters in Marietta, Georgia, but these were viewed as exceptions. For OEMs, investment in the Southeastern United States was not on the radar screen as they scurried to build low-cost production facilities in Asia, Eastern Europe, and Mexico. Yet in the 1990s the seeds were being planted for an aerospace cluster courtesy of the automotive industry (Fig. 14.10).

Automotive manufacturing in the United States was traditionally centered in the Midwest near Detroit with unionized labor forces, but the arrival of foreign automotive competitors altered the landscape. Honda was a pioneer in opening a nonunionized production facility in Marysville, Ohio, in 1982. Nissan (1983) and Toyota (1988) followed with their own nonunionized factories in Tennessee and Kentucky, respectively. Recognizing the altered

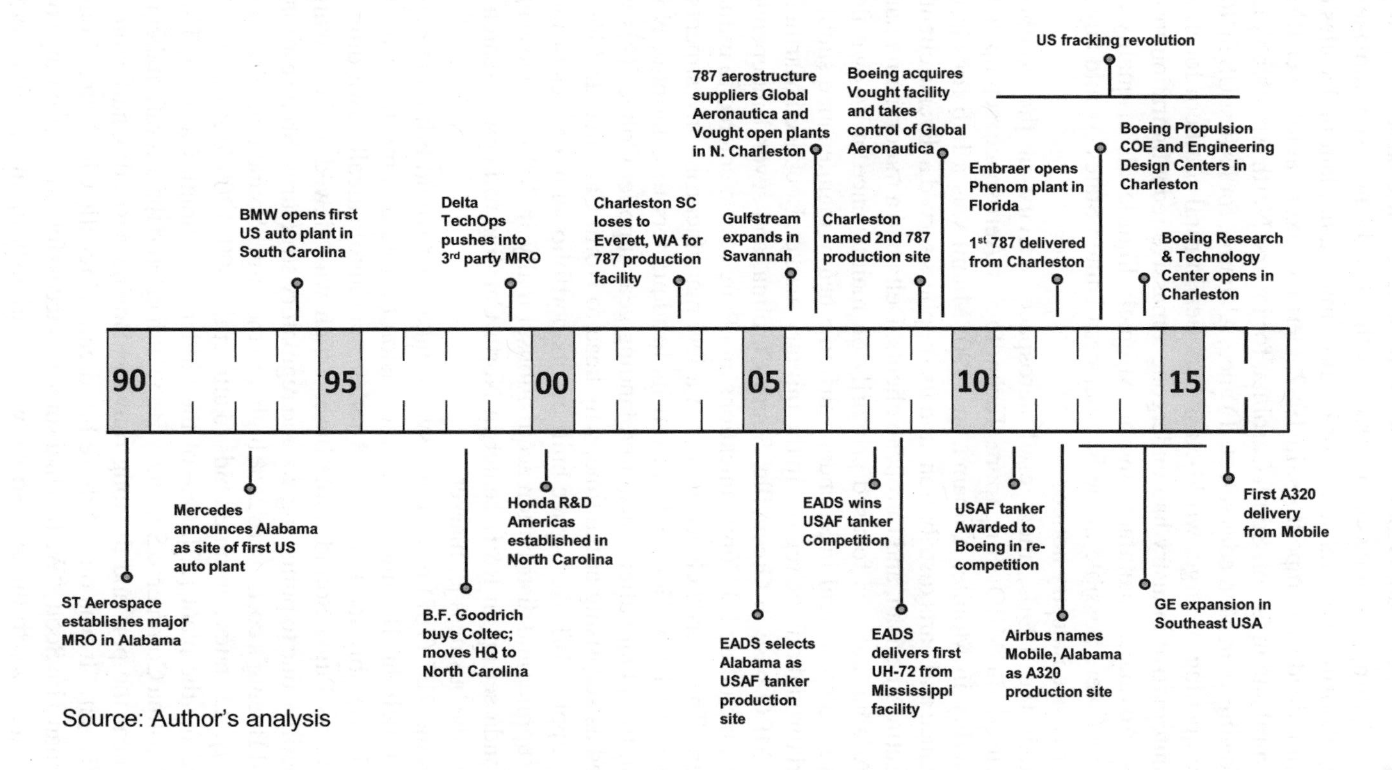

Fig. 14.10 Milestones in Southeastern US aerospace investment.

landscape, US auto giant General Motors announced it was starting a new car company, Saturn, in Tennessee in 1990. As the world's largest automaker at the time, choosing the rural south for its most important plant in decades did not go unnoticed by competitors. In 1992, German luxury automaker BMW chose a rural site in Greer, South Carolina, for its first North American plant. The following year, Mercedes selected Vance, Alabama, for its first US factory. With the opening of large, world-class factories in rural southern locations, the US automotive industry had undergone a massive reconfiguration in just 10 years. Advanced manufacturing at extremely high levels of quality and productivity were possible in the Southeastern United States. Would anyone in the aerospace industry notice?

Through the 1990s, most civil aerospace activity in the southeast was focused on MRO. Singapore Technologies Aerospace set up a US MRO facility in Mobile, Alabama, in 1990. Miami was a hub for aircraft maintenance and parts distribution, and its suppliers served a diverse clientele of US, Latin American, and European clients. Delta had a major maintenance hub in Atlanta, but was focused primarily on maintenance of its own fleet. There were also several independent airframe heavy maintenance suppliers scattered throughout the region. In the military market, Lockheed Martin had a major MRO base in Greenville, South Carolina, and several independent suppliers, including L3 Communications and Pemco, also catered to military customers. The decade closed with just a few manufacturing investments in the region. In 1998, Goodrich moved its headquarters to Charlotte, North Carolina, from Ohio after it acquired landing gear supplier Coltec. This was perceived as surprising at the time; why head to North Carolina? In Florida, interior supplier B/E Aerospace built out its portfolio with a series of niche interior supplier and distribution acquisitions. Finally, in 2000, Japanese auto giant Honda set up an R&D center in North Carolina to begin research on entry into the business jet market.

It was the 7E7 (787) production sweepstakes that would bring the region into the limelight. Boeing made the decision to hold a competition for the 787 final assembly facility in 2003 rather than automatically awarding the facility to its Puget Sound home base, which was viewed as increasingly uncompetitive due to punishing tax and regulatory structures. Some estimates indicated Boeing's costs were 25% higher in the Puget Sound than elsewhere. Ultimately, 22 states submitted bids to attract the 800–1200 jobs that Boeing promised for the plant [23]. One of the bidders was South Carolina. Then–South Carolina Commerce Secretary Bob Faith offered the aircraft maker two options near the port and the long runways Boeing desired: Charleston and Myrtle Beach. "It was one of those big dreams," recalls Charleston County Councilman Tim Scott, "We had had some unsuccessful bids in bringing some other big industry to the area so we were cautiously optimistic. We started looking at what we considered landing a whale" [24]. Fearing the loss of a crown jewel, Washington Governor Gary Locke and other state officials embarked on an eight-month battle to hold serve. In June, he signed into law a $3.2-billion, 20-year tax-break package that brought the gap between Everett and Kinston, North Carolina, the lowest-cost site competing for the 7E7, to less than $300 million [25]. In December 2003, Everett learned that it had won the competition. South Carolina felt it had the most competitive bid and was disappointed by the news. Was it a bridge too far given the fact that Boeing was engulfed in a corruption scandal that led to the resignation of CEO Phil Condit that month?

South Carolina's disenchantment was soon assuaged by the news that Vought Aircraft had selected North Charleston as the location for its 7E7 Dreamliner aft fuselage component manufacturing facility. It also announced that a 50/50 joint venture with Alenia North America, Global Aeronautica, would be established next door to the new facility in support of midbody fuselage production for the 787. As outlined in Chapter 8, both facilities caused major supply chain disruptions and were ultimately acquired by Boeing in 2008 and 2009. Thus, a new organization with several thousand employees was created that was later named Boeing South Carolina. Shortly after Boeing took control, employees represented by the International Association of Machinists (IAM) at the former Vought facility voted to decertify the union with a vote of 199 to 68, confirming South Carolina's right to work status; along the way, South Carolina continued to lobby for a 787 final assembly facility.

To South Carolina's good fortune, Boeing determined that it required a second production facility to reach 787 production rates beyond 10/month. A second 787 final assembly sweepstakes took place in 2009, and this time Charleston was no longer the underdog. Fresh with the memory of a 57-day strike at Everett the prior year, Boeing selected North Charleston as the site of its new 787 final assembly facility in October 2009 (Fig. 14.11). South Carolina offered some $450 million in state incentives to land its prize; the $750 million investment by Boeing was the state's largest initial investment in its history, easily exceeding the $500 million invested by BMW in its groundbreaking investment 15 years before [26].

While Boeing was grappling with the 787 development program, a strategy was brewing in Toulouse to transform Airbus's parent company EADS into a global aerospace powerhouse with a strong civil/military market balance. To execute the strategy, EADS needed greater access to the huge US military market, which comprised about half of all global military equipment spending. In 2002, it submitted an unsuccessful bid for a US Air Force tanker competition to replace aging KC-135 tankers with the A330-MRTT, a tanker design based on its A330-200 airframe. EADS lost the competition to Boeing's KC-767 bid. It also opened an engineering and design center in Wichita, its first outside of Europe. In 2003, it created an EADS North American division,

Credit: Copyright © Boeing

Fig. 14.11 Boeing South Carolina.

and in the following year, its US subsidiary Eurocopter set up a helicopter production facility in Columbus, Mississippi. This would pay major dividends in 2006 when the US Army selected the UH-145, a variant of the EC-145, for a purchase of 332 utility helicopters, with a potential total program life-cycle value of over $2 billion.

A massive corruption scandal surrounding the KC-767 lease deal erupted in 2003, when it was disclosed that Darleen Druyun, the former Principal Deputy Undersecretary of the Air Force for Acquisition, had passed information from EADS's bid to Boeing before she joined the company. This led to the resignation of Boeing CFO Michael Sears and the resignation of CEO Phil Condit. The US Air Force froze the tanker program and planned to run a new competition in the future.

EADS would require an in-country production facility (and a partner) to fulfill the tanker's contract obligations and evaluated four locations in the Southeastern United States for its production facility—Alabama, Florida, Mississippi, and South Carolina. In 2005, it selected Mobile, Alabama. This decision gave EADS access not only to a deep-water port, but also to one of the most powerful US congressional delegations. US Senator Jeff Sessions was a member of the Senate Armed Services Committee, and fellow senator Richard Shelby sat on the influential Senate Appropriations Committee. Both were Republicans during the George W. Bush administration who became forceful advocates for Airbus, offsetting some of Boeing's massive political influence in Washington, DC. Senator Shelby called the argument that the Airbus KC-330 entry would lack sufficient US content "bogus" and promised to "push every way we can" to make the case for Alabama and Airbus [27]. Alabama turned out to be a savvy political decision. Northrop Grumman would be the US partner. A new request for proposals (RFP) was issued in January 2007, and in June 2008 the EADS KC-35 entry won the competition. It was a huge, historic, yet short-lived victory for EADS. Boeing protested the award, the US Air Force reopened the bidding process, and Northrop Grumman pulled out of the program. The Boeing KC-46, a 767 variant, was selected in February 2011 in a $30 + -billion contract. EADS declined to protest.

The brooding in Alabama wouldn't last long. In 2012, EADS subsidiary Airbus shocked the industry when it announced that it would build an A320 final assembly facility in Mobile, its second outside of Europe. (The other was in Tianjin, China.) The $600-million facility would employ 1000 people and considerably expand its US footprint and credentials (Fig. 14.12). An incentive package worth $100 million helped to close the deal [28]. The entire Gulf Coast region, including Mississippi and Florida, cheered the move. Other benefits touted by Airbus were getting closer to its US customers and currency hedging. Industry guru Richard Aboulafia saw another motivation: "The real reason for doing it is to put leverage on the politicians and trade unions back home—but Airbus can't say that," he opined [29]. Regardless of its true motivation, Airbus had deftly executed a globalization strategy, and 19 years after Mercedes chose Alabama for its new auto facility, the state had brought home the world's largest jetliner manufacturers—the crown jewel of advanced manufacturing.

Credit: Airbus

Fig. 14.12 Airbus executives Barry Ecclestone and John Leahy at the Mobile grand opening.

While two final assembly anchors developed in South Carolina and Alabama, Florida—a state best known for its Space Coast—finally got into the act. In 2011, Embraer opened a final assembly facility for its Phenom business jets in Melbourne, Florida, taking advantage of the highly trained workforce and the many former employees from Cape Canaveral. In 2016, it would expand and add Legacy 450 and 500 business jet production to the facility.

Meanwhile, the momentum continued in South Carolina. In 2011, Nikki Haley assumed South Carolina's governorship and became an intense advocate for aerospace investment, acquiring the nickname "the senator from Boeing" along the way. Boeing doubled down on its commitment, adding interior manufacturing, propulsion, and research and technology facilities. GKN would establish a new aerostructures facility there in 2011, as well.

One more wave of aerospace investment would hit the Southeast in the 2010s, this one centered on aeroengines and advanced materials. The wave was aided by two shifts—one in global energy markets and the other a leading OEM's supply chain strategy. The fracking revolution discussed in Chapter 7 would lead to a significant energy cost advantage for US suppliers, particularly those depending on manufacturing processes related to natural gas. By January 2015, US natural gas prices were $2.59 per thousand ft^3 compared to $9.50 in Europe, $10.25 in China, and $14.00 in Japan.[11] The shift in supply chain strategy benefitting the region was led by GE Aviation, which decided the many new enabling technologies in its LEAP-X engines would stay primarily in the United States. This would lead to an array of investments in the Southeastern United States. Its famous LEAP-X fuel nozzles would be produced in a high-volume additive manufacturing facility in Auburn, Alabama. Asheville, North Carolina, would be the venue for a new ceramic matrix composite (CMC) engine component facility. "We have built eight new factories in the US in the last eight years," GE Aviation CEO David Joyce said in a 2016 interview. "Many of these facilities are our most advanced technologies."[12] Superalloy producer Carpenter Technology also invested in raw materials capability when it opened two facilities in Alabama to produce advanced alloys and powder metals in the 2014–2015 timeframe. To support burgeoning composite production, composites supplier Cytec also expanded its carbon fiber production footprint in Greensville, South Carolina.

In less than a decade, the Southeastern United States went from an aerospace manufacturing backwater to one of the globe's hotspots with Boeing and Airbus final assembly facilities (Fig. 14.13). Its pro-business environment, relatively inexpensive energy and land costs, committed political leadership,

[11]Chriss Street, "US Natural Gas Prices Collapsing After EPA Says Fracking Is Safe," Breitbart.com, 9 June 2015, http://www.breitbart.com/big-government/2015/06/09/u-s-natural-gas-price-collapsing-after-epa-says-fracking-safe. [retrieved 22 Aug. 2015].

[12]David Joyce (CEO—GE Aviation), interview with author, 16 March 2016.

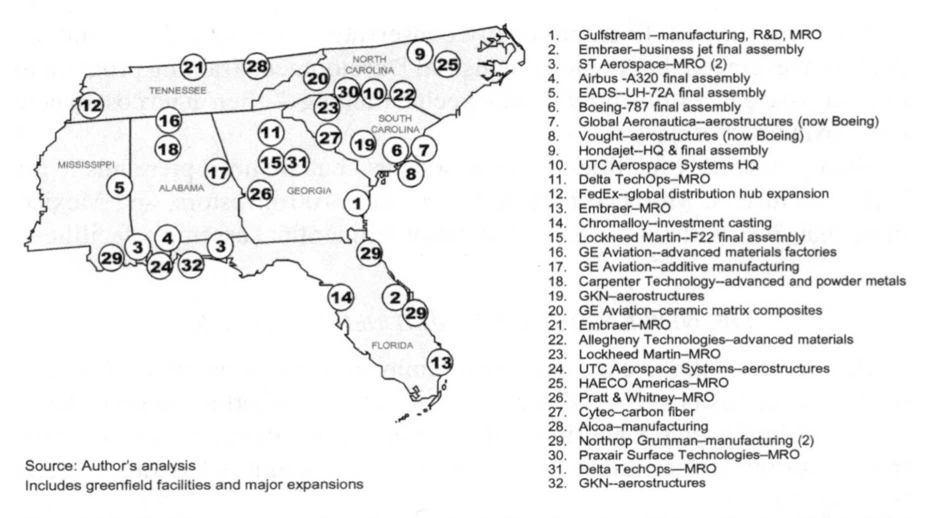

Fig. 14.13 Major aerospace investments in the Southeastern United States: 1990–2016.

and anchor investments by leading OEMs all underpinned its success. The success of the region proved that not all jetliner manufacturing was destined to move to emerging economies.

Four Clusters and a Funeral: The Takeaways

The stories of five jetliner clusters—four rising and one falling—highlight several realities for civil aerospace investment. Clearly, a positive business climate and infrastructure are crucial, as the contrast between California and the right-to-work states in the Southeastern United States demonstrates. Singapore is consistently rated as one of the most business-friendly countries, and Morocco and Mexico knocked down barriers to become investment-friendly locations. Decisions to build or relocate facilities don't happen frequently, but in aggregate they shift or create clusters over time.

Another takeaway is the importance of good government institutions, including intellectual property protection, economic development agencies, and the rule of law. Strong leaders aren't mandatory. It's true that Hamid Benbrahim El-Andaloussi played a pivotal role in Morocco, but one is hard-pressed to pick out a single, indispensable leader in the other three clusters that drove aerospace development. Strong institutions can be at the federal level, as in Singapore and Morocco, or at the local and state level as in Mexico and the Southeastern United States.

Training and development of human capital are crucial. Singapore aligned its training and university programs with the needs of its investors and created liberal immigration policies to ensure the ready supply of labor. Similarly, Morocco created IMA and Mexico created UNAQ to develop a skill base that each country lacked. Even politically conservative and limited-government Alabama leveraged the Alabama Industrial Development Training program to lure aerospace investors through job-specific training—often at no cost to new and expanding businesses.

Finally, assembly facilities are often magnets but are not a prerequisite for building clusters. Morocco attracted more than 100 investors, and Mexico more than 300, without the benefit of major jetliner final assembly facilities.

The New Landscape of Jetliner Production

The stories of the five clusters are a reminder that comparative advantage in the jetliner business is ephemeral. The unbundling of jetliner supply chains described in Chapter 8 and the shifting competitive dynamics created new manufacturing clusters since 2000—mostly to the south of existing clusters (see Fig. 14.14).

In the Americas, there were six major pre-2000 jetliner and civil aerospace manufacturing clusters. The largest was the Puget Sound ecosystem surrounding Boeing's Everett and Renton final assembly facilities, including Tier 1 suppliers and subtier manufacturers. The next largest was in Quebec and Ontario. Anchored by Bombardier final assembly facilities in Montreal and Toronto, it also included many component OEMs and subtier suppliers that also supported the region's automotive industry. Third was the Central United States cluster, stretching from Dallas-Ft. Worth to Wichita, Kansas, which included significant aerostructures and component capabilities, and was anchored by Cessna and Learjet final assembly facilities in Wichita, and Lockheed Martin and Bell Helicopter plants in Dallas-Ft. Worth. The Midwestern US cluster stretched from GE's aeroengine final assembly facility in Cincinnati through Indiana, Illinois, and Iowa to the west and Michigan to the north. The Northeastern US cluster was centered around the Connecticut operations of Pratt & Whitney and helicopter OEM Sikorsky. It was a cluster with falling employment, much like the shrinking Southern California cluster described previously in this chapter.

Three new clusters developed in the Americas after 2000. One was in Brazil and was concentrated around Embraer's Sao Jose dos Campos headquarters. It included its final assembly facility and some system OEMs, aerostructures firms, and subtier suppliers. The other two clusters, Mexico and the Southeastern United States, were described in this chapter.

In Europe, the Middle East, and Africa (EMEA), there were several major pre-2000 jetliner manufacturing clusters that were influenced heavily by Airbus and Rolls-Royce activity. The largest was in France—a very deep cluster with the world's second largest aerospace industry. It included Safran along with numerous aeroengine and aircraft systems suppliers in greater Paris, and the Aerospace Valley in Southwest France with hundreds of aircraft

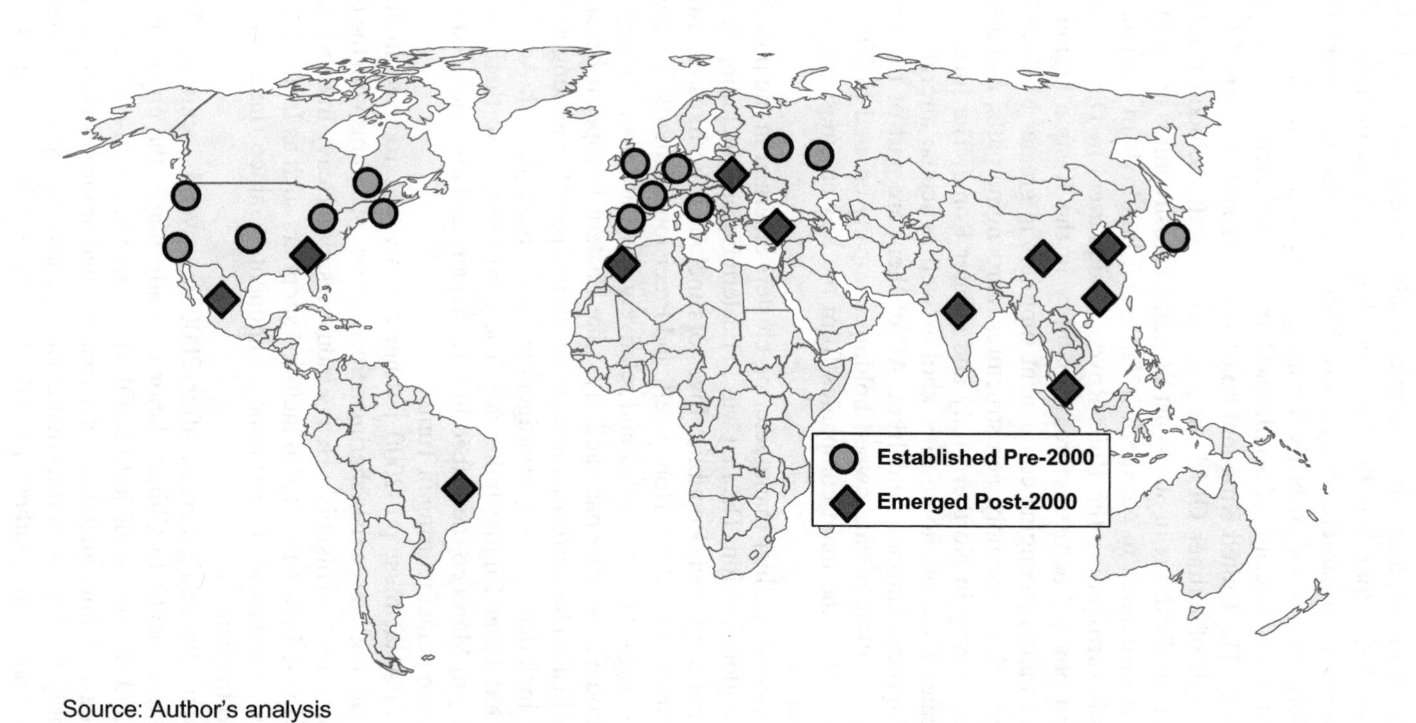

Source: Author's analysis

Fig. 14.14 Major civil aerospace manufacturing clusters.

system, aerostructures, and subtier suppliers surrounding Airbus's Toulouse headquarters. Next door was the German cluster, which included major concentrations of aerostructures, systems, MRO, and interiors suppliers in Northern Germany around Airbus's Hamburg final assembly facility, as well as advanced manufacturing in Southern Germany anchored by aeroengine supplier MTU. The United Kingdom boasted a large and diversified cluster despite the exit of jetliner OEMs in prior decades. It included aerospace component manufacturing in Southeast England, aerostructures in Southwest England, aerostructures in Northern Ireland, and hundreds of suppliers in the Midlands surrounding the Rolls-Royce headquarters in Derby. To the south, Spain boasted aerostructures capability in the Basque region, and aerostructures and aeroengine component suppliers in greater Madrid. Italy had helicopter final assembly, aerostructures, aeroengine parts, and aircraft components activity in Northern Italy and greater Rome. The final cluster was in Greater Moscow, home to Sukhoi as well as iconic aircraft OEMs Ilyushin, Tupolev, Antonov, and Irkut. All of them are part of the United Aircraft Corporation, a state-owned holding company created in 2006 that incorporates 30 of the main companies from the Soviet times, employing 100,000 people.

Three important jetliner manufacturing clusters emerged in the post-2000 era thanks to globalization. The first was in Eastern Europe, including Poland, Romania, and the Czech Republic. Investors included the likes of Airbus, United Technologies, MTU, Honeywell, and Precision Castparts. They were drawn to the region for its skilled and relatively low-cost labor. Turkey also became a hotbed for aerospace activity. Like Eastern Europe, it became a favored destination for aerostructures and machined parts for aeroengines and systems. It included two major indigenous companies: Turkish Aerospace Industries and Tusas Engine Industries. The final cluster, Northern Africa, was centered in Morocco (described in this chapter) and also included some labor-intensive manufacturing in Tunisia.

Finally, in Asia, most pre-2000 jetliner activity was focused in Japan because of Boeing's large risk-sharing aerostructures programs with the three Japanese "heavies"—Mitsubishi Heavy Industries, Fuji Heavy Industries, and Kawasaki Heavy Industries. It also included suppliers such as Ishikawajima-Harima Heavy Industries (aeroengine components), Jamco (interiors), and Tiejen-Seiki (actuators).

New Asian clusters emerged after 2000 in several locations. Three were in China, thanks to China's labor cost advantage, the emergence of COMAC, and the desire of jetliner OEMs to maximize in-country sales. Northern coastal China included firms concentrated around Airbus's Tianjin final assembly facility as well as manufacturing and engineering centers in Beijing and Harbin. In southern coastal China were indigenous firms and a string of foreign investments in the Guangdong and Fujian provinces. Major manufacturing centers included Guangzhou, Xiamen, and Zhuhai. Its focus was labor-intensive manufacturing, aircraft components and systems, and aerostructures. A cluster in central China, centered in Xi'an and Chengdu, included significant aeroengine manufacturing capability and was also home of turboprop OEM Xian Industrial Aircraft Corporation and AVIC Aviation Engine Corp—China's aeroengine national champion. The Shaanxi province alone, including Xi'an, had more than 100,000 aerospace workers. Although not yet a cluster, Shanghai was on its way in the mid-2010s as COMAC ramped up to produce the C919.

Another new cluster was Southeast Asia, including Singapore (described in this chapter), Malaysia, Thailand, Vietnam, and the Philippines. Investments focused on aeroengines, aircraft parts, aerostructures, and components. Finally, south-central India, including Hyderabad and Bangalore, gained the attention of investors seeking to leverage India's significant engineering capability and to gain access to its large civil and military aircraft market. Bangalore is anchored by national champion Hindustan Aeronautics. A wave of engineering investments after 2000 strengthened the cluster, with Bangalore alone boasting more than 20,000 aeronautical engineers.

The sum of these changes transformed the economic geography of aerospace. Twenty-five years after the end of the Cold War, jetliner manufacturing dispersed. The rise of Airbus, sourcing from low-cost countries, and the digitization of design transformed and dispersed jetliner supply chains. In the mid-1980s, nearly three-quarters of jetliner manufacturing activity was in the United States and Canada, with the balance in Europe. By 2015, the United States and Canada fell to just over half of all activity, with Europe comprising 35% and the rest of the world 11% (Fig. 14.15).

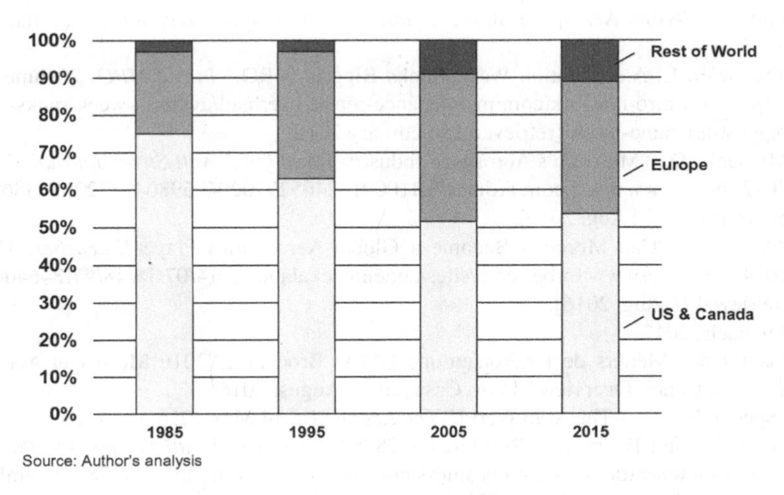

Fig. 14.15 Jetliner manufacturing by region.

REFERENCES

[1] Los Angeles Economic Development Corporation, *The Aerospace Industry in Southern California*, LAEDC, Los Angeles, 2012, p. 13.
[2] Mock, R., "1927 Aircraft Production," *Aviation*, 2 Jan. 1928, pp. 36–37.
[3] Hise, G., *Magnetic Los Angeles: Planning the Twentieth-Century Metropolis*, Johns Hopkins University Press, Baltimore, 1997, p. 117.
[4] Los Angeles Community Analysis Bureau, *The Economic Development of Southern California, 1920–1976*, Vol. 1: The Aerospace Industry as the Primary Factor in the Industrial Development of Southern California: The Instability of the Aerospace Industry, and the Effects of the Region's Dependence on It, Community Analysis Bureau, Los Angeles, 1976, p. 5.
[5] Westwick, P., *Blue Sky Metropolis*, University of California Press, Berkeley, CA, 2012, p. 30.
[6] Lotchin, R., *Fortress California 1910–1961: From Warfare to Welfare*, Oxford University Press, New York, 1992, p. 251.
[7] "California—An Aerospace Leader," *Aviation Week & Space Technology*, 4 Dec. 1989, p. S3 [sponsored supplement].
[8] Smith, B., "California Aerospace Firms Fighting on Three Fronts," *Aviation Week & Space Technology*, 9 Nov. 1992, p. 47.
[9] Velocci, A., "Hughes Chief Builds on Known Strengths," *Aviation Week & Space Technology*, 26 April 1993, p. 13.
[10] Roosevelt, M., "Aerospace Adapts to Survive in Orange County," *Orange County Register*, 1 Feb. 2013, https://www.ocregister.com/2013/02/02/aerospace-adapts-to-survive-in-oc. [retrieved 1 Feb. 2016].
[11] Anselmo, J., "Made in Mexico," *Aviation Week & Space Technology*, 2 April 2007, p. 67.
[12] Coffin, D., "The Rise of Foreign Aerospace Suppliers in Mexico," USITC Executive Briefing on Trade, February 2013.
[13] FEMIA, Mexican Aerospace Presentation, http://femia.com.mx/themes/femia/ppt/femia_presentacion_tipo_eng.pdf [retrieved 24 March 2017].
[14] ProMéxico, "Mexican Aerospace Industry: A Booming Innovation Driver," *Negocios*, June 2015, p. 17.
[15] Coffin, 2013.
[16] Fink, D., "Asian Aerospace Show," *Aviation Week & Space Technology*, 20 Jan. 1986, p. 13.
[17] Tegtmeier, L. A., "Aviation Week Ranks Biggest MRO," *Inside MRO*, 26 June 2013, http://www.mro-network.com/maintenance-repair-overhaul/aviation-week-ranks-biggest-airframe-mros. [retrieved 15 February 2018].
[18] Michaels, D., "Morocco's Aerospace Industry Takes Off," *Wall Street Journal*, 20 Nov. 2012, https://www.wsj.com/articles/SB1000142405297020405980457722676386826375 8. [retrieved 12 Aug. 2016].
[19] Alaoui, A., "Can Morocco Become a Global Aeronautics Player?," *Forbes*, 18 July 2014, https://www.forbes.com/sites/abdelmalekalaoui/2014/07/18/48/#7f8464dd6f36. [retrieved 16 Aug. 2016].
[20] Michaels, 2012.
[21] Institut des Métiers de l'Aéronautique (IMA) Brochure, "2016 Moroccan Aerospace Training Center Overview," IMA, Casablanca, August 2016.
[22] "Special Report—The Arab World," *The Economist*, 14 May 2016, pp. 9–10.
[23] Gates, D., and Turim, D., "Building the 787: A Timeline," *Seattle Times*, 15 Dec. 2009, https://www.seattletimes.com/business/boeing-aerospace/building-the-787-dreamliner-a-timeline. [retrieved 22 Aug. 2016].

[24] Shane, A., "How the Boeing Deal Was Done," Thestate.com, 2 Nov. 2009, http://www.thestate.com/news/local/article14355173.html. [retrieved 22 Aug. 2016].
[25] Gates and Turim, 2009.
[26] Shane, 2009.
[27] Fulghum, D., "Alabama Wins," *Aviation Week & Space Technology*, 27 June 2005, p. 34.
[28] Mecham, M., and Flottau, J., "The Right Time," *Aviation Week & Space Technology*, 9 July 2012, p. 50.
[29] Ibid.

Chapter 15

Horizons

This book has traced the development of a new, global jetliner ecosystem in the early 1990s that was shaped by five megatrends: shifting geopolitics, enabling technologies, globalization, airline deregulation and liberalization, and the rise of lessors. What are the trends from a mid-2018 vantage point that could shape the jetliner ecosystem over the next several decades?

The first trend is the rise of Asia. The region's demographics, infrastructure investment, and ongoing liberalization will drive a tripling of revenue passenger kilometers over the next 20 years, according to Airbus.[1] Asia will soon become the largest air travel market sometime and won't look back By 2036, according to Airbus, the Asian air travel market will comprise 41% of global demand, more than North America and Europe combined (36%). Domestic Chinese traffic will lead the way, with some 1.6 billion passengers expected by 2036.[2] This will translate into massive jetliner demand. Boeing anticipates that Asian airlines will spend $2.5 trillion on jetliners over the next 20 years—41% of global spending (Fig. 15.1).[3]

If the forecast is even remotely accurate, it portends a shift in where the future jetliner ecosystem will exist. Today Asia comprises less than 10% of jetliner manufacturing activity. This will grow as countries in the region ramp up aerospace investment and flex their political muscle—particularly China. The single-aisle C919 may never be a commercial success, but China will learn from the experience and is already planning a twin-aisle C929 in collaboration with Russia for the late 2020s. "Eventually China will be similar to what Airbus is," according to former Rockwell Collins CEO Clay Jones.

[1]Airbus, "2017 Airbus Global Market Forecast: 2017–2036," http://www.airbus.com/aircraft/market/global-market-forecast.html [retrieved 17 April 2018].

[2]Ibid.

[3]Boeing, "Current Market Outlook 2017–2036," www.boeing.com/cmohttp://www.boeing.com/cmo, [retrieved 17 April 2018].

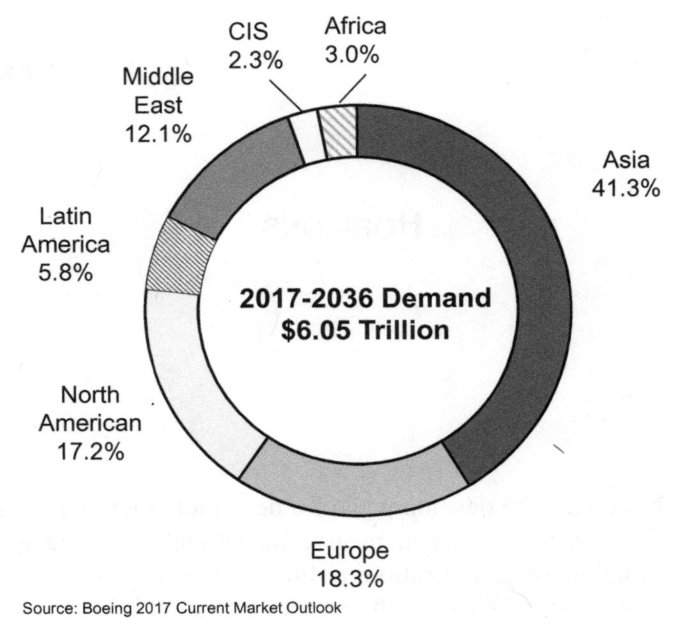

Fig. 15.1 Projected jetliner demand: 2017–2036.

"In 20 years, they might get a 4–5% share of the jetliner business. In 40 years, it could be 30%. No one has more political will and capital than China does."[4] The bottom line is that if anyone can break the Airbus–Boeing duopoly, it is China. And the competitiveness of the C929, and perhaps an update to the C919, will determine the pace of China's encroachment.

A second trend is adoption of new enabling technologies for jetliners. Additive manufacturing is slowly gaining traction, particularly in aeroengines, and should become mainstream for aerostructures and systems by the mid-2020s. The real payoff of additive manufacturing is when it is coupled with radical new design approaches. Geared turbofan architectures, pioneered by Pratt & Whitney, will spread to other original equipment manufacturers (OEMs). Rolls-Royce announced its intent to bring a geared turbofan to the market by 2025, and GE will likely follow suit. Another technology to watch is electric and hybrid propulsion. Airbus plans an electric regional jet demonstrator in the next decade, and eventually hybrid propulsion could play a role in larger jetliners. However, it is highly unlikely that we will see an all-electric jetliner for many decades, given the realities of energy density. Right now, the specific energy of batteries is roughly 2% that of liquid fuel. Factor in the efficiency of electric powertrains compared to conventional aeroengines

[4]Clay Jones (former CEO—Rockwell Collins), interview with author, 5 Nov. 2016.

and the figure grows to 7%. This means that 1000 lb of jet fuel yields about 14 times more energy than a 1000-lb battery [1].

The encroachment of autonomous aircraft should gradually impact jetliners. Just as Airbus eliminated the third member of the flight crew on the A310 and A300B, we may well see the elimination of the second crew member in the decades ahead. The technology to do this already exists; what is missing are regulations, procedures, training, and political will. The single-pilot jetliner will probably make its first appearance in cargo and military segments before it transitions to mainline passenger jetliners.

What about supersonic? It has been 15 years since the retirement of the Concorde, and several promising new supersonic designs are in process. These programs, however, will be more oriented towards business aviation than mainstream airlines. Aerion Corporation, for example, plans to bring an 8- to 12-seat supersonic business jet to the market in the mid-2020s. If Concorde taught us anything, it is that the demand for expensive supersonic travel is very limited and not applicable to airlines. Business aviation is its logical home for the foreseeable future.

Technology will also impact the way in which aircraft are designed and built. Highly automated "brilliant" factories leveraging the principles of Industry 4.0 will become more widespread. By combining cyberphysical systems, the Internet of things, cloud computing, and cognitive computing, productivity will continue to grow (good news for customers), while the role of touch labor will diminish (bad news for employees and governments).

A third trend to watch is sustainable aviation. Despite a growing focus on climate change, aviation has grown unfettered for decades and has operated outside of international and national emissions regulation. This changed in 2017 when the International Civil Aviation Organization approved a carbon offsetting strategy to cap international aviation emissions after 2020. The long-term goal is a halving of 2005 emissions levels by 2050. To approach this ambitious goal, five emissions reduction "levers" must be simultaneously attacked: aircraft technology, more efficient airline operations, improved aviation infrastructure and air traffic management, expanded use of biofuels, and market-based measures or carbon taxes.

OEM vertical integration is another trend that could reshape the jetliner ecosystem. Airbus and Boeing believe that the pendulum swung too far to outsourcing with the Tier 1 supply chain strategies, and they are intent on bringing more work in-house. Their supply chain decisions will be driven by the desire not only to save money, but also to control core technologies for aircraft of the future and expand their aftermarket revenue. This means that OEMs will likely play a greater role in avionics, mechanical and electrical systems, and even propulsion. There will be successes and failures, and eventually supply chain strategies will morph yet again. The massive 2017 United Technologies–Rockwell Collins merger announcement can be interpreted as a reaction to OEM vertical integration, and more mega-mergers are possible. Aeroengine OEMs could also become more vertically integrated, particularly as additive manufacturing gains acceptance. Some 35% of GE's Advanced Turboprop engine, which will enter service in 2019, will be printed via additive manufacturing. This leaves little room for outside suppliers, let alone a multitier supply chain. Suppliers will need to transform their portfolios and business strategies because of OEM vertical integration.

Beyond these trends, there are several wildcards that could shape the jetliner business. Terrorism is front and center, and the 9/11 attacks and subsequent recession underscore its negative impact on air travel. Airline consolidation, nearly complete in North America, could spread to other geographic regions if they continue to liberalize and deregulate. This would probably be bad news for jetliner OEMs and suppliers because customer bargaining leverage would increase. On the other hand, air travel could move towards protectionism.

Whether these predictions about trends are right or wrong, one thing is certain: The jetliner ecosystem will continue to evolve as it is shaped by economic, political, technological, and regulatory trends. It is dynamic. It is *aero*dynamic.

REFERENCE

[1] Adams, A., "The Age of Electric Aviation Is Just 30 Years Away," *Wired*, 31 May 2017, https://www.wired.com/2017/05/electric-airplanes-2/ [retrieved 6 Jan. 2018].

INDEX

Note: Page numbers followed by *f* or **t** (indicating figures and tables).

A300 Airbus, 16*f*, 17, 29
  aluminum fuselage, 248–249
  twinjet concept, 17
A300-600 series, 24
A300B Airbus, 17
A310 program, 21
  aluminum fuselage, 248–249
A310-300 (first jetliner), 245*f*
A320 Airbus, 2, 33*f*, 145
A320 *vs.* 737NG jetliner battle, 173–177
  jetliner deliveries, 176*f*
A320neo aircraft, 142–143, 145, 146, 147*f*
A330 Airbus
  avionics, 222
  twin-aisles, 2
  two-engine, 30
A340 Airbus, 109
  avionics, 222
  four-engine, 30
  twin-aisles, 2
A350 XWB, 166–168, 195–196, 208, 235, 253, 257, 285, 286, 307, 313, 329, 345
  cockpit, 234*f*
  shipset value, 234
  third moonshot, 59–62
  wing, 310*f*
A380 Airbus
  advanced materials, 249*f*
  first moonshot, 54–56, 55*f*
  wire harness issue, 192
AAR Corporation, 327, 347
Aboulafia, Richard, 162*f*
ACAC. *See* AVIC I Commercial Aircraft Company
ACARS. *See* Aircraft communications addressing and reporting system
ACSS. *See* Aviation Communications & Surveillance Systems
Additive manufacturing, 266, 382
Advanced turbine cooling technology, 103
Aero International Regional (AIR), 87
Aeroengines, 246
  development, 243
  materials evolution, 242*f*
  OEMs, 254, 384
  performance improvement, 123*f*
  program, 121
  reliability, 29
  technology and planning, 8
Aeroflot (Soviet Union), 9, 10
Aeropolitics, 18, 59
Aerospace, 239, 367
  historic moment in aviation and, 47
  titanium supply chain, 242
Aérospatiale, 17, 217
  from France, 17
Aerostructures, 194*f*
  suppliers (787 major)
Aerostructures supply chain, 261, 262*f*
AESS. *See* Aircraft environment surveillance system
AFDX. *See* Avionics Full-Duplex Switched Ethernet
Aftermarket, 322, 323
  percentage of profits from, 336*f*
Aftermarket (1980s), 326–328
Agency for Science, Technology and Research (ASTAR), 363

AIMS. *See* Airplane information management system
AIR. *See* Aero International Regional STAR
Air France, 9, 10, 26
Air France-KLM, 62
Air India, 10
Air transport
  avionics market shares, 236*f*
  passenger data channels, 282*f*
"Air-to-ground" approach (ATG approach), 281
Airborne platform computing systems, 235
Airbus, 2, 7, 17–18, 24, 30, 125, 217, 244–245
  bright spot, 166–168
  rarity, 197
  rise of, 24–29
  supply chain evolution, 196*f*
Airbus Industrie, 17, 25
Airbus–bombardier surprise, 177–179
  airbus takes control of C series, 178*f*
Aircraft, 39, 43, 239
  accessories, 212
  aluminum alloy, 240
  control and safety, 287
  electrical power-generating capability by, 309*f*
  equipment, 295–302, 302–307
    milestones, 317*f*
  functions, 211
  interior
    milestones, 274*f*
    segment, 270*f*
  manufacturers, 10
  manufacturing, 352
  navigation, 214, 215
  OEMs, 52
  production, 352
  programs, 149
  radars, 213
  safety, 215
  systems, 3, 189
    milestones, 304*f*
  wings, 249
Aircraft communications addressing and reporting system (ACARS), 219, 288
Aircraft environment surveillance system (AESS), 228
Aircraft on ground (AOG), 335–336
AiResearch, 297
Airline Deregulation Act, 19
Airline Regulation Act, 39–40
Airlines, 86, 359
  changing nature of, 48–51
  consolidation, 384
  crisis, 66–68
  maintenance alliances, 324–326
  operations, 288
  restructuring, 153–160
    cost of capital and oil prices, 158*f*
    lessors' share of fleet, 160*f*
    US airline consolidation, 154*f*
    world's largest airlines, 157*f*
  ticket prices, 4
Airplane information management system (AIMS), 222
Airport Noise and Capacity Act, 39
AJ Walter, 347
Al-Li. *See* Aluminum-lithium
Alaska Airlines, 219
Alcan, 259–260
Alcoa, 244, 259
Alcoa Davenport Works sprawls over 130 acres, 241*f*
Aleris, 260
Alitalia (Italy), 10
All Nippon Airways (ANA), 113
Allegheny Technologies, 260–261
Allied-Signal Aerospace, 224, 301–302
Allvac Metals Company, 243
Aluminum (Al), 252
  alloys, 255
  industry, 239
  monoplanes, 240
Aluminum Company of America, 266
Aluminum-lithium (Al-Li), 252
American Airlines, 10, 56, 71
ANA. *See* All Nippon Airways
Anglo-French working party, 17
AOG. *See* Aircraft on ground
APU. *See* Auxiliary power unit
Arbeitsgemeinschaft Airbus, 17
ARINC 429 twisted-wire standard, 223
ARINC 629 data bus, 223
ARINC 664 standard, 231
ARINC communications, 288
ARJ21 jet, 92, 140, 170
Armstrong, Neil, 106, 172

ASEAN. *See* Association of Southeast Asian Nations
Asian airlines, 45, 49, 381
Association of Southeast Asian Nations (ASEAN), 64
ATG approach. *See* "Air-to-ground" approach
ATG-4, 289
ATI, 258
ATR. *See* Avions de Transport Regional
Attrition, 90–92
  regional jet deliveries, 91*f*
  regional jet milestones, 92*f*
Automotive manufacturing in United States, 368
Autotune, 212
"Autotuned" airborne radios, 212
Auxiliary power unit (APU), 10, 296, 328, 357
  jetliner APU, 297*f*
Aviation, 211
  fuel, 132
Aviation Communications & Surveillance Systems (ACSS), 225, 233
Aviation Industry Corporation of China (AVIC), 90, 134–135, 169–170, 314
Aviation Week & Space Technology (AWST), 113, 125, 337–338
AVIC. *See* Aviation Industry Corporation of China
AVIC 2. *See* China Aviation Industry Corporation 2
AVIC I Commercial Aircraft Company (ACAC), 92
Avionics, 211, 219
  in Bustling (1990), 221–225
  early history, 211–214, 213*f*
  in jet age, 214–216
Avionics Full-Duplex Switched Ethernet (AFDX), 233
Avions de Transport Regional (ATR), 73
Avro C102, 8
Avro Lycoming ALF 502 turbofans, 87
Avro's CF-100 advanced fighter program, 8
AWST. *See Aviation Week & Space Technology*

BA. *See* British Airways
BAC 1–11 (British Aircraft Corporation), 12
BAC-111 program, 103
BAe. *See* British Aerospace
BAE Systems, 54, 55, 60, 91–92, 235, 339
Bandeirante, 77–78
BASA. *See* Bilateral Aviation Safety Agreement
Bateman, Don (inventor of GPWS and EGPWS), 224*f*
Beaudoin, Laurent, 74*f*
Bendix, Vincent, 213*f*
Bendix Aircraft, 299–300
Bendix radio division, 212
Béteille, Roger (Airbus president), 17
BFE market. *See* Buyer-furnished equipment market
Big Gamble, 111–113
Bilateral Aviation Safety Agreement (BASA), 357
"Black belt" manufacturing, 223–224
"Black boxes", 211
Bloomberg Business Week, 99
BOAC. *See* British Overseas Aircraft Corporation
Boeing
  airliner, 24
  B-47 Stratojet long-range bomber flew, 9
  B-52 aircraft, 103
  Boeing 2207 aircraft, 18
  Boeing 2707 Supersonic Transport programs, 104
  Boeing 717 aircraft, 119
  Boeing 727 aircraft, 11*f*
  Boeing 747 aircraft, 14*f*, 29, 103
  Boeing 757 aircraft, 108, 112
  Boeing 767 aircraft, 21, 112, 26 108
  Boeing 777 aircraft, 2, 247
  Boeing 7J7 aircraft, 109
  launch of 737 MAX, 143–147
  one-two punch, 23*f*
  supersonic program, 19
  Supply Chain Nightmare, 192–196
Boeing Commercial Airplanes, 24
Boeing Technical Research Center, 1
Bombardier, 186–189, 253
  747 aircraft, 94–95
  C-series, 134–137
Boom–bust pattern, 119
Borman, Frank (Apollo astronaut former), 20

Boullioun, E. H. "Tex", 24
Brainiacs, 235
  avionics in Bustling (1990), 221–225
  avionics in jet age, 214–216
  Brussels bombshell, 225–227
  early history of avionics, 211–214
  evolution of air transport avionics market shares, 236
  GE's return to avionics, 234–235
  goodbye third pilot, 216–219
  momentum shift, 227–232
  supplier consolidation, 219–221
  Thales rising, 232–234
British Aerospace (BAe), 24, 54, 72, 72*f*, 87, 146
British Aircraft merger (1977), 24
British Airways (BA), 4, 24, 113
  yin-yang layout, 277*f*
British Overseas Aircraft Corporation (BOAC), 8, 10
British Petroleum, 24
Brussels bombshell, 225–227
"Business withhold" list, 198
Bustling (1990), avionics in, 221–225
Buyer-furnished equipment market (BFE market), 290
Bypass ratio, 101
Bypass turbojet, 101

C-130 military aircraft, 15
CAAC. *See* Civil Aviation Administration of China
CAB. *See* Civil Aeronautics Board
California aerospace, 354
California-based OEM, 29
Canadair Regional Jet (CRJ), 76, 76*f*, 79, 221–222
  CRJ 700, 89
  CRJ 900, 89
Canadian Pacific Airlines Comet, 9
Canadian revolution, 73–77
Cape Town Convention, 159–160
Caravelle, 12
  SE 210 Caravelle, 9
Carbon fiber reinforced plastics (CFRPs), 244
"Carbon-carbon" graphite matrix, 299
CARE program, 345
Carter, Jimmy (US President), 19
Carter Administration, 19
CASA. *See* Construcciones Aeronauticas SA
Cathode ray tubes (CRTs), 30, 219
CATIA, 42
Cavity magnetron, 213
CCS. *See* Common core system
Ceramic matrix composites (CMCs), 138, 373
CF6-50 powerplants, 17
CFD. *See* Computational fluid dynamics
CFIT. *See* Controlled flight into terrain
CFM, 106, 138–139
  CFM56-2 engine, 106
  CFM56-3 engines, 107, 118
  CFM56-5 engine, 27, 118
  engineering task, 118–119
  LEAP, 137–140
  LEAP-1B, 146*f*
CFRPs. *See* Carbon fiber reinforced plastics
Chênevert, Louis (Pratt & Whitney President), 122
China, rising, 314–315
China Aviation Industry Corporation 2 (AVIC 2), 190–191
CIS. *See* Commonwealth of Independent States
CIS/MS. *See* Crew information system/maintenance system
Civil Aeronautics Board (CAB), 19
Civil aviation, 213–214
Civil Aviation Administration of China (CAAC), 63, 170, 172
Cleveland-based Parker Appliance Company, 352
CMCs. *See* Ceramic matrix composites
Cockpit, 14
  A350 XWB cockpit, 234*f*
  commonality, 221–222
  European Dream, 14
  787 cockpit with Collins displays, 230*f*
Cockpit voice recorders (CVRs), 215
Cold hearth melting process, 247
Cold War, 102, 351–352
Collins, Art, 213*f*
Collins Aerospace Systems, 318
Collins Radio Company, 212, 214, 220
COMAC. *See* Commercial Aircraft Corporation of China
Comet, 8, 12

accidents, 215
aluminum fatigue, 242–243
de Havilland Comet 4, 101
de Havilland DH 106 Comet, 7
fuselage, 9
jetliner, 242–243
Commercial Aircraft Corporation of China (COMAC), 140–142, 141*f*, 170, 314
Common core system (CCS), 229–230
Commonwealth of Independent States (CIS), 185
"Community air carrier" concept, 42
Commuters, 71
Competitive dynamics of 1970s, 19–24
Composites, 55, 244–247
Computational fluid dynamics (CFD), 248
Concorde program, 18–19, 43, 99, 105, 298–299, 383
Consolidation, 315–319
Consolidation protagonist, 303*f*
Consolidators, unlikely, 272–276
Constellium, 259–260
Construcciones Aeronauticas SA (CASA), 17–18, 44
Consumers, 211
Continental Airlines, 153
Controlled flight into terrain (CFIT), 215–216
Convair, 12
Convair 880 jetliners, 103
Convair 990 jetliners, 103
Convair F-102, 103
Corrosion, 250
resistance, 239
Cost reduction, 330
Council on California Competitiveness, 354
Crew information system/maintenance system (CIS/MS), 230
CRJ. *See* Canadair Regional Jet
CRTs. *See* Cathode ray tubes
Cuddington, Charles, 111
CVRs. *See* Cockpit voice recorders

d'Estaing, Valéry Giscard, 21
Dart engine, 101–102
DASA. *See* Deutsche Aerospace AG
DC-10 airplanes, 15, 31
DC-8 airplanes, 10
de Havilland Comet 4, 101
de Havilland DH 106 Comet, 7
de Havilland Ghost jet engines, 8, 8*f*
"Deferred production cost" category, 12
Delta Air Lines, 10, 24, 26, 37
Densification, 283
777-300ER densification example, 284*f*
Deutsche Aerospace AG (DASA), 44, 54, 81
Deutsche Airbus (West German company), 17
Directional solidification (DS), 123
Donegan, Mark, 263, 263*f*
Double-vacuum process, 247
Douglas, McDonnell, 109
Douglas A3D, 103
Douglas Aircraft, 9, 11, 353
Douglas X-3 Stiletto, 241–242
Draper, Charles, 214
Dreamlifter, 66
electric architecture, 194
Drift rate, 215
DS. *See* Directional solidification
Dual band radios, 211–212
Dun's Review, 21
Duralumin, 240

E2 aircraft, 147*f*
EA. *See* Engine Alliance
EADS. *See* European Aeronautic Defence and Space Company
Eagle Services Asia (ESA), 334
Earnings before interest and taxes (EBIT), 196–197, 265
EASA. *See* European Aviation Safety Agency
Eastern Airlines, 10, 24
Eaton Aerospace, 356
EBIT. *See* Earnings before interest and taxes
EC. *See* European Commission
Ecclestone, Barry, 373*f*
Economic Development Board (EDB), 360–361
EDB. *See* Economic Development Board
EDI. *See* Electronic data interchange
EFISs. *See* Electronic flight instrumentation systems
EHA. *See* Electrohydrostatic actuators
Eichner, Ira, 326*f*

El Al (Israel), 10
El-Andaloussi, Hamid Benbrahim, 364*f*
Electrical power management, 235
Electrohydrostatic actuators (EHA), 307–308
Electronic data interchange (EDI), 186
Electronic flight instrumentation systems (EFISs), 219
Electronic information management device, 233–234
Electronics, 211
Embraer, 77–80, 88, 186–189
  EMB-145, 78*f*
  regional jet, 115
Embraer. *See* Empresa Brasileira de Aeronáutica
EMEA. *See* Europe, Middle East, and Africa
"Emergence" program, 365
Emirates, 68, 115, 154–155
  April 2015 route map, 156*f*
Emirates Air, 49
Empresa Brasileira de Aeronáutica (Embraer), 77
Engine Alliance (EA), 121
Engine OEMs, 345
Engine Validation Noise and Emissions Reduction Technology program (EVNERT program), 130
Engineering service providers (ESPs), 205–206
Engines Turn Or People Swim (ETOPS), 329
"ERJ", 79
ESA. *See* Eagle Services Asia
ESPs. *See* Engineering service providers
Ethiopian Airlines, 164
ETOPS. *See Engines Turn Or People Swim;* Extended-Range Twin-Engine Operational Performance Standards
EU. *See* European Union
Europe, Middle East, and Africa (EMEA), 376
European "Air bus" project, 17
European Aeronautic Defence and Space Company (EADS), 54, 192, 372
European aerospace production, 365
European Aviation Safety Agency (EASA), 173, 329, 340
European Commission (EC), 47, 226
European Dream, 7, 34
  accelerated innovation, 10–12
  Airbus A320, 33*f*
  Boeing 747, 14*f*
  Boeing's one-two punch, 23*f*
  Boeing's supersonic program, 19
  cockpit, 14
  competitive dynamics of 1970s and US deregulation, 19–24
  early years of jet age, 7–10
  Jetliner deliveries, 13*f*, 22*f*, 32*f*
  Kolk machine, 15
  late-1980s growth, 31–33
  Plowden Report, 17
  revolutionary Boeing 727, 11*f*
  rise of airbus, 24–29
  twin-aisle contenders, 16*f*
  Twin-Aisle Refresh, 29–31
  UDF, 33–34
  World's largest airlines, 18*f*
European restructuring, 53–54
European suppliers, 256
  Constellium, 259–260
European Union (EU), 42, 60
EVNERT program. *See* Engine Validation Noise and Emissions Reduction Technology program
eXConnect service, 289
Extended-Range Twin-Engine Operational Performance Standards (ETOPS), 29, 39, 110, 222, 302
  influence, 328–330
  certification, 2

FAA. *See* Federal Aviation Administration; US Federal Aviation Administration
FACC. *See* Fisher Advanced Composites Corporation
FADEC engine. *See* Full authority digital electronics control engine
Farnborough Air Show (1978), 24
"Father of inertial navigation". *See* Draper, Charles
"Father of the A380". *See* Thomas, Jürgen
FCC. *See* Flight control computer
FDR. *See* Flight data recorder
Federal Aviation Administration (FAA), 43, 173, 216, 248, 305, 323, 357
FedEx, 50

Fee-for-service dream house, 86–87
FHS. *See Flight Hour Services*
Fiber optic gyroscope technology (FOG technology), 215
Fisher Advanced Composites Corporation (FACC), 246
Flight control computer (FCC), 222
Flight data recorder (FDR), 215
Flight Hour Services (FHS), 345
Flight International, 110
Flight management system (FMS), 216
Flight-control software, 27
Fly-by-wire flight controls, 27, 217
FMS. *See* Flight management system
FOG technology. *See* Fiber optic gyroscope technology
Fokker, 27, 80–82
  Fokker F70, 81*f*
Force of conviction, 113–116
France, jetliner business in, 9
France-based Zodiac Aerospace, 273
Fuji Heavy Industries, 193
Full authority digital electronics control engine (FADEC engine), 108
"Functional silos", 115

GA Telesis, 347
Galvanic corrosion, 252
Gamesa, 256
Garrett Aviation, 220
GDP. *See* Global gross domestic product; Gross domestic product
GE. *See* General Electric
Geared turbofan, 109
Geared turbofans (GTFs), 128, 130
  milestones, 144*f*
  Pratt & Whitney's GTF architecture, 131*f*
Gearheads, 143–147
  Boeing's launch of 737 MAX
  bombardier C-series, 134–137
  breakthrough, 132–134
  CFM's LEAP, 137–140
  COMAC, 140–142
  landscape changing, 149–150
  Pratt & Whitney, 148–149
  Regional Reengining, 147–148
  Shot Heard 'Round World, 142–143
General Electric (GE), 100, 103–104, 170, 254, 267
  acquiring British Airways' engine maintenance facility, 333*f*
  Aviation, 373
  CF-6
    CF6-45B2, 21
    engine, 15
    turbofans, 108
  CF6-80C2, 30
  GE-136 unducted fan on MD-80 aircraft, 109*f*
  GE36, 33
  GE90
    composite fan blade, 248*f*
    design strategy, 120
    GE90-115 engine, 120*f*, 123
    models, 120
  GE9X aeroengines, 162–163
  return to avionics, 234–235
General Motors (GM), 201
GEnx, 122, 123
  aeroengines, 61
German cluster, 376
Germanwings, 17
GIMAS. *See* Groupement des Industries Marocaines d'Aéronautique et Spatiales
Glass laminate aluminum reinforced epoxy (GLARE), 249
Global Aeronautica, 193
Global air transportation system, 57
Global airline restructuring, 62–64
Global economy, 227
Global gross domestic product (GDP), 45
Global jetliner ecosystem, 381, 382*f*
Global positioning system (GPS), 217–218, 218*f*
Globalization, 3, 227
  engineering, 204–208
  harnessing, 189–191
GM. *See* General Motors
Goodrich Aerostructures and Rohr, 257
GPA. *See* Guinness Peat Aviation
GPS. *See* Global positioning system
GPWS. *See* Ground proximity warning system
Green Revolution, 99
Gross domestic product (GDP), 1, 26, 172, 359
Ground proximity warning system (GPWS), 216

Groupement des Industries Marocaines d'Aéronautique et Spatiales (GIMAS), 365
GTFs. *See* Geared turbofans
Guinness Peat Aviation (GPA), 31
Gulf Air, 49
Gyrocompasses, 212
Gyroscope, 212, 215
Gyrostabilizers, 212

HAECO. *See* Hong Kong Aircraft Engineering Company
HAESL. *See* Hong Kong Aero Engine Services Limited
Hard alpha inclusion, 247
Hawker Siddeley Aviation, 24
Hawker Siddeley Dynamics, 24
Hawker Siddeley HS 121 Trident, 12
Head-up display (HUD), 219
Hecho En México (Eaton aerospace), 355–359
  aerospace facilities, 357*f*
  Mexican aerospace cluster growth, 358*f*
Herzner, Fred (GE Aviation executive), 11, 112
Hexcel, 260
HF radios. *See* High-frequency radios
Hickton, Dawne, 265*f*
High pressure (HP), 254
High-bypass "geared turbofan" architecture, 30
High-frequency radios (HF radios), 214
High-pressure compressor (HPC), 139
High-pressure turbine design (HPT design), 139
Higher-temperature superalloys, 244
Hong Kong Aero Engine Services Limited (HAESL), 334
Hong Kong Aircraft Engineering Company (HAECO), 327, 334
HP. *See* High pressure
HPC. *See* High-pressure compressor
HPT design. *See* High-pressure turbine design
HUD. *See* Head-up display

IAE. *See* International Aero Engines
IAG. *See* International Consolidated Airlines Group
IAM. *See* International Association of Machinists
IATA. *See* International Air Transport Association
ICAO. *See* International Civil Aviation Organization
ICBM warheads. *See* Intercontinental ballistic missile warheads
IDS. *See* Integrated Defense Systems
IFE. *See* Inflight entertainment
IFEC. *See* Inflight entertainment and connectivity
ILFC. *See* International Lease Finance Corporation
ILSs. *See* Instrument landing systems
IMA. *See* Institut des Métiers de l'Aéronautique; Integrated modular avionics
Indian Airlines, 17
Inertial navigation system, 214, 215
Inertial reference suite (IRS), 230
Inflight entertainment (IFE), 269, 270, 273*f*, 282
Inflight entertainment and connectivity (IFEC), 280
Information technology systems (IT systems), 50
Institut des Métiers de l'Aéronautique (IMA), 365–366
Instituto Tecnológico de Aeronáutica (ITA), 77
Instrument landing systems (ILSs), 212–213
Integrated Defense Systems (IDS), 161
Integrated modular avionics (IMA), 222, 228
Integrated product team (IPT), 205
Intercontinental ballistic missile warheads (ICBM warheads), 299
Interior designers
  evolving connectivity, 287–290
  interiors in 2010s, 283–285
  interiors in new millennium, 276–278
  interiors segment comes of age, 290–292
  internet cafes in sky, 279–282
  jetliner interiors, 269–272
  operations headaches, 286–287
  supplier consolidation, 278–279
  unlikely consolidators, 272–276

Zodiac and B/E Aerospace interior acquisitions, 287*f*
International Aero Engines (IAE), 27, 107
board, 30
International Air Transport Association (IATA), 153
International airlines, 10
International Association of Machinists (IAM), 370
International Civil Aviation Organization (ICAO), 25, 39, 213, 328, 383
International Consolidated Airlines Group (IAG), 155
International Lease Finance Corporation (ILFC), 55
International Society of Transport Aircraft Trading (ISTAT), 163
conference, 60–61, 142
International Traffic in Arms Regulations (ITAR), 183
Internet
bubble, 339
cafes in sky, 279–282
IPT. *See* Integrated product team
Iran–Iraq war, 37
Iridium communications, 288
IRS. *See* Inertial reference suite
Ishikawajima Ne-20, 100
ISTAT. *See* International Society of Transport Aircraft Trading
IT systems. *See* Information technology systems
ITA. *See* Instituto Tecnológico de Aeronáutica
ITAR. *See* International Traffic in Arms Regulations
ITC. *See* US International Trade Commission

J79 engine, 102
JAA. *See* Joint Aviation Authorities
JAL (Japan), 10, 162
Japan Aircraft Development Corporation, 245, 256
Japan Airlines, 271
"Japanese Heavies", 193
JET. *See* Joint European Transport
Jet age
avionics in, 214–216
early years of, 7–10
Jet engines, 99–103
Jetliners, 19, 214, 327
age, 7
avionics
innovations, 214
suppliers, 220*f*
business, 12
component and system suppliers, 311*f*
deliveries (1980s), 31, 32*f*
ecosystem, 3*f*, 184*f*, 239, 384
fleet growth, 330*f*
interiors, 269–272
manufacturing by region, 378*f*
manufacturing clusters, 377
materials, 240, 242*f*
Alcoa Davenport Works sprawls over 130 acres, 241*f*
composites, 244–247
emergence of tier 1 aerostructures suppliers, 255–257
evolution of aeroengine materials, 242*f*
innovation golden age, 247–251
new frontiers, 266–267
precision castparts' acquisitions, 264*f*
revolution from below, 261–266
787 shocker, 251–255
supplier competition, 257–261
VIM, 243
new landscape of production, 375–378
civil aerospace manufacturing clusters, 375*f*
OEMs, 256
orders, 159
supply chains, 184–186, 225
Joint Aviation Authorities (JAA), 43
Joint European Transport (JET), 26
Joint venture (JV), 139–140, 191, 258, 311, 334, 361
Jones, Clay, 228*f*
JSA Research analyst Paul Nisbet, 56
JT3 turbojet engine (1950), 103
JT3D engine, 106
JT8D engine, 104, 106
JT9D-3A engine, 123
Jumo 4 (1944), 100
Just-in-time production system, 200–201
JV. *See* Joint venture

Kaiser, Henry, 241
Kaiser Aluminum, 260
Kawasaki Heavy Industries, 193
Kelleher, Herb (Charismatic CEO), 20, 48*f*
Key performance indicators (KPIs), 191
King Radio, 215
KLM airlines (Netherlands), 10, 271
Kolk, Frank (American Airlines president), 15
"Kolk machine", 15
Korean War, 8
KPIs. *See* Key performance indicators
Kroll process, 241

L-1011 airplanes, 15
L-188 Electra, 15
"La Década Perdida", 355–356
La Guardia Airport's runways (New York), 15
Laker, Freddie, 20, 24
Laker Airways, 20
Landing-Gear Advanced Manufacturing Co., Ltd. (LAMC), 314
Landscape changing, 149–150
LAX. *See* Los Angeles International Airport
LCCs. *See* Low-cost carriers; Low-cost countries
LCDs. *See* Liquid crystal displays
Leading Edge Aviation Propulsion (LEAP), 138
  fuel nozzle, 140*f*
Leahy, John, 175*f*, 373*f*
Lean electronics, 228–229
Leaning operations, 200–204
LEAP. *See* Leading Edge Aviation Propulsion
LEAP-X fuel nozzles, 139, 373
Lear, Bill, 213*f*
Learoscope, 212
Legacy aircraft, 275
Legacy US airlines, 20
Lessors, 4
LHT. *See* Lufthansa Technik
Lie-flat decision, 276
Lifeblood, 326–328
  Aftermarket (1980s)
  airline maintenance alliances, 324–326
  birth of MRO sector (1990s), 330–338
  ETOPS influence, 328–330
  MRO, 321–324, 338–340
  PMA parts, 340–342
  return on assets, 343–348
  return to growth, 342–343
Lightweight alloys, 240
Line replaceable units (LRUs), 222
Line-of-sight communications, 213–214
Line-of-site communications, 219
Line-replaceable modules, 233
Line-replaceable units (LRUs), 194, 231
Liquid crystal displays (LCDs), 223
Lockheed, 25
  F-104 fighter, 102
  tri-jet program, 104
  U-2, 103
Lockheed Martin's C-5, 12
Los Angeles City Hall, 351
Los Angeles International Airport (LAX), 352
LOT Polish Airlines, 164
Low-cost carriers (LCCs), 4, 20, 26, 43–44, 143–144, 154, 157*f*, 289, 342
Low-cost countries (LCCs), 189–191
Low-pressure compressor (LPC), 127
Low-pressure turbine (LPT), 127
LPC. *See* Low-pressure compressor
LPT. *See* Low-pressure turbine
LRUs. *See* Line replaceable units; Line-replaceable units
Lufthansa (Germany), 10, 12, 17, 18, 21, 30, 42, 59, 62, 134–136, 142, 145, 161, 163, 167, 172, 176, 280, 338
Lufthansa CityLine, 76, 79, 82, 90, 93
  CRJ700, 88*f*
Lufthansa Group, 155
Lufthansa Technik (LHT), 337, 339–342, 347
  Wolfgang Mayrhuber foresaw globalization, 332*f*

Machine shops, 261
Maintenance, repair, and overhaul (MRO), 5, 321–323, 330, 356
  activity cost structures, 337*f*
  birth of MRO sector (1990s), 330–338
  conferences, 338*f*
  facilities, 190–191
  in new millennium, 338–340
  origins, 323–324
  sector development, 335*f*

Maintenance Steering Group (MSG), 323–324
Malaysia Singapore Airlines (MSA), 359
Marshall, Colin, 114
Massachusetts Institute of Technology (MIT), 77, 201
Material innovation golden age, 247–251
Material supplier competition, 257–261
  leading aerospace raw material suppliers, 259*f*
Matis aerospace. *See* Morocco Aero-Technical Interconnect Systems aerospace
Mbps. *See* Megabyte per second
McCune, Michael, 128*f*
McDonnell Aircraft Corporation, 12
McDonnell Douglas Company, 12, 15, 20, 25, 28, 31, 34
  aircraft factory in Columbus, Ohio, 1
  McDonnell F-101, 103
  McDonnell F-4 fighter, 102
  MD-11 aircraft, 2, 31, 109
  MD-12X, 40
  MD-80 aircraft, 25
  MD-83 aircraft, 28
  MD-87 aircraft, 28
  MD-88 aircraft, 29
  MD-89 aircraft, 34
  MD-90 aircraft in China, 1
ME262 aircraft, 100
Mechanical actuation systems, 235
Mega-merger, 47, 318
Mega-systems award, 311–314
Megabyte per second (Mbps), 280
Mid-air collisions, 219
MIDPARC industrial park, 366
Military aeroengine programs, 207
Military aircraft, 103, 242–243
Minibus, 87
MIT. *See* Massachusetts Institute of Technology
Mitsubishi Heavy Industries, 193
Mitsubishi Regional Jet (MRJ), 132–134, 170–171
Modern miracles
  Brian Rowowe's Big Gamble, 111–113
  changing fortunes, 106–110
  jet engines, 99–103
  large-thrust firewoworks, 119–123
  module reuse, 124–125
  new aeroengine market, 116–117
  Pratt & Whitney, 103–106, 125
  revolutionary technology, 123
  Rolls-Royce force of conviction, 113–116
  single-aisle roller coaster, 118–119
  Trent, 110–111
Module reuse, 124–125
Momentum shift, 227–232
  787 cockpit with Collins displays, 230*f*
  avionics comparison, 231*f*
Monocoque design, 240
Monti, Mario, 227*f*
"Mood lighting" on production aircraft, 283
Moog, Bill, 301*f*
Moonshots
  A350 XWB (third moonshot), 59–62
  A380 (first moonshot), 54–56
  airline crisis, 66–68
  changing nature of airlines, 48–51
  debacles, 164–166
  drive innovation (2000s), 307–311
  early 1990s bust, 39–44
  European restructuring, 53–54
  global airline restructuring, 62–64
  jetliner deliveries, 46*f*
  largest jetliner fleets, 38*f*
  late 1990s boom, 51–53
  mega-merger, 47
  moonshot delays, 64–66
    787 rolled out in 2007, 66*f*
    jetliner deliveries, 67*f*
  new super-jumbo, 44–45
  787 (second moonshot), 56–59
Morocco (unlikely cluster), 364–368
  Moroccan aerospace companies, 367*f*
Morocco Aero-Technical Interconnect Systems aerospace (Matis aerospace), 364
Moscow–Omsk–Irkutsk route, 9
Motoren-und Turbinen-Union (MTU), 327
MRJ. *See* Mitsubishi Regional Jet
MRO. *See* Maintenance, repair, and overhaul
MSA. *See* Malaysia Singapore Airlines
MSG. *See* Maintenance Steering Group
MSG-3 methodology, 324
MTU. *See* Motoren-und Turbinen-Union

NAFTA. *See* North American Free Trade Agreement
Neeleman, David, 48*f*
Neumann, Gerhard, 102, 105*f*
New aeroengine market, 116–117
New ecosystem, 1–5
90-minute rule, 29
No Fly List. See "Business withhold" list
Nonrecurring engineering (NRE), 59, 187
North American F-100, 103
North American Free Trade Agreement (NAFTA), 190, 356
North American market, 324
Northrop Grumman B-2 Spirit stealth bomber, 245
Northwest Airlines, 30
NRE. *See* Nonrecurring engineering

O'Leary, Michael, 48*f*
OEM. *See* Original equipment manufacturer
OEMs. *See* US original equipment manufacturers
Open rotor program, 129
Operations headaches, 286–287
Orange County Register, 355
Original equipment manufacturer (OEM), 3, 17, 37, 160, 171, 183, 220, 239–240, 261, 295, 322, 351, 382
  alternative, 340–342
  and supplier profit margins, 197*f*
  vertical integration, 383–384

PAIG. *See* Premium Aircraft Interiors Group
PAL. *See* Philippines Airlines
Pan Am. *See* Pan American Airways
Pan American Airways (Pan Am), 10, 14, 28
"Paper engine", 30
Paris Air Show (1973), 19, 105, 109
Partnering for Success initiative, 198
Parts manufacturer authority parts (PMA parts), 340–342
Passenger control units (PCUs), 272
Passenger services, 288
PCC. *See* Precision Castparts Corporation
PCUs. *See* Passenger control units
People Express, 20
Philippines Airlines (PAL), 339
Phoenix, 223
Pierson, Jean (Airbus CEO), 28, 28*f*
Pierson, Jean (Airbus leader), 47
Piston engines, 328
Piston-engine aircraft, 9
Plowden Report, 17
PMA parts. *See* Parts manufacturer authority parts
"Pod" configuration, 277
"*Power to Fly, The*" (Rowe), 114
Pratt & Whitney, 101, 102, 113, 122, 125, 148–149
  JT8D turbofan jet engines, 10
  JT9D turbofans, 14
  JT9D-7R4, 21
  PW-2037, 24
  PW4056, 30
Pre-preg, 260
Precision Castparts Corporation (PCC), 262, 264–265
Premium Aircraft Interiors Group (PAIG), 279
Premium economy, 278, 285*f*
Primus integrated modular avionics system, 225
Profitability, 124–125
Program accounting, 11
Puget Sound ecosystem, 376
PW4084 engine, 123

Qatar Airways, 166–167*f*

R&D. *See* Research and development
Radiation Laboratory (Rad Lab), 213
Rapid plasma deposition process (RPD process), 267
Ravaud, René, 105*f*
Raw materials, 258–259
RB211 engines, 111
Reagan, Ronald, 25
Reengining twin-aisles, 162–164
Regional feeds, 82
Regional jets, 116–117
  competition, 168–171
  deliveries, 168*f*
Regional Reengining, 147–148
Regional revolution, 116
  attrition, 90–92
  Bombardier's 747, 94–95
  British aerospace responds, 87

Canadian revolution, 73–77
Embraer, 77–80
exiting Fokker, 80–82
fee-for-service dream house, 86–87
regional jet aeropolitics, 85–86
regional jet demand growth, 82–85
Comair route map, 84*f*
regional jet deliveries, 83*f*
regional jet transition, 93–94
US regional airline statistics, 94*f*
70-seaters, 87–90
turboprop dominance (1980), 71–73
Request for proposals (RFP), 15, 372
Research and development (R&D), 56, 244, 361
Resin transfer infusion method (RTI method), 135, 266
International Metals, 257
Resin transfer molding technology (RTM technology), 138
Retrenchment
A350 XWB, 166–168
airbus–bombardier surprise, 177–179
moonshot debacles, 164–166
mother of all jetliner battles, 173–177
new regional jet competition, 168–171
reengining twin-aisles, 162–164
restructuring airlines, 153–160
single-aisle hijinks, 171–173
twin-aisle delays, 161–162
Return on assets, 343–348
Return on invested capital (ROIC), 159, 283, 346–347
Return on net assets (RONA), 51, 159, 192–193
Return to growth, 342–343
Revenue passenger kilometers (RPKs), 4, 18, 49, 155
Revenue stream, 322
Revolutionary aeroengine technology, 153
Revolutionary technology, 123–124
RFP. *See* Request for proposals
Rhenium, 254
Ring-laser gyro technology (RLG technology), 215
Risk and revenue sharing partner (RRSP), 191–192
RLG technology. *See* Ring-laser gyro technology
Robins, Ralph, 114*f*
ROIC. *See* Return on invested capital
Rolls-Royce, 100, 105, 361–362
force of conviction, 113–116
module reuse, 125
phenomenon, 344
Rapprochement, 148–149
RB207 powerplant, 17
RB211 engines, 15, 24
RB211 turbofans, 15
RB211-524G/H, 30
RB211-535C powerplants, 24
Rolls-Royce Avon engines, 9
Rolls-Royce Conway, 101*f*, 103
Rolls-Royce Nene aeroengines, 8
Seletar campus, 363*f*
RONA. *See* Return on net assets
Rowe, Brian, 102, 107, 112
Big Gamble, 111–113
Royal Air Force, 100
RPD process. *See* Rapid plasma deposition process
RPKs. *See* Revenue passenger kilometers
RRSP. *See* Risk and revenue sharing partner
RTI method. *See* Resin transfer infusion method
RTM technology. *See* Resin transfer molding technology
Ruffles, Philip, 111

Saab 340 aircraft, 71, 73
Sabena (Belgium), 10
SAESL. *See* Singapore Aero Engine Services Pte. Ltd
San Diego–based Convair, 10
SAS. *See* Scandinavian Airlines System
SC production methods. *See* Single crystal production methods
Scandinavian Airlines System (SAS), 9, 10, 11, 324–325
Schmidt, Helmut (German Chancellor), 21
Scottish Aviation, 24
SE 210 Caravelle, 9
Seattle, 223
70-seaters, 87–90
Embraer E170, 89*f*
Lufthansa CityLine CRJ700, 88*f*
707 jetliner, 242–243
737 aircraft, 12
737 MAX aircraft, 147*f*

737 Next Generation development programs (737NG development programs), 52
737-300 aircraft, 25
737NG aircraft, 146
747.ETOPS aircraft, 2
777-200LR aircraft, 53, 120
777-300ER aircraft, 51, 59, 120, 264, 283
777-9X, 163
787 cockpit with Collins displays, 230*f*
787 shocker, 251–255
  aircraft structural material composition, 252*f*
  composites penetration, 251*f*
7E7 (787) production sweepstakes, 369–370
Sextant Avionique, 221
SFE. *See* Supplier-furnished equipment
SFR. *See* Société Francaise Radio-électrique
Shanghai Aircraft Industrial Corporation, 29
“Sharklets”, 142
“Shed”, 71, 72*f*
Shortwave radios, 214
SIAEC. *See* Singapore Airlines Engineering Company
Siddeley, Hawker, 17
Singapore (Asia’s Überclustster), 359–364
  aerospace investments, 362*f*
Singapore Aero Engine Services Pte. Ltd (SAESL), 334
Singapore Airlines, 46
Singapore Airlines Engineering Company (SIAEC), 361, 363
Singapore’s Paya Lebar airport, 359
Single crystal production methods (SC production methods), 123
Single crystal turbine blade, 246
Single-aisle hijinks, 171–173
Single-aisle roller coaster, 118–119
Single-pilot jetliner, 383
SITA communications, 288
Six Sigma principles, 223–224, 228–229
“Sleeperette” seats, 271
Smiths Aerospace, 229
  flight management systems, 235
Snecma, 105
Société Francaise Radio-électrique (SFR), 211–212
Southeastern United States (Aerospace Manufacturing Hotspot), 368–374
  aerospace investments, 374*f*
  Boeing South Carolina, 371*f*
  milestones in Southeastern US aerospace investment, 369*f*
Southern Californian aerospace, demise of, 351–355
  aerospace companies in, 353*f*
  aerospace manufacturing in, 354*f*
Southwest Airlines, 20, 153
Sparaco, Pierre (Aviation journalist), 7
SPDB process. *See* Super-plastic diffusion bonding process
Sperry, Elmer, 213*f*
Sperry Flight Systems, 216
Spinning mass technology, 215
Spirit AeroSystems, 257
Sporty Game, The, 116
SR-71 military aircraft, 15
SSJ100. *See* Sukhoi Superjet 100
ST Aerospace, 361, 363–364
Steam gauges, 219
Stonecipher, Harry, 58*f*
Sud Aviation, 17
Sukhoi Superjet 100 (SSJ100), 169
Super Guppies, 185
Super-jumbo Airbus, 44–45
Super-plastic diffusion bonding process (SPDB process), 253
Superalloy producer Carpenter Technology, 374
Superalloys, 243, 260–261
SuperFan engine, 30
Supplier consolidation, 219–221, 220*f*, 278–279
Supplier-furnished equipment (SFE), 231
Supply chain 2.0
  Boeing’s Supply Chain Nightmare, 192–196
  Bombardier and Embraer, 186–189
  globalizing engineering, 204–208
  harnessing globalization, 189–191
  jetliner supply chains, 184–186
  leaning operations, 200–204
  more for less era, 196–198
  rebalancing, 198–200
  tier 1 expands, 191–192
Supply chains, 3
Surveillance capability, 216

Sutter, Joe, 12
Swissair, 10
Swissair Flight 111, 273–275

TAC. *See* Taiwan Aerospace Corporation
TAESL. *See* Texas Aero Engine Services LLC
Taiwan Aerospace Corporation (TAC), 40–41
TCAS. *See* Traffic Alert and Collision Avoidance System
Terrorism, 384
Texas Aero Engine Services LLC (TAESL), 334
Thales rising, 232–234
Thatcher, Margaret, 24, 110
Thatcher–Reagan deregulation, 27
"Third Package" of liberalization, 42
Thomas, Jürgen, 55
Tier 1 aerostructures suppliers emergence, 255–257
  aerospace raw materials demand (2015), 255*f*
  aerostructures outsourcing, 256*f*
Tier 1 supply chain
  models, 3
  strategies, 5
Tier 1s, 255
Tier 2s, 255
Tier 4, 261, 266
TIMET. *See* Titanium Metals Corporation of America
Tinidur, 243
Titanium, 241, 261
  sponge, 257
Titanium Metals Corporation of America (TIMET), 241
Total quality management, 223–224
TotalCare approach, 334–335
Toyota Motor Company, 186
Toyota Production System (TPS), 200–201
TPS. *See* Toyota Production System
Traffic Alert and Collision Avoidance System (TCAS), 219
Trans World Airlines, 29
Trent 600, 111
Trent 800 development program, 115
Trent engines, 110–111
Trent XWB, 99–100, 123
Trident, 103
Triple-vacuum process, 247
Trippe, Juan, 12
TRUEngine, 345
Trunkliner program, 1
Tupolev 104, 9
Tupolev TU-144, 19
Turbine blades, 244
Turbofans, 101, 115
  engine, 102*f*
Turboprop dominance (1980), 71–73
Twin-aisle
  aeroengines, 120*f*
  contenders, 16*f*
  delays, 161–162
  engines, 123
Twin-Aisle Refresh, 29–31
Two-spool design, 103

U-2 military aircraft, 15
UAC. *See* United Aircraft Corporation
UAE. *See* United Arab Emirates
UDF. *See* Unducted fan
Udvar-Hazy, Steven F, 60*f*
UltraFan, 168
Unducted fan (UDF), 33–34, 109
Unical Aviation, 347
Union de Transports Aeriens (UTA), 325
United Aircraft Corporation (UAC), 137
United Airlines, 10, 153
United Airport in Burbank and Mines Field, 352
United Arab Emirates (UAE), 342
United Kingdom
  aircraft and aeroengine manufacturers, 8
  Hawker Siddeley, 17
United Parcel Service (UPS), 50
United States (US)
  aircraft, 212
  airlines, 90, 155
  airspace, 57
  carriers, 11
  deregulation, 19–24
  jet aircraft, 100
  long-range bombers and transport aircraft, 7–8
  military aircraft, 240–241
  US Central Intelligence Agency, 242
  US Congress, 19
  US Federal Trade Commission, 47

United States (US) (*Continued*)
US Navy radar system, 216
US-based Allegheny Technologies, 257
United Technologies Corporation (UTC), 149, 202
UPS. *See* United Parcel Service
Ural Boeing Manufacturing, 258
US Air Force, 9, 103
cargo aircraft, 12
US Federal Aviation Administration (FAA), 29, 271–272
US International Trade Commission (ITC), 177
US original equipment manufacturers (OEMs), 73
Used and serviceable material (USM), 334, 340, 347
USM. *See* Used and serviceable material
UTA. *See* Union de Transports Aeriens
UTC. *See* United Technologies Corporation
UTC Aerospace Systems, 257, 316

V2500 engine, 27
Vacuum induction melting (VIM), 243
Variable pitch stators, 102
VAS Aero Services, 347
Verkhnaya Salda Metallurgical Production Association (VSMPO), 240, 257–258
Very large commercial transport (VLCT), 44
Very-high-frequency (VHF), 213–214
VHF. *See* Very-high-frequency
VHF omnidirectional range (VOR), 215
Vickers VC-10, 12
Vickers Viking turboprop aircraft, 8
"Victor highways", 215
VIM. *See* Vacuum induction melting
VLCT. *See* Very large commercial transport
von Ohain, Hans, 100
VOR. *See* VHF omnidirectional range
VSMPO. *See* Verkhnaya Salda Metallurgical Production Association

Warren, David, 215
Weather radar, 216
Welch, Jack, 227*f*
"White-sheet" approach, 106
Whittle, Frank, 100
"Wide chord" fan, 108
World Trade Organization (WTO), 85
World War II, 77, 100, 101, 352

Xtra-Wide-Body (XWB), 61

Ziegler, Bernard, 27, 216, 217
Zodiac Aerospace, 315

About The Author

Kevin Michaels, Ph.D., has 32 years of aerospace experience, and is a globally recognized expert in the aerospace manufacturing and MRO sectors. He is Managing Director and founder of AeroDynamic Advisory, and has a broad range of consulting expertise developed from engagements with leading aerospace companies across the globe. A contributing columnist to *Aviation Week & Space Technology*, Michaels holds B.S. Aerospace Engineering and M.B.A. degrees from the University of Michigan, and M.Sc. and Ph.D. degrees in International Relations from the London School of Economics.

Supporting Materials

A complete listing of titles in the Library of Flight series is available from AIAA's electronic library, Aerospace Research Central (ARC), at arc.aiaa.org. Visit ARC frequently to stay abreast of product changes, corrections, special offers, and new publications.

AIAA is committed to devoting resources to the education of both practicing and future aerospace professionals. In 1996, the AIAA Foundation was founded. Its programs enhance scientific literacy and advance the arts and sciences of aerospace. For more information, please visit www.aiaafoundation.org.

WITHDRAWN